Studies of Brain Function Vol. 16

Jos J. Eggermont

The Correlative Brain

Theory and Experiment in Neural Interaction

With 81 Figures and 2 Color Plates

Springer-Verlag
Berlin Heidelberg New York
London Paris Tokyo
Hong Kong Barcelona

Professor Dr. Jos J. Eggermont
Faculty of Social Sciences
Department of Psychology
University of Calgary
2500 University Drive N.W.
Calgary, Alberta, Canada

ISBN 3-540-52326-X Springer-Verlag Berlin Heidelberg New York
ISBN 0-387-52326-X Springer-Verlag New York Berlin Heidelberg

Library of Congress Cataloging-in-Publication Data. Eggermont, Jos J. The correlative brain: theory and experiment in neural interaction / Jos J. Eggermont. p. cm. – (Studies of brain function; vol. 16) Includes bibliographical references. Includes index. ISBN 0-387-52326-X (U.S.: alk. paper) 1. Neural circuitry. 2. Neural networks. 3. Neuroplasticity. I. Title. II. Series: Studies of brain function; v. 16. [DNLM: 1. Brain – physiology. 2. Models, Neurological. 3. Neural Transmission – physiology. 4. Neurons – physiology. W1 ST937KF v. 16 / WL 300 E29c] QP363.3.E34 1990 612.8′2 – dc20 DNLM/DLC for Library of Congress 90-9811 CIP

Printed in Germany

Typesetting: K+V Fotosatz GmbH, Beerfelden
2131/3145 (3011)-543210 – Printed on acid-free paper

Preface

This monograph is the result of a course given to graduate students and to the faculty of the Dept. of Medical Physics and Biophysics of Nijmegen University, Nijmegen, The Netherlands, in the fall of 1984 and 1985. The course was intended to put together experiment, theory, and analysis methods in order to study neural interaction and coding in the brain. The following pages give a survey of neural interaction and its experimental substrate: correlated neural activity. The basic reason for restricting myself to vertebrate brains was to keep the material concise. As the text developed, however, it became more of a review, than a research monograph, in the attempt to balance theoretical and experimental aspects in brain research. Consequently, the book can be read from various points of view: that of requiring an overview of theories and theoretical principles, or an overview of experimental studies in neural interaction and the methods that can be used, or with the conviction that theory and experiment cannot be separated. In the latter case the book should be read from beginning to end.

A way to read through the theoretical sections and the experimental sections of the book is presented in the following flow chart;

Theory: → Chap. 2 – Chap. 4 – Chap. 5 ↘
Introduction → Chap. 1 — Chap. 10 → Chap. 14
Experiment: → Chap. 3 – Chap. 6 – Chap. 7 – Chap. 8 ↗ ↘ Chap. 9

Thus the common aspects for both approaches being in the Introduction and Chapters 1 and 10 to 14, theory and experiment merging in the latter chapters. The first nine chapters deal with a stationary brain that has constant parameters and is basically memoryless, so that systems theory can be applied. Chapter 9 gives a compact description of what is theoretically possible in such a case. Chapters 10 to 14 discuss the changing brain: development and learning in its various aspects. Unless one assumes that all these changes are slow compared to the experiment time, systems theory cannot be applied in these cases. In another sense also the book consists of two parts. Part one covering the first nine chapters dealing with a static brain, and part two comprising the final five chapters dealing with plasticity and memory.

The text contains a number of 'Boxes', indicated in italics in the Contents, which detail some of the more technical aspects of procedures and concepts used in the main text. Readers who are not completely familiar with the topics discussed in these 'Boxes' may find them helpful. If the topic is completely new to the reader consultation of the references given in the 'Boxes' is strongly recommended.

The reader will probably become very quickly aware of the fact that in nearly every chapter there are references to papers written by George Gerstein. His activities in this field range from models for spontaneous activity in single neurons to experiments and methods designed to probe the workings of neural assemblies, and cannot be overemphasized. I hope that this monograph can do justice to the many contributions he has made to brain research up to now.

This monograph could not have obtained its final form without the meticulous editorial and scientific rigor used by Dr. Ad Aertsen when reading through the manuscript, and the subtle ways he used in telling me that parts were wrong, out of proportion, not clear or not important. If the text is clear and readable it is due to him, if it is not I am to blame for not incorporating all of his advice.

Calgary, Spring 1990 JOS J. EGGERMONT

Contents

Introduction

The functional aspects of the brain can be studied in many ways. It is possible to monitor its uptake of metabolites by imaging techniques displaying the course of radioactive labeled air, salts, or sugars with gamma cameras or positron emission tomography. However, to date these methods lack sufficient resolution in the temporal as well as spatial domain. More traditional techniques use the electrical activity produced by the myriad of nerve cells in the brain. Pasting gross electrodes to the human scalp allows us to monitor the ongoing activity of the more superficial cortical layers via the electroencephalogram as well as stimulated or voluntary activity through the various evoked potentials. The results thus obtained always reflect the activity of a relatively large group of nerve cells occupying some cm^3 of brain tissue. Evoked potentials with better spatial resolution, originating from a few mm^3 of brain tissue can be recorded by using microelectrodes with relatively large exposed tips that are pushed into the brain. Reducing the electrode tip to exposed lengths of only a few μm allows one to register the activity of only a few nerve cells; very small tips of around 0.1 μm even make it possible to penetrate a single nerve cell: the classical (intracellular) single-unit recording.

1 Single-Unit Analysis Versus Population Response Analysis

Much of our knowledge of the central nervous system has been derived from single-unit recordings in various parts in the brain, from the sensory nerve fibers up to the primary cortex, via the association cortex to the motor cortex and then down again to the motor units. The streamlining of sensory messages into motor action shows that "communication and control are the essence of brain function" (Moore 1980). We will, however, encounter other views of the brain which demonstrate that only about 1% of the input to the cortex is extracortical. One of these is postulated by Braitenberg (1974), that "the cortex to a large extent is a thinking machine working on its own output".

The single-unit view of brain function implicitly assumes that a sufficiently large sample of neural units represents all there is to measure in a given nerve tract or nucleus. Furthermore one often assumes that there are meaningful single-unit correlates of behavioral actions. When one advances to higher levels of the brain, the neural properties become more and more complex as compared to those of more peripheral neurons. In the single-unit view, neurons become tuned to certain "information-bearing parameters" (Suga et al. 1981), i.e., neurons become feature extractors. Some neurons in the bat's auditory cortex selectively respond to target distance, others to relative target velocity. Neurons in the visual cortex may selectively respond to bars of a certain length and specific orientation. Neurons in the HVc nucleus of songbirds are reported to respond only to the birds' own specific vocalization. One can carry this view to the extreme and assume that perceptual processes are localized in individual neurons or one may believe that there are single cell correlates of perceptual phenomena (e.g., a discussion in Rowe and Stone 1980). There has been a great amount of emphasis on feature extraction analysis by single neurons, especially in the visual system (e.g., Hubel and Wiesel 1968). This can be considered an artifact, albeit a very stimulating one, of the tendency of neurophysiologists to study one neural cell at a time; a tendency which in turn was determined by technical constraints.

It has of course been realized that single-unit analysis has its limitations, but there have been so many important questions for which single-unit analysis provided an answer that little attention was paid to the functional interaction between neural units. Ideally on should study a very large number of nerve cells simultaneously, observe their responses for every feature that the brain might use, and then try to conceptualize the results obtained. Since this is obviously impossible, one generally attempts to record sequentially from a number of units, one neuron at a time, the responses to more or less restricted sets of stimuli. This generally takes the form of mapping experiments whereby the changing tuning of the neurons to a particular stimulus parameter such as sound frequency, line orientation, localization of a stimulus in space, and stimulus velocity, is studied as a function of the location of the neurons in the structure under observation (e.g., Merzenich and Reid 1974; Hubel and Wiesel 1968).

As has been remarked above, this type of analysis introduces a conceptual bias: one is encouraged to assume that the tuning of the neurons to the experimental variables actually determines the function of the structure under investigation. When only pure tones are used in the study one obviously will find a tonotopic organization (if present) and not a preferred mapping of species-specific vocalizations. Furthermore, there is inevitably a bias in the recording because the sample will contain more large cells since these are likely to produce stronger electric fields (Rall 1962) and therefore more stable recordings. The early studies in single cell electrophysiology were all intracellular studies (e.g., Eccles 1957). The bias in these studies was of course toward relatively large cells, with the advantage that the recording is unambiguous and that information about many converging cells might be obtained when postsynaptic potentials are identified. The investigator is then able to identify functionally related neurons and even estimate the strength of synaptic couplings on the basis of the size and shape of postsynaptic potentials.

Most cells in the vertebrate central nervous system (CNS) are too small to allow prolonged intracellular recording so that most recordings are obtained extracellularly. This method lacks the unambiguous character of an intracellular recording and a stable large action potential waveform is actually the only criterion for deciding that the record is from a single unit. If the recording microelectrode does not actually penetrate the cell wall, one cannot be totally sure that the result is from a single cell and not from some sort of stable combination of contributions from several synchronously firing cells, or from (not necessarily synchronous) units with the same spike waveform, the first possibility occurring especially in structures with very homogeneous response properties. However, if all contributing units have the same tuning property, single-unit properties can be correctly deduced.

Microelectrode tip size determines to a large extent whether one will sample single-unit activity or multiunit activity. Multiunit activity is always recorded extracellularly and comprises a group of neurons whose cell bodies lie within a small area around the electrode tip. Typically a spherical or ellipsoidally shaped volume with a radius of around 15 μm yields, for a cell density of 2×10^5 per mm^3, a small group of nerve cells numbering three to five, and having action potential amplitudes in the range of 50–200 μV, but occasionally large spikes also occur. With this type of recording one has some knowledge about the spatial arrangement of the contributing units, but gains no insight regarding their possible interconnections unless one is able to recognize the individual cell's contribution to the firing behavior of the other neurons.

Multiunit recordings can be performed using either multielectrode arrays (e.g., Krüger 1983 for a review) or single electrodes (e.g., Gerstein 1970). Although the first method may cause damage to the brain if the structures to be studied are located deep in the brain, it does permit a large spatially extended region of the brain to be sampled simultaneously. The second method requires the separation of a multiunit spike train into single-unit spike trains (e.g., Abeles and Goldstein 1977). Intermediate recording situations using a few microelectrodes – each of which picks up the activity of a few units – open the possibility of studying short-range as well as long-range neuronal connectivities (Gerstein et al. 1983).

A promising alternative technique which allows the simultaneous recording of electrical activity in a relatively large area of the brain is the optical recording of the membrane potential changes (e.g., Grinvald 1984 for an introductory review). By applying fluorescent dyes to the brain in vivo and illuminating the brain with a laser, changes in fluorescence will accompany the generation and propagation of action potentials and postsynaptic potentials. The technique results in intracellular evidence of population activity, and may reach a temporal resolution in the millisecond range and a spatial resolution in the μm domain. At present the resolution is about $50\times50\times50$ μm^3 and the record clearly represents multiunit activity.

The struggle to understand the brain started with the application of the method most appropriate at that time: the morphological investigation. Ramon y Cajal's pioneering work resulted in the neuron concept: neurons are the structural entities that constitute the brain. New experimental techniques were developed which allowed a probing into the functioning of the individual neural elements (Adrian 1928; Eccles 1957) by recording the action potentials that were generated by the neurons and subsequently studying their relation to the synaptic potentials. Present technical and computational developments promote the investigation of how neurons are functionally connected. This has already been done on the local circuit level (Shepherd 1979) and an increasing effort is being put into the theoretical approach toward the understanding of neural assemblies (e.g., Palm 1982a). The brain cannot be studied as an isolated structure. Whether we view it as a thinking machine or as a communication system, its functioning will determine behavior, at the same time being dependent on it. Understanding brain and behavior are mutually dependent processes. Can one understand behavior by knowing the properties of all the neurons in the brain, i.e., from an analysis of individual neurons in isolation, or does one require knowledge about how functional neural groups, or neural networks act? In the single-unit approach, the relation to behavior is given by or limited to the following points:

1. Neurons are tuned to information-bearing parameters (feature extraction).
2. Single-cell correlates of perceptual processes include the idea of pontifical or command cells.
3. Neurons are mutually independent and have well-defined receptive fields.

If one favors the neural network approach these points become:

1. Spatiotemporal patterns of firing activity determine the status of the network.
2. There is collective decision-making and there are no command neurons.
3. The receptive field concept loses its meaning; cause and effect become obscure.

The phenomenon of interaction determines whether individual neurons connected into a network give rise to emerging properties that supersede those of its parts. The brain can be viewed as a democracy of nerve cells, and as in any democracy, it will be governed by cooperation and competition. An inquiry into these concepts will be the main theme of this monograph.

2 Outline of this Book

In this monograph we will deal with correlation between the activity of different neurons and will try to show that the study of the central nervous system is incomplete without taking cooperative effects into account. First a survey of the single neuron properties – spontaneous activity, receptive fields, and spectrotemporal response properties – will be given with emphasis on how these properties change from (sensory) periphery to cortex. This will provide us with a basis for understanding the formation of feature detectors or feature extractors through neural interaction. Small neural nets generally are sufficient to explain this. More global properties of neuronal interaction lead to models for large neural nets and for mass action in the nervous system. After a discussion of the static and dynamic aspects of neural interaction, including habituation, long-term facilitation, and memory, we will in considerable detail describe ways of measuring neural interaction. Methods to visualize the results obtained and ways to quantify neural correlations are an important aspect in the study of the connections between neurons. For the major sensory systems and other areas of the vertebrate CNS we will give a survey of the experimental results from simultaneously recording from more than one neural unit. The consequences thereof for our view of nervous system action will be extended to aspects of plasticity, context-dependent behavior, and stimulus-dependent neural interaction. The many similarities with the study of associative memory and cooperative effects will be illustrated.

In many ways the outline of this book follows historical lines. One may view the 1960's as the decade of statistical analysis of spontaneous nervous activity and of the emergence of theories explaining the nature of this activity. The advent of the computer era has certainly been an important aspect in this. Increasing computational power in the 1970's resulted in the emergence of nonlinear analysis techniques applied to central nervous system activity and to stimulus-response relations. The 1980's showed as the decade of spatial and spatio-temporal analysis through the use of multielectrode recording techniques and multiunit analysis. The emphasis on parallel processing in computation is both stimulating and necessary to elaborate on the enormous amount of data that the new experimental techniques yield. This book features experimental and theoretical chapters alternately, but as the text advances these aspects become more and more intermingled. This reflects the view of the author that experimental and theoretical approaches should not be dealt with independently but in an integrated way. Both approaches should learn from each other: theory should become more aware of the boundary conditions set by the experimental results, and experiments should be guided by theoretical considerations.

1 Information Coding

1.1 Information Coding and Transmission by Single Cells and Cell Populations

We consider the brain as a communication system. In its simplest form such a system consists of a transmitter sending a message, a channel in which the message may be perturbed by noise and distortion, and a receiver which decodes the message. To be able to decode the message, the code must be known. While obviously it is known to the brain, or intrinsically used to control various actions, this code is not known in principle to the experimenter. The experimenter knows only the carrier of the neural code: the single-unit spike trains. The communication process can be studied at different levels: one may want to put numbers on the capacity of the transmission channel, one may want to decipher the meaning of the transmitted symbols, or one may ask if the message elicits the desired conduct or behavior. For an early review on the issue of neural coding, see Perkel and Bullock (1968).

Information theory offers a method to quantify the transmission processes in the brain, but only in a statistical sense and under, generally not rigidly fulfilled, conditions. In 1949 Shannon published his *Communication in the Presence of Noise* in which he derived optimal values of symbol transmission rates: the channel capacity. A few years later MacKay and McCulloch (1952) used information theory to calculate the limiting information capacity of a neuronal link assuming either a binary code (time is quantized in bins in which a spike does or does not occur) or a time interval code (information is transmitted as a string of symbols representing the length of successive intervals) and assuming that successive impulse positions are statistically uncorrelated. They concluded that their calculations supported interval modulation codes over binary codes. A binary or frequency code could be based on the assumption that the carrier is an irregular pulse train, e.g., a Poisson pulse train in which the intensity conveys the message. An interval code suggests that interspike intervals should be independent symbols and the message be analyzed as a Markov chain. It will be demonstrated later that most neural spike trains are neither Poisson nor simple Markovian in character. One may therefore rightfully suggest that spontaneous activity is only evidence that the neuron functions in a complex computing task caried out by an assembly of neurons in which it participates (Braitenberg et al. 1965).

Before considering in more detail the advantages and disadvantages of various codes we should consider how the channel that transmits the information is built up. It consists ideally of axons and synapses and for the sake of simplicity we incorporate most of the cell body and dendrite properties in the synapse. Thus the synapse incorporates the release of transmitter, the formation of postsynaptic potentials, and the transformation of psp's into conducted action potentials (ap).

Box A. Information Theory
A general communication system (Fig. A.1) consists of an information source which provides a message, a transmitter which sends the message modulated upon a carrier signal, a transmission channel in which the signal may be perturbed by noise or distortion, a receiver which collects the (perturbed) signal and decodes it in order to obtain the message. For the CNS the information source may be the sensory world, the message is commonly called the stimulus; sensory receptors function as transmitters giving off spike trains as carriers of the information. Noise in the form of spontaneous activity may be added, higher-order neurons receive and decode the spike trains. In the central nervous system the symbols might be the length of interspike intervals, the average firing frequency over fixed time intervals, a favored pattern in a spike train, etc.

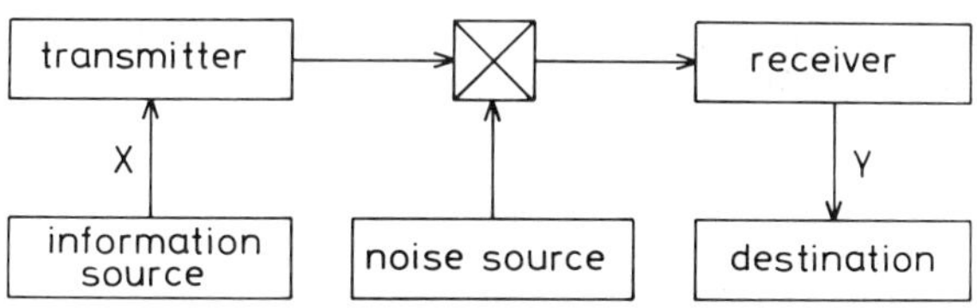

Fig. A.1. Schema of a general communication system

Information can be treated as a measurable quantity: one can put a number to it. Let a message consist of a set of symbols $X = \{x_i\}$, each x_i occurs with probability $p(x_i)$. It is customary to define the *information content* of symbol x_i as (e.g., Eckhorn and Pöpel 1974)

$$I(x_i) = -{}^2\log p(x_i) \text{ [bit]} . \quad (A.1)$$

From this we learn that the more likely x_i is to occur to our prior knowledge, the less will be the information content of such an occurrence. When x_i is equally likely to occur or not to occur, $p(x_i) = 0.5$, and $I(x_i) = 1$ bit. By taking a weighted average over all possible symbols that the source can deliver, we obtain the average information content or the uncertainty associated with the information source. This is also called *information entropy* and defined as

Box A (continued)

$$H(X) = \sum_i p(x_i)\ I(x_i)$$
$$= - \sum_i p(x_i)\ {}^2\log\,[p(x_i)]\ \text{[bit/symbol]}\ . \qquad \text{(A.2)}$$

Analogously one can define the information entropy H(Y) in the output signal which consists of a set of symbols $Y = \{y_j\}$.

The noise added in the transmission channel will give rise to an additional uncertainty, the dissipation entropy, $H(Y|X)$:

$$H(Y|X) = - \sum_{i,j} p(y_j, x_i)\ {}^2\log p(y_j|x_i)\ , \qquad \text{(A.3)}$$

where $p(y_j, x_i)$ is the joint probability of occurrence of y_j and x_i; $p(y_j|x_i) = \frac{p(y_j, x_i)}{p(x_i)} \leqslant 1$ is the conditional probability of occurrence of symbol y_j when x_i has been sent.

The information T from the message that is transmitted is given by

$$T = H(Y)\ -\ H(Y|X)$$
$$= \sum_{i,j} p(y_j, x_i)\ {}^2\log \frac{p(y_j, x_i)}{p(y_j)\ p(x_i)}\ . \qquad \text{(A.4)}$$

The rate at which this transinformation T is transmitted is given by

$$T^1 = \frac{1}{\Delta t} T\ [\text{bit s}^{-1}]\ ,$$

where Δt is the duration of a symbol in the signal.

The *information capacity* (or channel capacity) is defined as (Stein 1967)

$$C = \max_{\{p(x)\}} T\ \text{[bit]}\ . \qquad \text{(A.5)}$$

In the special case that the signal has a Gaussian distribution with variance s^2 the information entropy is

$$H = {}^2\log \sqrt{2\pi\, e\, s^2}\ . \qquad \text{(A.6)}$$

It can be shown that in the limit of large t $C = -{}^2\log \sqrt{2\pi\, e\, s^2}$ (Stein 1967).

In the case that the signal has an exponential distribution $H = {}^2\log(e \cdot \frac{1}{\lambda})$ and in this case $\frac{1}{\lambda}$ takes the place of the standard deviation s for the Gaussian distribution (Nakahama et al. 1972).

The channel capacity of myelinated nerve fibers in the frog has been studied by Lass and Abeles (1975) and Abeles and Lass (1975). It appeared that there was a small but distinct jitter in the interval transmitted along the axon. The channel capacity, equal to the maximal firing rate, was observed to be inversely proportional to the standard deviation of the jitter in the intervals. For a standard deviation of about 10 μs and a refractory period of 3 ms the channel capacity was about 1000 bit s^{-1}. Introducing a synapse with a jitter of a few ms immediately reduces the channel capacity to about 100 bit s^{-1}. If a number of synapses are connected in series in the transmission link, the precision of timing diminishes with about the square root of the number of synapses. We may expect therefore that a single neuron chain from periphery to cortex containing 4 synapses will have a channel capacity of about 50 bit s^{-1} at most.

Estimations of channel capacity cannot be made without specifying the code, and the above studies assumed an interval code. Mountcastle (1967) described a variety of neural codes in single axons such as a frequency code (the intensity of a stimulus is reflected in firing frequency), the interval code, and the pattern code. The latter code may be appropriate when there are strong serial correlations between successive intervals. Eckhorn et al. (1976) calculated the so-called transinformation content for three different codes and three different physiological preparations. The codes were the frequency code, a time-weighted average frequency code and a so-called signal code. In the latter case no assumptions about a code are made and the information content was calculated on the basis of the entire spike train structure. The preparations studied were the transmission from γ-efferents to the muscle spindle afferents, muscle spindle afferents to dorsal spino-cerebellar tract neurons, and retinal ganglion cells to the principal cells of the lateral geniculate nucleus. It appeared that for all preparations the weighted-average code yielded results ranging from 90–99% of those of the signal code. This demonstrates that nearly all information is transmitted by the linear part of the coupling. A brain acting with such a code will operate in a linear manner upon the input provided by the first-order sensory nerve fibers [Mountcastle 1967]. The results for the rate code did not approximate the signal code closely: only 5–92% (the actual number depending on the rate of input information), suggesting that in most situations the frequency code is unreliable for single neuron transmission. Although the frequency code may be unreliable for transneuronal information processing, most peripheral sensory nerves appear to code stimulus intensity in the frequency of firing. Stein (1967) investigated the influence of the distribution of the firings in the spike train upon the information capacity of a neuron for a frequency code. When intervals between successive nerve impulses are identically and independently distributed random variables, with a probability distribution having mean μ and variance σ^2, then the distribution of the number of nerve impulses in a fixed interval t is asymptotically normal with mean $x \simeq t/\mu$ and variance $s^2 \simeq \sigma^2\, t/\mu^3$. The uncertainty of a normal distribution is as we have seen: $H = \log[\sqrt{2\pi e} \cdot s]$. The asymptotic information capacity for large t in such a case is given by

$$C^0(1) \sim \log\left[\int_{x_{min}}^{x_{max}} \frac{dx}{\sqrt{2\pi e}\cdot s(x)}\right]. \tag{1.1}$$

In terms of the interval distribution this reads

$$C^0(1) \sim \log\left[\sqrt{\frac{t}{2\pi e}} \int_{\mu_0}^{\mu_1} \frac{d\mu}{\sqrt{\mu}\cdot\sigma}\right], \tag{1.2}$$

where $\mu_0^{-1} = x_{max}$, the maximum firing frequency, μ_1^{-1} is the minimum firing frequency. For most interval distributions $\sigma = b\,\mu^m$ so that

$$C^0(1) \sim \log\left[\sqrt{\frac{t}{2\pi e}} \cdot \frac{\mu_0^{-k} - \mu_1^{-k}}{bk}\right], \tag{1.3}$$

where $k = m - 0.5$. Serial correlations between the (successive) intervals have an effect on the variance of the frequency distribution, which becomes $s^2 \simeq a\,\sigma^2 t/\mu^3$
with $a = 1 + 2\sum_{i=1}^{\infty} \varrho_i$ and ϱ_i is the ith order serial correlation coefficient. Equation (1.2) then becomes

$$C^{\varrho}(1) \sim \log\left[\sqrt{\frac{t}{2\pi e}} \int_{\mu_0}^{\mu_1} \frac{d\mu}{\sqrt{\mu a}\cdot\sigma}\right]. \tag{1.4}$$

Under the assumptions made for the standard deviation of the distribution Eq. (1.3) now reads

$$C^{\varrho}(1) \sim \log\left[\sqrt{\frac{t}{2\pi e}} \cdot \frac{\mu_0^{-k} - \mu_1^{-k}}{\sqrt{a}\cdot b\cdot k}\right]. \tag{1.5}$$

A more compact version of this equation, taking into account Eq. (1.3) reads

$$C^{\varrho}(1) = C^0(1) - \frac{1}{2}\log a\ , \tag{1.6}$$

which indicates that only the summed serial correlation is important, and not the configuration of the ϱ_i.

Finally the case is considered that there are N independent, identical neurons conveying the message, then

$$C^{\varrho}(N) = C^0(1) - \frac{1}{2}\log a + \frac{1}{2}\log N\ , \tag{1.7}$$

i.e., the asymptotic capacity for information transmission is increased by $1/2 \log N$ bit. One may interpret this also in the way that the same amount of information can now be transmitted in a fraction $1/N$ of the original time. Such situations might occur, e.g., in the auditory nerve where typically one inner haircell is innervated by 10–20 (probably independent, Johnson and Kiang 1976) afferent nerve fibers of which in general three types can be considered in terms of spontaneous activity and threshold (Liberman and Oliver 1984).

If the N parallel nerve fibers were completely correlated in their firing patterns, one obviously would not gain anything in terms of capacity for information transmission, and with incomplete correlation the gain will be less than in the case of independence. In the case of N identical nerve fibers, the pooled mean firing rate will be reduced by a factor N, and the pooled variance will be

$$S = N \cdot s^2 + s \sum_{i,j} C_{ij} ,$$

where C_{ij} is the covariance between the firing of neurons i and j. By writing $C_{ij} = \sigma_i \cdot \sigma_j \cdot r_{ij}$ where r_{ij} is the correlation coefficient and introducing the factor $d = 1 + \frac{2}{N} \sum_{i,j} r_{ij}$, we arrive at

$$C^r(N) \sim C^0(N) - \frac{1}{2} \log d , \tag{1.8}$$

where we have to take the $\sum$ over $\frac{1}{2} N(N-1)$ pair combinations. The most general situation, N correlated neurons with serial correlation between the intervals of the individual units, has a transmission capacity of

$$C^{\varrho r}(N) \sim C^0(N) + \frac{1}{2} \log N - \frac{1}{2} \log a - \frac{1}{2} \log d . \tag{1.9}$$

A distribution quite often encountered in the description of interspike intervals is the gamma distribution for which $\sigma = \mu/\sqrt{n}$, where n is the order parameter of the distribution. In the above formulas then $m = 1$ and $b = \frac{1}{\sqrt{n}}$, e.g.,

$$C_\gamma^0(1) \sim \log \left[\sqrt{\frac{tn}{2\pi e}} \cdot \frac{\mu_0^{-0.5} - \mu_1^{-0.5}}{0.5} \right] . \tag{1.10}$$

To put some numbers on the reducing effects on information transmission by serial and parallel correlation – still assuming a frequency code – let $\mu_0^{-1} = 100\,s^{-1}$, $\mu_1^{-1} = 10\,s^{-1}$, $n = 4$, $\varrho_1 = 0.6$, $\varrho_2 = 0.2$, $\varrho_3 = 0.4$, $\varrho_{4 \to \infty} = 0$; $r_{ij} = 0.6$ for all i, j and $N = 10$. Then for $t = 10$ s we obtain that $C_\gamma^\varrho(1)$ is about 15% less than C_γ^0. The parallel correlation does not have much effect either since it reduces $C^0(N)$ in Eq. (1.7) by about 6%. Even a very strong correlation of $r_i = 0.9$ produces for the situation considered a "loss" of only about 9% as compared to the independent case, and is thus much better than the single neuron capacity and far more reliable.

Although we have paid much attention to frequency code or weighted-average frequency codes, there are some indications that interval codes might be operative. Sanderson et al. (1973) consider transmission from retinal ganglion cells to principal cells in the lateral geniculate nucleus and argue that the lateral inhibition mechanisms in the LGN are able to extract information from the distribution parameter n in the case that the overall firing frequency does not change.

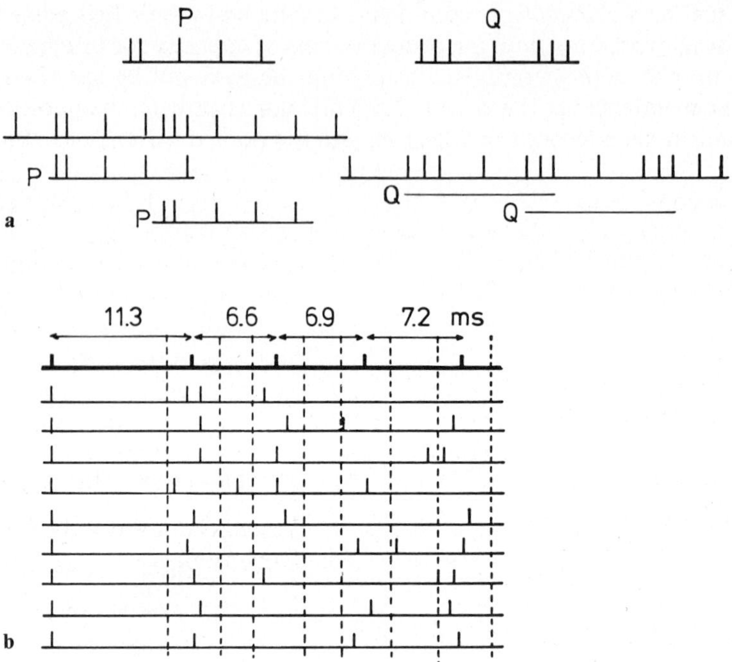

Fig. 1.1 a, b. Occurrence of favored patterns in single-unit spike trains (after Dayhof and Gerstein 1983 a, b), patterns may be found overlapping each other. **a** Illustrates the concept of a pattern and certain forms of overlap. **b** Shows a set of patterns found in spike trains from the cat striate cortex that more or less conform to the template shown in the first row

More elaborate single neuron codes depending on the serial properties of neuronal interspike intervals are likely to exist. Dayhoff and Gerstein (1983 a, b) observed in the crayfish claw control system favored patterns of spikes in 23 out of 30 spike trains investigated. These favored patterns consisted generally of 3–7 spikes and occurred more often than chance alone would predict. In the cat striate cortex they found in 16 out of 27 spike trains patterns that contained up to 7 spikes. This suggests that such favored patterns (cf. Fig. 1.1) are relatively common. They may have an important property in that such a code may be much less vulnerable to noise than interval or frequency codes (Sherry and Klemm 1980; Legendy and Salcman 1985).

When we combine these serial correlations in the favored patterns with correlation between spike trains of individual neurons, it seems likely that there are favored patterns in a more general sense. Patterns may manifest themselves in a spatiotemporal way, i.e., across a population of neurons. Such patterns can be found from cross-correlation functions between spike trains by taking only those spikes that are present in the correlation peaks. Frostig et al. (1984) took only the events in the primary complex of the correlation histogram and obtained what they called "information trains". A similar procedure has been used by Epping and Eggermont (1987), but it may well be worthwhile to incorporate the firings due to the autocorrelation within the single unit spike trains to arrive at more general spatiotemporal codes, e.g., by calculating higher-order correlation functions.

In the CNS there is generally a large degree of convergence, i.e., higher-order neurons may be innervated by relatively large numbers of afferents. The response of such higher-order neurons may therefore be based on a weighted average of the incoming activity. Such ensemble averages might be able to obtain the same reliability in a short time, as compared to the long-term averages that we have discussed so far, to arrive at the information transmission capacities for single neurons. When the average over n neuronal spike trains of duration t is equivalent to one single unit train of length nt, then we call this process an ergodic process. This, however, is hard to prove rigidly since ergodicity requires that the neurons behave in an identical way (in distribution sense), that the discharges of the neurons are uncorrelated, and that transient effects are avoided, i.e., the spike trains are stationary. In such cases then the CNS is able to average out unwanted noise in the relatively short time that it generally has to reach a decision. Dormont et al. (1982) have shown that the ensemble activity of only eight neurons recorded successively in the ventrolateral thalamus of the behaving cat was much better correlated with the cat's reaction time than was the case for any of the neurons considered individually. Simultaneous recordings from neurons in the cat striate cortex using microwire bundles led Gerstein and Michalski (1981) to suggest that the activity pattern among several neurons in response to a stimulus (in this case an oriented light bar) is more coincident that the similarity in the response pattern of each of the single units to successive stimuli. It might therefore be that actually fewer neurons are needed for detection, i.e., yield reliable PST histograms, than would be expected on the basis of a set of independent neurons and on the basis of the number of stimuli needed to give a reliable single-unit PST histogram. This was also suggested by John (1972) in saying that cooperative processes involving large numbers of neural elements might generate physical consequences and concomitants not to be predicted from the characteristics of individual elements.

In most real life situations, the extent to which information transmission can be enhanced by summation or averaging over the activity of numbers of cells will be limited by any underlying correlation between neurons. In such cases not all rates or interval values are equally likely and therefore some prior knowledge as to when the firings will occur is available. As we have seen, the capacity for information transmission drops in such cases. One of the central issues to investigate will be whether the CNS transmits information in arrays of more or less independent neurons (e.g., Erickson 1974, 1982) or in coordinated neural populations composed of local circuits that participate in several distributed systems (e.g., Mountcastle 1978). However, in both cases sensory information is mediated by simultaneous activity in ensembles of neurons, i.e., in spatiotemporal activity patterns. Understanding of the brain, or parts thereof, thus requires that the relationship between such spatiotemporal activity patterns and the behavior of the animal be established. Such multichannel codes of information will require some form of coincidence detection, neurons that are sensitive to temporal differences in the arrival of action potentials from various other neurons. One noteworthy example is found in the auditory system. Time differences as small as 10 μs between the arrival of sound at the two ears produces differences in the firing behavior of cells in the auditory midbrain of the animal that receive afferents from both ears (e.g., Rose et al. 1966).

In later chapters we will discuss aspects of brain function related to this communication aspect, e.g., sensorimotor control, as well as aspects that pertain to its action upon its own output, such as learning and memory. These aspects can be only poorly understood on the basis of the single-unit concept and will require more flexible and adaptable units of brain function such as neuron ensembles.

1.2 Cooperative Effects and Ensemble Coding

Cooperative effects can be studied by the simultaneous recording from groups of neurons and by studying correlation in their firing behavior. On the basis of the outcome of the measurements, one might conclude, for instance, a functional independence between neurons, in which case population encoding is operative, e.g., as a population rate coding. Especially in structures where most of the neurons are broadly tuned there is a necessity for a population coding in a so-called across-fiber pattern (Erickson 1982). As an example we may consider the superior colliculus, a midbrain structure playing an important role in the control of eye movements. In monkeys the visual receptive fields in the superficial layers are quite large and so are the motor fields in the deeper layers. Still (reflex-) saccadic eye movements are quite accurate (Ottes et al. 1985a, b). In order to explain this, a population coding and encoding model has been devised (Ottes et al. 1986; van Gisbergen et al. 1987) which accurately describes the generation of the correct saccade from a pattern of motor activity. The across-fiber pattern is therefore utilized in the form in which it is present. At no point is it aimed at a pontifical or command neuron, but transformations occur to make the neuron pattern to appear in a form which facilitates comparison with other stimuli (see also Georgoupolos et al. 1986).

Alternatively, one might obtain strong temporal coordination between the individual neurons, in which case coding via distributed networks may be more appropriate. These networks are termed assemblies (Hebb 1949; Palm 1982a), synfire chains (Abeles 1982), or repertoires (Edelman 1978). At this point it may be remarked that the assembly definition of John (1972) and that of Hebb (1949) do not necessarily coincide. We will take up this point in Chapter 13. An important factor in these networks is the strength of the interneuronal coupling. Braitenberg (1963) attempted to derive an expression for a normalized coupling coefficient on the basis of histological measures of cell body volume V_T of neuron type T, the density of neurons N_T of type T, the volume of the axonal field A, and the volume of the dendritic field D. He arrived at a coupling between receiving neuron h and sending neuron k:

$$\tilde{\tilde{w}}_{hk} = \frac{V_k}{\Sigma_T N_T V_T} \cdot \frac{D_h \cap A_k}{D_h \cdot A_k} \cdot E \ ,$$

where E is some unit of excitation produced in a unit volume of gray substance by the arrival of an impulse in each of the axon terminals contained in it. A coupling coefficient in more restricted sense would be

$$w_{hk} = \frac{D_h \cap A_k}{D_h \cdot A_k} \ , \tag{1.11}$$

since the other terms mainly deal with the ultimate effect on the output firing behavior on the basis of the available input. Thus the relative amount of overlap between the receiving dendritic field and the sending axonal field would on average determine the strength of the neuronal coupling. Stark et al. (1968) elaborated along these lines to define a net transfer function on the basis of convergence and divergence factors of the axonal fields. Taken together with the coupling coefficient, this might give a conceptual starting point for evaluating histological data. Meek (1983) elaborated on a combined Golgi-electron microscope analysis of the goldfish optic tectum. He estimated for each cell type the number and distributions of synaptic contacts. Together with the known target layers for the various afferent tracts into the optic tectum and the assumption that "within a certain presynaptic zone all types of presynaptic elements distinguished are presumed to make equal total numbers of synaptic contacts", he arrived at a connectivity matrix giving the relative contribution of different types of presynaptic elements to the total population of receptive synaptic contacts on each cell type. This connectivity matrix, besides the afferent contacts, also features the interneuronal connectivity and the tectal efference. These are obviously starting points; no distinction has been made between excitatory and inhibitory synapses and between relative importance for spike generation of somatic synapses versus those on distal dendrites.

On the basis of the coupling strength, some principal differences in network behavior can be distinguished. In the limit of weak coupling no single connection will determine the firing probability of a neuron, for strong coupling cause and effect might be confused as to what neuron determines the behavior of another neuron in the network. Sejnowski (1976a) describes that in the case of weak coupling local interactions will dominate the firing behavior in the ensemble, while in the case of strong coupling neurons can no longer be investigated individually but only as a coordinated ensemble. Long-range interactions will become significant, i.e., neurons several synapses removed from each other may still be coordinated in their activity (Abeles 1982). In the case of weak coupling one can still analyze the neural ensemble in terms of feature detectors, and population codes such as population firing rate will be a perfect replica of the input stimulus (e.g. Knight 1972). Receptive fields can still be defined for the individual neurons in the ensemble although they may no longer be stimulus-invariant (Eggermont et al. 1983a; Johannesma and Eggermont 1983). In the case of strong coupling, cause and effect may become obscure and standard systems analysis, which is at the basis of the receptive field concept, becomes useless. When this happens the neuronal circuit under study will be something different from the sum of its parts (Johannesma et al. 1986). If one studied the neuronal circuit element in isolation, i.e., as a single unit, its properties determined in that way might have little relevance to the operation of the entire circuit. The neuron population will behave in a context-dependent way and depend on the stimulus as well as on the history of the neural net because also the correlation strengths may become time varying (e.g. Shaw 1978).

Time-varying correlations find their basis in time varying connectivities in the neural net. This can be illustrated with models for associative memory (Kohonen 1984). Such models can be summarized in two equations:

$$\eta_h(t) = \xi_h(t) + \sum_k w_{hk}\,\eta_k(t) \tag{1.12}$$

$$\frac{dw_{hk}}{dt} = \alpha\,\eta_h(t)\,[\eta_k(t) - \eta_{kb}]\ , \tag{1.13}$$

where $\xi_h(t)$ are input firing rates to cell h, $\eta_h(t)$ are output firing rates of cell h. In these models there is a coupling with strength 1 on the input from cells outside the population and a coupling $w_{hk} \leqslant 1$ for inputs derived from the outputs of cells in the same population. The w_{kh} depend on the correlation between η_h and η_k, η_{kb} is the invariant background activity of neuron k; $\alpha > 0$ for excitatory and $\alpha < 0$ for inhibitory synapses. Time-varying correlation is hypothesized to be the basis of nervous system plasticity, thus neural correlation studies should be combined with experiments on plastic changes in the nervous system. Behavioral paradigms that are appropriate for such experiments must be capable of producing changes in the neuronal couplings in the assembly structure within the time during which stable recordings can be obtained. Gerstein et al. (1980) describe two such paradigms. Provided that they are powerful enough it will be possible to measure whether the strength or the nature of the coupling between the observed neurons is varied by the conditioning procedure.

2 Spontaneous Activity

2.1 Noise or Information Carrier?

Spontaneous activity is an important object of study because it is the basis of two ideas regarding information processing in the CNS. One viewpoint considers spontaneous activity as *neuronal noise* (Fitzhugh 1957, 1958; Siebert 1965; McGill 1967), the presence of which sets a limit to the detection capabilities and sensitivity of our senses and obscures the stimulus-related activity. The problem of the analysis of messages carried by the nerve fibers for the brain is then similar to the engineering problem of detecting a signal in a noisy communication channel. An increase in the number of independent noisy channels which all carry the same stimulus message but with uncorrelated spontaneous activity will then increase the amount of information transferred at a given intensity or decrease threshold intensity. This can be called ensemble coding. Fitzhugh (1957) calculates that for N independent nerve fibers the threshold intensity is proportional to $N^{-0.62}$. An example of such a situation might be present in the auditory system where one inner haircell is innervated by 10–20 afferent nerve fibers with greatly varying spontaneous activity properties and slightly varying thresholds (e.g., Geisler et al. 1985). The other viewpoint reflects the other extreme: spontaneous activity is considered to reflect the *main process* in the CNS upon which the sensory input either acts as a modulator or is regarded as a mere perturbation of the spontaneous activity (e. g., Rodieck et al. 1962). The modulator view is implicitly held also by Fitzhugh (1958) when he considers a stimulus to act mainly by distorting the spontaneous interval statistic. This distortion can be considered as a multiplication of the interspike intervals by a time-dependent frequency function induced by the stimulus. Evidence for a modulator view of stimulus action also comes from the auditory system. Auditory nerve fibers with high spontaneous activity display a much lower threshold to stimuli (approximately 20 dB) when synchronization to the stimulus is considered than when a certain increase in firing rate is taken as a criterion. In the vestibular nerve there is a very regular spontaneous discharge, stimulation can increment as well as decrement the firing rate (e. g., Precht 1974). In retinal ganglion cells the dark discharge in off-center cells is larger than in on-center cells. Stimulation decreases activity in off-center cells and increases them in on-center cells (Barlow 1969). Essentially the difference between both viewpoints may be characterized as spontaneous and induced neural activity being additive or multiplicative respectively (Aertsen pers. commun.).

Since most of the evidence goes to spontaneous activity as an important aspect in the transmission of information, we have to consider whether modulation of an in general stochastic pulse train is a faithful method to transmit signals. A completely unstructured series of impulses can be described by a *Poisson process*. This is characterized by three properties:

1. The probability of an event in the time interval $[t, t+\Delta t]$ is equal to $\lambda(t)\Delta t$.
2. The probability of no events in that time interval is equal to $1-\lambda(t)\Delta t$, and
3. The joint probability of n events in interval [s, t] and m events in interval [u, v] is equal to the probability of n events in [s, t] times the probability of m events in [u, v].

This implies that interspike intervals are statistically independent. If $\lambda(t) = \lambda$ the process is called a homogeneous Poisson process, otherwise an inhomogeneous Poisson process. A deterministic signal can be conveyed by a Poisson process if the intensity $\lambda(t)$ is a unique function of the signal. However, because the Poisson process is a stochastic process the communication channel is inherently noisy. Here we see that both extreme viewpoints of spontaneous activity merge again. In the practice of nervous system activity Poisson processes do not occur since neurons have refractory properties: there is a certain time after an impulse in which they cannot fire again and/or a time in which the firing probability is reduced. Neural spike trains showing these refractory properties can be described by renewal processes (e.g., Johnson and Swami 1983) for which the intervals are still statistically independent, but property 3 no longer holds. It can be shown that non-Poisson pulse trains, like renewal process-governed spike trains, can be used to transmit signals but with a certain distortion due to the refractory properties. If the receiver knows the recovery function then the signal can be obtained (Johnson and Swami 1983; Jones et al. 1985) in a faithful way. In general, spontaneous activity is most pronounced (in terms of mean rate) in the periphery of sensory systems and is commonly attributed to the stochastic transmitter release from the receptor cells (Koerber et al. 1966; Rodieck 1968; Sewell 1984). Arguments in favor of this are that destruction of cochlear function destroys spontaneous activity in auditory nerve fibers. Sewell (1984) demonstrated that log spontaneous rate, R_s, is linearly related to the endolymphatic potential V, or:

$$R_s = R_0 \exp(kV) . \tag{2.1}$$

Only the integrity of the inner haircells is required for normal spontaneous activity, as shown by Dallos and Harris (1978) in kanamycin-treated cochleas. The view that synaptic transmission contributes to spontaneous activity at more central levels has also been stressed (e.g., Shaw and Vasudevan 1974). In addition it has been shown that cortical areas isolated from sensory input exhibit spontaneous activity (Smith and Smith 1965). Isolated slabs of cat cortex (15×4×4 mm) show the emergence of spontaneous activity about 1 week after deafferentation; this spontaneous activity is no longer influenced by the cat's behavior or by stimulation (Burns and Webb 1979). Thus there are both extrinsic and intrinsic sources of spontaneous activity in the CNS, the extrinsic source is the transmitter release in the receptor cells. Intrinsic sources may be any agglomerate of nerve cells since, for example, for the second-order vestibular neurons which lose spontaneous activity after vestibular nerve section a recovery of some spontaneous activity is reported some time after acute surgery (Precht 1974).

The extrinsic sources of spontaneous activity dominate, since at the cortical end of the sensory processing chain the spontaneous discharge rate has dropped considerably. For example, in the visual system of the cat the mean spontaneous rate is about 35 spikes per second in the retinal ganglion cell, 14 spikes per second in the next sensory station, the lateral geniculate nucleus, and only about 6 spikes per second in the striate cortex (Hertz et al. 1966).

In the auditory system a clear distinction can be made between three groups of spontaneous rates in auditory nerve fibers. These rates are well correlated with the morphology of their dendritic terminals as well as with threshold for auditory stimulation (Liberman and Oliver 1984) although the extent of the latter is disputed (Geisler et al. 1985). The group of fibers with spontaneous rates less than 0.5 spikes per second has the higher thresholds and small terminals that synapse with the inner haircells. The group with spontaneous rates higher than 18 spikes per second has the lowest thresholds and synapses with large terminals, the third group has properties in between. Small and large terminals are found on all haircells, generally one haircell gives input to 10–20 auditory nerve fibers. These findings support the idea that the amount of spontaneously arriving transmitter at the postsynaptic membrane is a major determinant of the spontaneous rate. This also supports the modulation theory since it is likely that small modulations are more easily detected on high spontaneous rate carriers than on low ones. Spontaneous rates of up to 150 spikes per second are not unusual in the cats auditory nerve (Kiang et al. 1965), can also be found in the cats cochlear nucleus complex (Pfeiffer and Kiang 1965; Koerber et al. 1966) and in the superior olivary complex (Goldberg et al. 1964) but are absent in the auditory cortex, where about 50% of the fibers discharge with less than one spike per second and the remainder in between 1–35 spikes per second (Goldstein et al. 1967). An intermediate position is taken by the medial geniculate body where rates are between 1 and 20 spikes per second (Aitkin et al. 1966). There exists great variability in spontaneous activity of the auditory nerve among various vertebrate classes. While mammals usually show a clear distinction between fibers with low and high spontaneous rates (a bimodal distribution with for the cat a median rate of about 40 spikes per second), bird's in general show a unimodal distribution. In general, all nerve fibers in the birds auditory nerve show spontaneous rates above 5 spikes per second and with a median rate of about 90 spikes per second (in the blackbird and pigeon). This higher rate was attributed to a higher body temperature (Sachs et al. 1980). In various reptiles spontaneous rates were either bimodally distributed (Caiman; Klinke and Pause 1980) with rates between 0.5 and 80 spikes per second or unimodal (Tokay gecko, Eatock et al. 1981) with rates between 1.6 and 40 spikes per second. Temperature for the caiman was 27 °C and for the gecko 24 °C. In frogs (*Rana pipiens* and *Hyla cinerea*) Megala and Capranica (1981) found a unimodal distribution with spontaneous rates less than 25 spikes per second (temperature ≃ 22 °C). Still lower on the phylogenetic scale, the goldfish auditory nerve showed spontaneous rates of up to 200 spikes per second. Thus there is no uniformity in the various vertebrate classes although body temperature seems to limit the upper boundary for the spontaneous rates.

In the vestibular system the spontaneous discharge is quite regular. Precht (1974) reports that in the freely moving cat the rate depends on the level of wakefulness. In awake cats cells in Deiters nucleus and in the superior nucleus fire on average 40 spikes per second. In the vestibular nerve the resting discharge is between 2 and 30 spikes per second.

Somatosensory receptors (e.g., the cutaneous mechanoreceptors) are specialized nerve endings and their resting discharge rate rarely exceeds 1–2 spikes per second (type I receptors) or can occasionally reach 20 spikes per second (type II receptors), which again indicates the role of transmitter release in the receptors (Burgess and Perl 1973). The somatosensory thalamus can exhibit spontaneous rates of up to 50 spikes per second.

Thus, with the notable exception of the somatosensory system, the spontaneous rate in general decreases when going from sensory periphery to sensory cortex.

2.2 Quantification and Representation

Since spontaneous activity is a stochastic process, a description by mean rate alone is, except in case of a Poisson process, insufficient and a quantification by means of interval densities has come into use. An estimator for the interval density is the interval histogram. The interval density is defined as:

$$I(t) = \lim_{\Delta t \to 0} \frac{1}{\Delta t} \text{Prob}\{\text{spike in } (t, t+\Delta t) \mid \text{previous spike at zero}\} .$$

The first use of the interval histogram in this field was made by Rodieck et al. (1962) and it appeared that various well-known distributions could be fitted to interval histograms in, e.g., the cochlear nucleus or the retinal ganglion cells. The basic problem of those days was which probability distribution would provide the best fit; this estimation had implications for the deduction of the underlying process.

The most simple statistical process is the Poisson process for which events are indistinguishable and independent. In the case of a homogeneous Poisson process, the interval probability density function (pdf) is exponential in shape. The inverse conclusion, namely that when the interval pdf is exponential therefore the process is Poisson, is not justified; it also needs to be shown that the intervals are independent. A graphical way of showing this is through the joint-interval histogram, which is an estimator of the joint interval density:

$$J(t_1, t_2) = \lim_{\Delta t \to 0} \frac{1}{\Delta t} \text{Prob}\{\text{spike interval in } (t_2, t_2+\Delta t)/\text{previous spike interval in } (t_1, t_1+\Delta t)\} .$$

When $J(t_1, t_2)(\Delta t)^2 = J(t_1)\Delta t \cdot J(t_2)\Delta t$, i.e., the probability of finding interval t_1 followed by interval t_2 is equal to the product of the interval densities then the intervals are independent. In practice this is shown by plotting the column-means vs. row and row-means vs. column; each set of means in the case of independence will fall on straight lines parallel to the axis. This is a necessary but not sufficient test of independence; a sufficient test follows from the serial correlogram or from higher-order joint-interval histograms or from evaluating the equation:

$$J(t_1, t_2) = J(t_1) \cdot J(t_2)$$

in a direct way as, e.g., in Eggermont et al. (1981). A sample of interval and joint-interval histograms is shown in Fig. 2.1. Only one of the neurons quali-

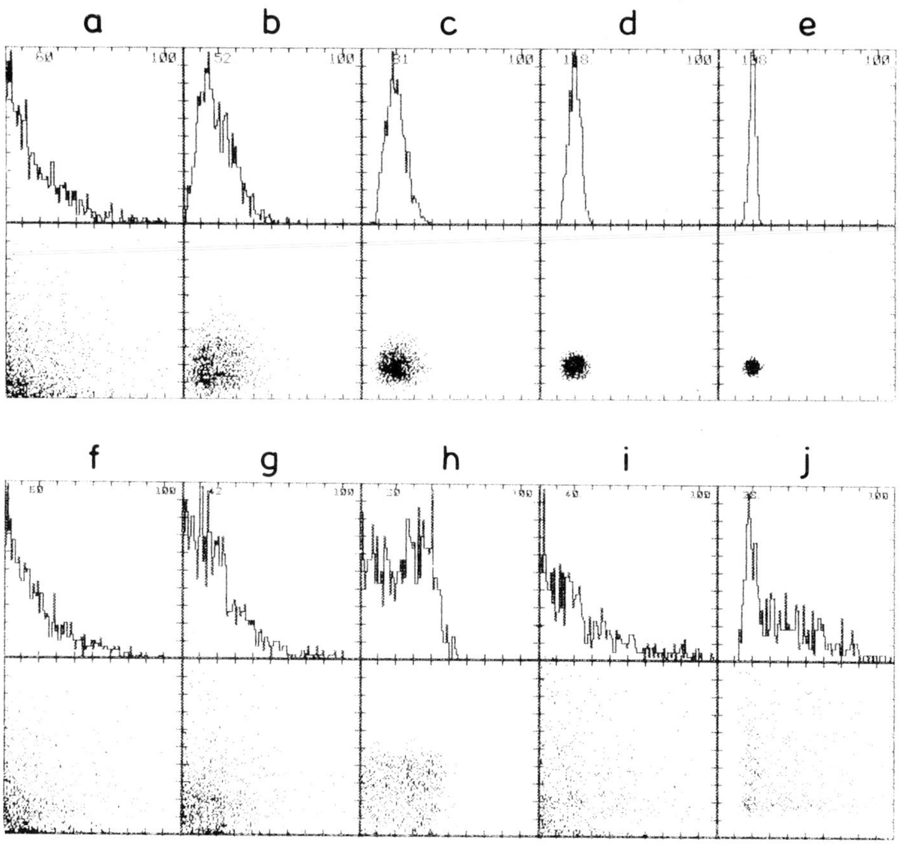

Fig. 2.1 a–j. Survey of interval histograms and joint-interval scattergrams. In **a–e** the process changes from Poisson to more regular properties. In fact all these distributions are the result of simulating a gamma-function interval distribution (see Box B) with $\alpha = 1, 3, 10, 30$, and 100. The distributions in the *lower part* (**f–j**) are more complex and resulted from simulating interacting spike trains

fies as a Poisson process, namely unit a, of which the interval histogram is exponential and the joint-interval histogram indicates independence of consecutive intervals.

When the Poisson model is applied to neural spike trains, intervals smaller than the refractory period of the cell do not occur. It has been suggested (e. g., Goldberg et al. 1964) that for certain classes of neurons the interval histogram (IH) can be described by a Poisson process with dead time R giving an interval distribution of the form

$$I(t) = \lambda\, e^{-\lambda(t-R)} \;. \tag{2.2}$$

Box B. Distribution Functions

For positive stochastic variables a variety of interval distributions can be used to fit the interval histogram. The most simple one relating to a Poisson counting process is the exponential distribution

$$I(t) = \lambda\, e^{-\lambda t} \;. \tag{B.1}$$

Related to the exponential distribution is the hyperexponential distribution, describing the time between events in a process where events are generated with probability p_1 from an exponential distribution with intensity λ_1 and with probability $(1-p_1)$ from another exponential distribution with intensity λ_2:

$$I_h(t) = p_1 \lambda_1\, e^{-\lambda_1 t} + (1-p_1)\lambda_2 e^{-\lambda_2 t} \;. \tag{B.2}$$

This is a mixture of exponential distributions and applies when processes giving rise to exponential interval distributions are operating in parallel.

When "exponential" processes are operating in series, a more complex class of interval distributions is found. These are the gamma distributions

$$\gamma(t) = \frac{\lambda^{\alpha} t^{\alpha-1}\, e^{-\lambda t}}{\Gamma(\alpha)} \,, \tag{B.3}$$

where $(\alpha-1)$ is called the order of the γ-distribution, α indicates the number of exponential processes in series. When $\alpha = 1$ an exponential distribution results, for α is an integer $\Gamma(\alpha) = (\alpha-1)!$ In case α is a natural number, the distribution is usually called the Erlang distribution and then often written as

$$E_\alpha(t) = \frac{(\alpha\mu)^{\alpha} t^{\alpha-1}\, e^{-\alpha\mu t}}{\Gamma(\alpha)} \;. \tag{B.4}$$

The Erlang distribution describes the probability of observing an interval of length t between every α^{th} element of an exponential process; the

Box B (continued)

distribution therefore results from deletions of a Poisson process. The Erlang distribution is the sum distribution of α random variables each exponentially distributed with intensity μ.

The relative spread or coefficient of variation defined as standard deviation related to the mean, for the γ-distribution is $\sqrt{\alpha}$, the mean is given by $\mu = \frac{\alpha}{\lambda}$.

The χ^2-distribution often used in statistical testing is also a special case of the γ-distribution with $\lambda = 1/2$; $t = \chi^2$ and $\alpha = 1/2\, \nu$ with ν the number of degrees of freedom.

There are other distributions that are very hard to distinguish from the gamma distribution. One of these is thc log normal distribution:

$$L(t) = \frac{1}{\sigma\sqrt{2\pi}}\, t^{-1} \exp\left\{\frac{(\ln t - \mu)^2}{2\,\sigma^2}\right\}, \qquad \text{(B.5)}$$

with μ and σ^2 the mean and variance of the normal distribution.

Another distribution resulting from the diffusion approximation of a random walk model (e. g., Cattaneo et al. 1981) is the inverse Gaussian distribution:

$$Y(t) = K\, t^{-W} \exp\left\{-\phi\left(\frac{\mu}{t} + \frac{t}{\mu}\right)\right\}. \qquad \text{(B.6)}$$

For the Wiener process $W = 3/2$ and ϕ/μ is constant.

In Fig. B.1, the distributions (B.3), (B.5), and (B.6) are shown for some parameters such as to provide the same mode. The close resemblance of all distributions illustrates that a model-free approach to deduce the underlying process is doomed to fail in nearly all cases.

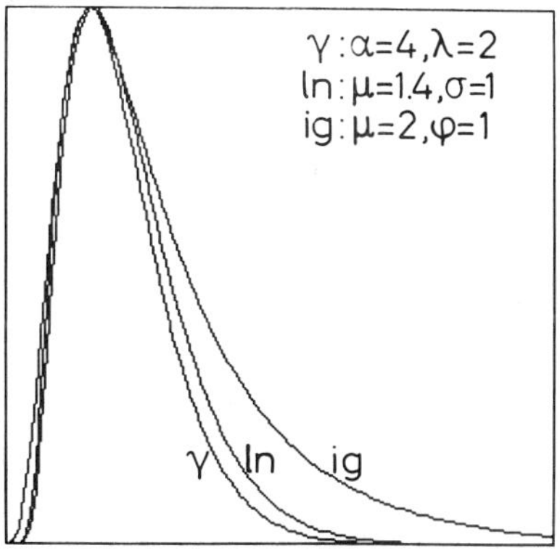

Fig. B.1. Comparison of a gamma (γ) distribution [Eq. (B3)], a log-normal (*ln*) distribution [Eq. (B5)], and an inverse gaussian (*ig*) distribution [Eq. (B6)] with identical modes. Only on basis of the tail portion is some distinction possible

The gamma distribution [Eq. (B.3)] was first used by Kuffler et al. (1957) to fit interval histograms from retinal ganglion cells. By using serial correlation coefficients it was shown that the impulses did not occur at random but that there was a slight but significant tendency for short intervals to be followed by long ones and vice versa. Therefore an interval distribution resulting form a Poisson process with a dead time was ruled out.

When α is large the gamma distribution becomes more and more symmetrical in shape and nearly Gaussian, but with a rather large relative spread. The quite regular interval distributions occurring in the vestibular nerve usually have a relative spread of only a few percent, they therefore require fits with truncated (no intervals $\leqslant 0$) Gaussian distributions.

Recently, Wise (1981) and Cattaneo et al. (1981) considered another class of distribution functions denoted as the "inverse gaussian distribution" [Eq. (B.6)]. It is argued that random walk models lead to interspike interval distributions of this kind. This type of distribution suggests a plot of log Y(t) versus log t. Wise investigated whether this type of distribution could be used to fit interval distributions for spontaneous activity from somatosensory cerebral cortex of the rabbit and respiratory neurons in the cat. It was found that in general a single curve did not fit the data and it was concluded that a single distribution was seldom present.

A property of some interval histograms is the so-called scaling invariance: the IH retains its shape when the spike rate increases by a factor a (e.g., by steady-state stimulation or temperature increase) and the interval axis is reduced by a factor a. This property is also found when second- or higher-order IH's are constructed, giving the distribution of intervals formed by joining two adjacent intervals, e.g.,

$$I_2(\tau) = I(\tau) * I(\tau) = \frac{1}{2} I\left(\frac{\tau}{2}\right) , \tag{2.3}$$

where $*$ denotes convolution. This is exactly how gamma distributions are constructed in a situation with independent intervals in the primary process, i.e., a renewal process. Distributions following Eq. (2.3) are called stable distributions of order 1/2. The inverse gaussian is stable of order 1/2 also in case the random walk includes a drift term (ϕ/μ) as shown by Gerstein and Mandelbrot (1964). A continuous version of this model was elaborated upon by Pernier and Gerin (1971), resulting in improved fits to interval distributions. Recently, more emphasis has been placed on the relation between pulse number distributions (PND) and pulse interval distributions (PID) and this actually has led to the notion that most of the descriptions based on PID's alone are basically untenable. For an exponential PID the corresponding PND might be represented by a Poisson counting process. In that case the underlying process is straightforward. In most cases however, namely in all nonrenewal processes, the PND cannot be predicted from the PID (Levine 1980). For a renewal process, Levine showed that the standard deviation for the firing rate is related to the sampling period and the mean and variance of the interval distribution:

$$\sigma_{N/T} = \sigma_{PID}\,\mu_{PID}{}^{-3/2}T^{-1/2} \,. \tag{2.4}$$

This means that $\sigma_{N/T}$ vs. log T should be a straight line with slope $-1/2$. It was evident that for the actual PND obtained the standard deviation was much lower – in the goldfish optic nerve – than predicted. The underlying process is therefore nonrenewal and rules out a gamma distribution as the proper description.

2.3 The Auditory System

In the auditory nerve nearly all fibers show spontaneous activity that can be described by a Poisson process with dead time: the interval distributions are exponential and of the form described by Eq. (2.2) with dead times ranging from 0.8–1.6 ms in the guinea pig (Manley and Robertson 1976). The intervals appeared to be independent of each other (Kiang et al. 1965); thus all necessary conditions for a Poisson process are fulfilled (cf. Fig. 2.2). For the cat (Kiang et al. 1965) and guinea pig (Manley and Robertson 1976), the mode of the interval histogram is generally 4–7 ms. Geisler et al. (1985) found for cat auditory nerve fibers that the coefficient of variation (ratio of standard deviation and mean) of the interval histogram was very close to one for mean intervals between 10 ms and 2000 ms and noticed that this was consistent with a Poisson process modified with brief refractory periods. The source of spontaneous activity is in the cochlea, since destruction thereof abolishes all spontaneous activity; destruction of only the outer haircells has no effect (Dallos and Harris 1978). The relation between terminal size and mean spontaneous rate (Liberman and Oliver 1984) suggests strongly that the synaptic release process of the inner haircells is the source of the spontaneous activity. The observation of spontaneous postsynaptic potentials in the goldfish auditory nerve dendrites (Furukawa et al. 1978) supports this. In the *cochlear nucleus* complex a distinction must be made between the ventral part (VCN) and the dorsal part (DCN), which both receive direct input from the same auditory nerve fibers, since all of these bifurcate, one branch going to the AVCN and the other branch entering through the PVCN into the DCN (Kiang 1984). Disruption of the cochlea abolishes spontaneous activity in the VCN while that in the DCN remains. There is, however, also a distinct change in the type of interval histograms found, units with modes less than 12 ms are observed only in animals with intact cochleas, while both intact and cochlear-destroyed animals show interval histograms with modes larger than 12 ms. It was also observed that the changes in spontaneous activity occur instantaneously after cochlea destruction and that after up to 77 days no change in spontaneous activity was noted (Koerber et al. 1966), unlike for second-order vestibular neurons (Precht 1974) or isolated cortex (Burns and Webb 1979). In the cochlear nucleus complex, approximately 25% of the interval histograms can be described by an exponential distribution, about 25% by a quasi-Gaussian process, 25% had bimodal interval histograms or was "scaling invariant", and the remaining 25% could not be

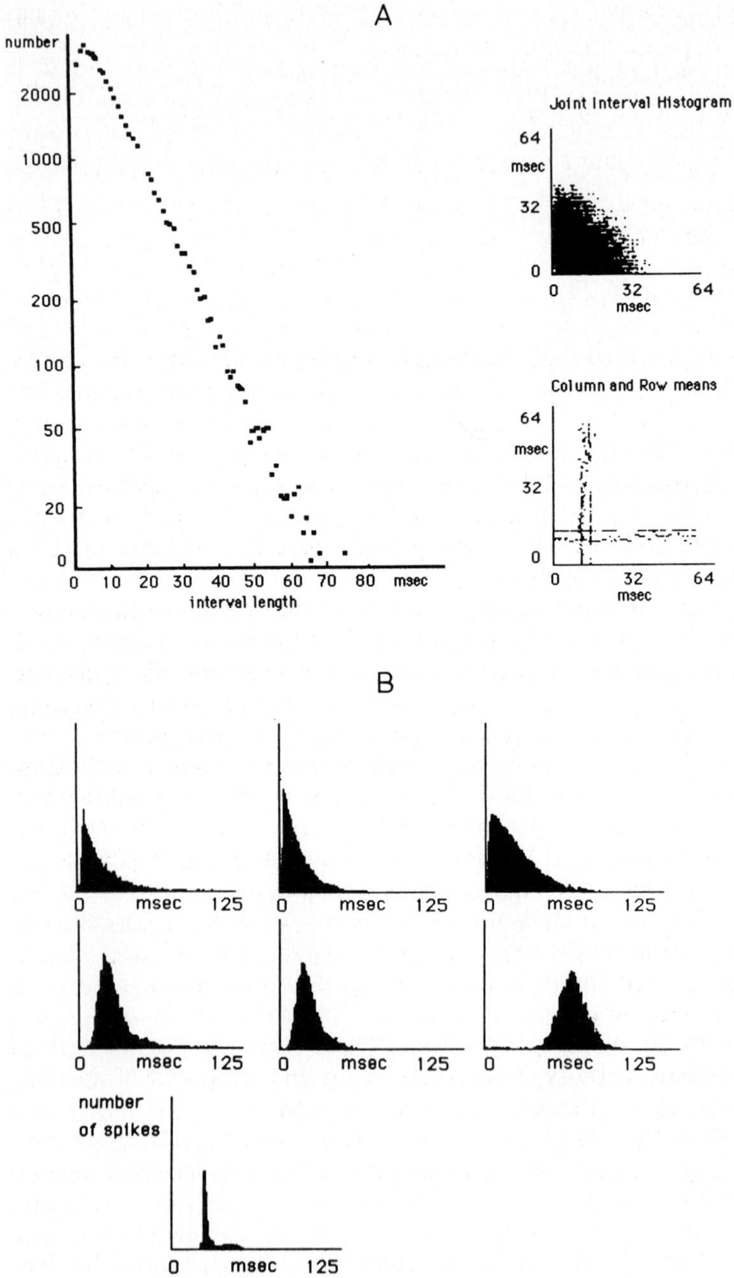

Fig. 2.2 A, B. Spontaneous activity in the auditory nerve and cochlear nucleus. The spontaneous activity of the auditory nerve is quite variable between fibers, but all interval histograms can be described by an exponential interval distribution, and show independence of subsequent intervals (**A**). In the cochlear nucleus a wide range of interval distributions are found ranging from exponential to nearly Gaussian (**B**). (After Kiang et al. 1965; Koerber et al. 1966)

classified (cf. Fig. 2.2) (Rodieck et al. 1962; Pfeiffer and Kiang 1965; Koerber et al. 1966). In an earlier study, Grossman and Viernstein (1961) reported that 30 out of 31 cochlear nucleus units showed exponential interval histograms.

In the *olivary complex* (Goldberg et al. 1964) neurons were classified according to their responsiveness to tones. One group of high discharge rate units (>150 spikes per second) showed "Poisson" properties, another group of high discharge rate units had more regular interval histograms but with serial correlation between the intervals. Low-discharge rate (<50 spikes per second and no intervals less than 25 ms) units also had regular interval histograms. Spontaneous activity for higher auditory centers and auditory cortex has been poorly described and only figures for mean discharge rate are available, although Aitkin et al. (1966) described gamma-type interval histograms for units in the medial geniculate body with mean rates between 8–20 spikes per second. For *vestibular* neurons (Walsh et al. 1972), a fairly regular discharge type is found, strongly contrasting the Poisson-like auditory nerve fiber discharges. A model was advanced linking irregular discharge to the presence of relatively few synaptic terminals on one fiber, such as for the auditory nerve, and regular discharge as due to the summed influence of many haircell synapses innervated by the same vestibular nerve fiber.

Teich and Khanna (1985) studied the mean-to-variance ratio of the number distribution for auditory nerve fibers in the cat and found that for low-, medium- and high-spontaneous rate fibers, regardless of characteristic frequency, this ratio was between 1 and 2 (for a Poisson process the ratio should be equal to 1). The fact that this ratio is larger than 1 points to more regularity than expected on the basis of a renewal process. This excludes homogeneous and inhomogeneous Poisson processes and doubly stochastic Poisson processes with dead time as a description of the spontaneous activity of the auditory nerve. In addition, it was discovered that spike pairs occurred frequently in the activity pattern; this again does not support a renewal process.

When we consider the spontaneous activity in the goldfish saccular nerve as reported by Fay (1978), it is found that this nerve is not as homogeneous as the mammalian one. Poisson-like interval histograms were found in only 19% of the cases and bimodal ones in 41% of the cases, 8% showed very regular firings and was not auditory, the remainder could not be classified. Thus in goldfish auditory nerve, properties as far as spontaneous activity is concerned are more like those for the mammalian cochlear nucleus, regardless of the fact that transmitter release in the haircell-neuron synapse was reported to follow a binomial process, hardly to be distinguished from a Poisson process (Furukawa et al. 1978).

2.4 The Visual System

In the visual system the statistical properties of *retinal ganglion cells* have been studied extensively (Kuffler et al. 1957; Hertz et al. 1964; Fusters et al. 1965; Heiss and Bornschein 1966; Ogawa et al. 1966; Rodieck 1968; Levine and

Shefner 1977) although the results are not very consistent and probably caused by differences in preparation and anesthesia. Kuffler et al. (1957) studied six units in the decerebrated cat extensively and demonstrated that none of the spike trains could be described by an exponential interval distribution but all could be fitted with gamma distributions with order parameter between 1 and 7. They found the first serial correlation coefficient to be negative, suggesting a tendency for short intervals and long intervals to alternate. In contrast, Hertz et al. (1964), studying 13 units in decerebrated cats, found all of them to have independent successive intervals and an interval distribution that could as a rule be described by an exponential distribution (cf. Fig. 2.3). Heiss and Bornschein (1966) studied 116 ganglion cells and found in 18 cases a bimodal distribution and in 29 cases a multimodal interval histogram, while the remainder had monomodal histograms. An exponential curve could be fitted through the peaks in these multimodal histograms. Ogawa et al. (1966) studied 29 units in ten cats and found seven units with a mean rate of 69 spikes per second to have an exponential interval histogram. Forty-five percent of the units could be fitted by a gamma-type distribution; these units had a mean rate of 25.8 spikes per second. The remainder showed either bimodal or multimodal interval histograms with mean rates of 38.3 and 27.2 spikes per second. In 31% of the units, and increased tendency to repetitive firing under barbiturate anesthesia was noted. Rodieck (1968) found a linear relation between mean interval and standard deviation of the intervals with slope of unity, which argues for a Poisson process; from the intercept a dead time of around 20 ms was estimated. Levine and Shefner (1977) recorded from 36 different-ganglion cells in the goldfish and observed no distribution approximating an exponential distribution; approximately 50% of the cells observed displayed multimodal interval histograms. They explained this multimodal distribution on the basis of independent inputs to control the generation of spikes. On the basis of the finding that the coefficient of variation remained constant under conditions that changed the firing rate (e.g., stimulation), they rule out gamma distributions as adequately describing the interval distributions. On the basis of a model that makes the variability of the intervals intrinsic to the ganglion cell and conceived as a membrane leakage resistance that has normally distributed values, they derive a so-called hyperbolic normal distribution for the intervals t:

$$H(t) = \frac{1}{\beta\sqrt{2\pi}}\, t^{-2} \exp\left[-\frac{(\alpha - t^{-1})^2}{2\beta^2}\right], \tag{2.5}$$

where α and β are related to membrane resistance mean and variance and to the longitudinal axon resistance. This distribution could very well be fitted to all the unimodal interval histograms while weighted sums [as in Eq. (B.2)] were fitted with the bi- and multimodal distributions. Smith (1979) argued that this distribution has nonexisting first and second moments, and proposes a slight modification based on gamma-distributed changes in the membrane leakage resistance that resulted in a slightly modified "hyperbolic" gamma distribution

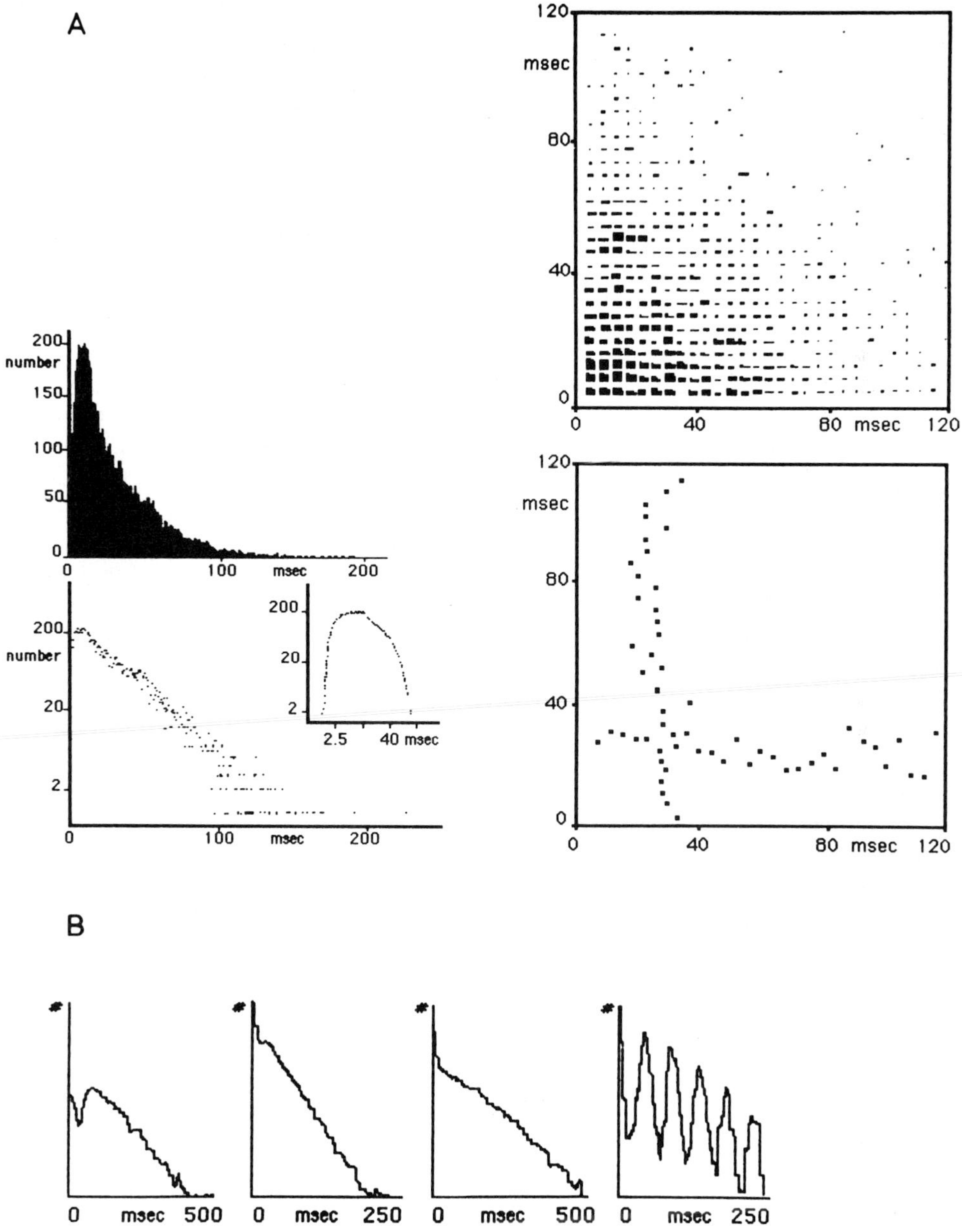

Fig. 2.3 A, B. Spontaneous activity in the optic nerve, and lateral geniculate nucleus (LGN). In this particular example the optic nerve fiber discharges resemble those of a Poisson process: negative exponential interval distribution and independence of subsequent intervals (**A**) (After Hertz et al. 1964). Other investigators find more gamma-like distributions as well as bimodal distributions (see text). In the LGN quite complex interval histograms are found, mostly bimodal or multimodal (**B**). Note hat all *vertical axes* are on a logarithmic scale

$$G(t) = \frac{b^{-\alpha} t^{-(\alpha-1)} e^{-1/bt}}{\Gamma(\alpha)} , \qquad (2.6)$$

that fits the interval histograms equally well and has the desired mathematical existence properties. This once more indicates that many distributions can be fitted to the same interval histogram. Some, however, are more justifiable than others, especially when a convincing model generates them. It is noted that Levine and Shefner's model in the limit approaches random walk models and then predicts an interval histogram best fitted by a gamma distribution of an order equal to the rate of change of the noise variable over the firing rate. This model assumes that the variability in the ganglion cell multiplies a steady input signal, thereby discarding any effect left of noisy transmitter release peripheral to the ganglion cell. Barlow and Levick (1969) review evidence for statements that the variability is extrinsic to the receptor cells and conclude that "there is no evidence against the view that maintained activity arises in the receptors". The isomerization process as a source was rejected by Frishman and Levine (1983), who instead propose – for the cat – a distal noise source proximal to the low-pass filtering of the receptor bipolar synapse. In this paper they also reject the possibility that the spike-generating mechanism is a source of the variability. This then basically leaves two sources for the variability in the interspike intervals of retinal ganglion cells: the receptor (but not the isomerization process and therefore most likely the transmitter release) and the membrane noise in the ganglion cell. A similar model for the auditory nerve fiber has been described by Eggermont (1973).

In the *rostral superior colliculus* (Radil-Weiss et al. 1968; Skvaril et al. 1971) about 1/3 of the 31 neurons studied had an exponential type interval histogram, five were Gaussian in shape, only one histogram could be fitted with a gamma distribution, one was bimodal, and the others were left unclassified.

For the *lateral geniculate nucleus* more data are available. Levick and Williams (1964) studied 233 single units in the LGN and optic radiation and found stable patterns, irregularity, as well as regularly unstable patterns. Interruption of the retinal activity caused in 61% of the units a powerful suppression of the dark discharge in LGN units. Bishop et al. (1964) distinguished four types of interval histogram on the basis of the properties of the long intervals, and two types based on the properties of the short intervals. Long and short intervals are distinguished by a well-marked depression in the interval histogram. All combinations of the various short and long interval types were found to occur, and no simple interval histograms were found. Eysel and Gaedt (1971) basically confirmed the findings of Bishop et al., but in addition observed an L-type histogram showing a very short latency mode and also a wide distribution of longer intervals although with very low frequency of occurrence. They also studied joint probability histograms and found that comparable bimodal interval histograms from optic tract and LGN differed in their joint properties. In the optic tract short and long intervals most likely alternate, while in the LGN especially short intervals tend to follow each other more frequently ("bursts"). There was also a higher probability for optic tract neurons that the longer intervals followed each other than in the LGN. Overall, however, a study of the transition probabilities indicated that also in LGN neurons alternations of short and long intervals regularly occur.

On the basis of an assumed gamma distribution for the intervals of ganglion cells and the independence thereof Bishop et al. (1964) considered two models to account for the observed interval histograms in the LGN. The first model assumed excitatory convergence of gamma-distributed independent inputs, the second model assumes inhibitory deletions from gamma-distributed inputs. We will deal with these models in more detail in Chapter 4. Hertz et al. (1964) found most of their 25 LGN units to possess transition types between the exponential and the bi- and trimodal distributions (cf. Fig. 2.3). Arnett (1975) distinguished LGN units with burst components from optic tract neurons which were burst free, although for both types of units rhythmically discharging cells were also found. Munemori et al. (1984) studied 47 LGN cells in detail and classified the spontaneous discharges into three main groups on the basis of the shape of the interval histogram and the autocorrelation function. They arrived at the gamma cell whose intervals fit a gamma-type distribution and whose autocorrelation function has some periodic properties which dampen within a time less than 10 ms. The second type was called the burst cell, whose interval histogram has a peak at interval value < 8 ms and of which the autocorrelation function displayed several large peaks. The third type, the multimodal cell, has a complex interval histogram shape with three or more peaks and an autocorrelation function with periodic properties. Gamma cells were found in laminae A 1 and C both for X- and Y-cells, the other cell types were observed in laminae A and A 1. Generally the burst cells were Y-cells and the multimodal cells were X-cells. Nakahama et al. (1971, 1972) studied the statistics of LGN interval distributions with a variety of techniques: serial correlation coefficients, coefficient of variation, joint-interval histogram, χ^2-test for Markov processes and Shannon's information entropy (see Box A). They found virtually no or very small negative serial correlation, coefficients of variation which were very close to 1/2, statistical independence of neighboring intervals both from joint-interval histograms and the χ^2-test, while also information entropy measures indicated that the state of a neuron depended little on its past activity.

A severe problem in studying thalamic neuron properties, however, is posed by their dependence on the state of rest of the brain. The dark discharge of LGN cells is on average greater during REM sleep than during slow-wave sleep and also greater than in the aroused state (Levick 1973).

In the *striate cortex* the maintained discharge of neurons does not have the spike clustering typical of LGN cells during slow-wave sleep. Horn (1962) found in the unanesthetized unrestrained cat very regularly firing cells having a mean interval of 9.42 s. Griffith and Horn (1966) analyzed 69 sections of spontaneous activity (each consisting of 2000 spikes) from 52 cells. In 43 sections the running means test resulted in stationarity of firing. In all these cells the interval histograms deviated significantly from that expected for an exponential one (cf. Fig. 2.3). However, if a less strict criterion was applied, they found that eight cases could be adequately described by a Poisson distribution. As the simplest explanation for the nonstationary cases, they proposed an inhomogeneous Poisson process. Hertz et al. (1964) studied 145 units and characterized most of them by either a bimodal or trimodal distribution, and found that continuous stimulation had hardly any influence. They found this to confirm the idea that the specific afference has only "perturbing" influence upon the activity of the cells. However, after blocking the mesencephalic reticular formation, they found that the retinal input determines the cortical activity. Levick (1973) indicated that most visual cortical neurons have monomodal interval histograms. A more differentiated view is given by Sanseverino et al. (1973), who found in the striate cortex units that displayed a sequence of very irregularly occurring impulses with a mean rate of 6.77 spikes per second and an interval histogram resembling an exponential. In areas 18 and 19 they found neural elements discharging with bursts of spikes (doublets, triplets, or quadruplets) separated by highly variable periods of silence or very low activity. The interval histograms resembled a bimodal one with exponential decay but mostly the L-shaped type previously described for the LGN by Eysel and Gaedt (1971). Average firing rates in A18 was 7.8 spikes per second, and only 2.76 spikes per second in A19. They found for the 134 units studied that grouped activity was not associated with that observed in sleep in LGN. Cattaneo et al. (1981) used an inverse Gaussian distribution [cf. Eq. (B.6)] to fit the interval histograms for simple cells. For complex cells they found clusters of spikes to occur during maintained stimulation; the intervals of spikes in the clusters did not change with stimulation although the frequency of occurrence did. The short-interval peak (2–4 ms) in the interval histogram distinguishes the complex from the simple cells.

More evidence of spike bursts in the striate cortex of awake cats comes from the chronic recordings of Legéndy and Salcman (1985). They recorded simultaneously from up to five fine wire electrodes implanted in the striate cortex. Burst activity was found in the majority of cells, bursts containing typically 10–50 spikes. It appeared that bursts were generally temporally close for different units, especially between pairs of neurons recorded by the same electrode. In addition, they found evidence for recurring spike patterns across multi-units, which suggests an extension of the single unit temporal-pattern procedures developed by Dayhoff and Gerstein (1983a, b).

2.5 Other Brain Areas

For the *somatosensory* system only a few detailed studies of spontaneous activity are available. We have already mentioned that primary cutaneous sensory fibers have a resting discharge of only 1–2 spikes per second. Amassian et al. (1964) found in the paralyzed cat that most neurons in the dorsal column had either quasi-Gaussian interval histograms with a mean around 18 ms or slightly bimodal interval histograms with modes around 40 and 50 ms. Interval histograms in the cuneate nucleus were found to be bimodal with a short mode less than 4 ms and a long mode around 20–30 ms, very similar to those of neurons in the lateral geniculate nucleus. In the dorsal spino-cerebellar tract Knox et al. (1977) observed in anesthetized cats that 52% of the neurons had quasi-Gaussian interval histograms, the remainder had higher-order gamma distributions. Mean rates were 3–48 spikes per second. Only occasionally was a bimodal distribution observed.

In the medullary-, pontine-, as well as mesencephalic region three basic patterns of spontaneous activity are found: a low-rate irregular discharge, more regular continuous discharges at rates of 50–100 spikes per second, and bursts of unit discharges with rates in the burst of 50–70 spikes per second up to 500 spikes per second occasionally and separated by silent intervals (Pompeiano 1973).

In the *thalamus* of the macaque monkey, Werner and Mountcastle (1963) found for 15 spontaneously discharging units that the slope of the regression line between standard deviation and mean firing rate was between 1.20 and 1.96, but with intercept values ranging from −12 to −51 ms. This suggests a varying coefficient of variation depending among others on the mean rate. Poggio and Viernstein (1964) found for the ventrobasal thalamus in the unanesthetized monkey in only 2 out of 119 samples from 26 neurons evidence for Poisson properties of the spike train. In 16 of 20 cells a positive serial correlation coefficient was found and the process could not be described by a simple Markov process. Mostly very periodic autocorrelation functions were found, suggestive of bi- and multimodal interval histograms. The authors suggested that the periodic character of these variations reflected the neural network in which the neurons participated. In the nonspecific thalamus it was found that 70% of the neurons had an exponential interval histogram and that 20% could be described with a gamma distribution (Radil-Weiss et al. 1968; Skvaril et al. 1971). These conclusions, however, were based upon graphical considerations only. The same authors studied spontaneous activity in the caudate nucleus and found 60% exponential distributions, 15% gamma distributions and about 5% quasi-Gaussian distributions.

In the lateral thalamus of humans, Bates (1973) describes spontaneous activity with rates of 1–5 per second, bursts occur with rates of 1–10 per second with spike rate in the burst as high as 200 per second. It was noted that bursts could come and go without any apparent alterations in the patient's state. This takes in the effects of the state of sleep or wakefulness upon the level and type of spontaneous activity. Nakahama et al. (1966) considered the firing in the ventrobasal thalamus in immobilized cats as a simple Markov process and defined on that basis a statistical dependency between adjacent intervals, defined as

$$D = 1 - \frac{H_1}{H_0}, \tag{2.7}$$

where H_1 is the information entropy of the transition matrix and H_0 is the information entropy of the probabilities calculated for the various interval classes. Note that H_1 is the same as $H(Y \mid X)$ in Eq. (A.3), H_0 is equivalent to $H(Y)$, and that D is actually a normalized version of T [Eq. (A.4)]: the information transmitted by the message. For an n-dimensional normal distribution $H_0 = \ln\{(2\pi e)^{n/2} |a_{ij}|^{1/2}\}$ and $H_1 = {}^{e}\ln[(2\pi e)|a_{ij}|/\sigma]$ where $|a_{ij}|$ is the determinant of the covariance matrix and σ^2 is the variance (Nakahama et al. 1971). For interspike interval class values τ, it was observed that for 16 out of 22 neurons a power relation of the form

$$D = \left(\frac{\tau}{a}\right)^{-b} \tag{2.8}$$

was found. Surprisingly, arousal or anesthesia had no influence upon a (the dead time of the dependency) but only upon b: arousal caused b to decrease and anesthesia caused b to increase. It was found that $0.23 < b < 1.02$ and it was concluded that the orderliness of the spike train could not be totally due to the sequential structure thereof. In addition it was found that b was linearly related to the mean firing rate both for a group of neurons and for individual neurons during various states (Nakahama et al. 1966) of anesthesia or arousal. In a subsequent paper, Nakahama et al. (1971) indicated that the difference between the D value for the original spike train and the shuffled spike train constitutes a measure of statistical independence of intervals. If these D-values differ significantly the intervals are not independent. In addition it was found, using various measures to test for independence of intervals, that neurons in the mesencephalic reticular formation and in the red nucleus showed positive first-order serial correlation coefficients, had a coefficient of variation larger than 1/2 and were of 2nd to 4th order Markov type indicating that the state of these neurons depended highly on their past activity (Nakahama et al. 1972). This was associated with the presumed task of these neurons: to regulate the general state of the brain and to associate with posture respectively.

Dormont (1972) compared 46 cells in the ventro-lateral thalamus in the awake cat for slow-wave and fast-wave cortical activity. It was found that the pattern of spontaneous activity of a given cell is correlated with the state of the electrocorticogram. Rhythmic burst activity was found only if the level of anesthesia was not too deep; during deep anesthesia no spontaneous activity is present. During fast-wave activity the underlying process was nearly renewal, indicating that the various inhibitory feedback mechanisms that are assumed during the slow-wave period, producing more or less rhythmical burst discharges, are rather inoperative during fast-wave activity. It was found that during fast-wave activity the interval histograms could be fitted with gamma distributions.

Spontaneous activity of *cortical neurons* is also very much dependent on the state of the animal. Evarts et al. (1962) noticed that 153 out of 242 neurons recorded form in visual cortex, brainstem and suprasylvian cortex of unrestrained unanesthetized cats had lower spontaneous rates during waking than during sleep. To be more precise, it was found that the median values were lower in the waking state, but no significant difference in the mean discharge rates could be demonstrated. Evarts et al. also noticed that the population variance was larger during waking than during sleep and they suggested that these states differ especially with respect to the organization of spontaneous activity. In contrast to these observations, Murata and Kameda (1963) reported for the same preparation that most of the cortical neurons increased their rate of firing following arousal. These observations were performed for more than 300 neurons in auditory, visual, and motor cortices.

In the isolated forebrain of the unanesthetized cat Smith and Smith (1965) studied a population of units with spontaneous activity of at least one spike per second. For cells in the median suprasylvian gyrus the interval histogram was double exponential [i.e., a hyperexponential distribution, see Eq. (B.2)]. In order to determine whether the separate processes generating this compound distribution could be described by Poisson processes the authors constructed "joint occurrence matrices". They distinguished intervals within bursts (short interval – short interval transitions), intervals between bursts (long interval – long interval transitions), entry to bursts (long interval – short interval transitions), and exits from bursts (short interval – long interval transitions). From the study of these transitions they concluded that they had sampled a rather uniform population of neurons and that the spike train could be viewed as a Poisson shower (burst process) gated on and off by some other stochastic process also Poisson-like in character. A corroboration of this model can be concluded from the data of Pernier and Gerin (1975) obtained for 71 neurons in the neocortex of the unanesthetized rabbit. They observed unimodal as well as bimodal interval histograms. All observed distributions could be described by a "hyper inverse Gaussian" distribution. It was noted that each cell could present several types of interval histograms even under a constant vigilance level. The individual interval distributions of the combined process could be described by the inverse Gaussian distribution [Eq. (B.6) with $W = 3/2$] quite appropiately for all 71 neurons. A second-order analysis of the intervals suggested a two-state semi-Markov model to describe the transitions from short intervals to long intervals etc. The analysis bore great resemblance to the joint occurrence matrix analysis of Smith and Smith (1965).

Motor neurons in the cat cortex (O'Brien and Fox 1969), however, have interval histograms, which vary between the Poisson and the normal distribution, that could well be fitted with a gamma function. They observed that in the more superficial layers the interval histograms were more Poisson-like and at larger cortical depths were more like normal. They advance the following model to explain this: larger cells (at greater cortical depths) require more synaptic input to produce a spike than smaller cells (in the more superficial layers) and would also provide a greater surface of integration of various synaptic inputs and be therefore less sensitive to single inputs. The finding that the Gaussian interval histogram neurons also had long duration spikes and therefore were likely to come from large cells is consistent with this suggestion.

To quantify the changes in the temporal pattern of spontaneous discharges in cat association cortex during sleep and wakefulness, Noda and Adey (1970) used the autocorrelation function. During deep sleep the autocorrelation histogram showed an early peak at approximately 20 ms, this peak was not observed in REM sleep and during wakefulness where instead a rather flat autocorrelation histogram was obtained. This suggests that only during deep sleep clustered firing is found. To further quantify the irregularity of firing the coefficient of variation for the interspike intervals was calculated. All regression lines between standard deviation and mean interval passed through the origin in the four states considered: arousal and wakefulness, deep sleep, intermediate sleep, and REM sleep. The slopes of the regression lines (coefficients of variation) were smaller in arousal and REM sleep. The means of the coefficients of variation were 1.11 in arousal, 1.09 in REM sleep, 1.27 in wakefulness, 1.47 in intermediate sleep, and 1.56 in deep sleep. This underscores the evidence for clustered firing during deep sleep as found from the autocorrelation histogram.

Instead of the gamma-function description for interval histograms of cortical neurons, several other descriptions have been advanced. Burns and Webb (1976) favor the log normal distribution [see Eq. (B.5)] which is characterized by the modal interval and a geometric standard deviation and which seems to fit better. Webb (1976a, b), using this description, observed a shortening of modal interval when the animal fell asleep and an increase of modal interval when the animal was alarmed. Their results thus corroborate the findings of Evarts et al. (1962) and contradict those of Murata and Kameda (1963). Burns and Webb (1982) further tried to quantify the changes that occurred when the animal's state changed from sleep to awake to alarmed. They noticed that the firing rate during sleep was linearly related to that during the awake state and that again the "awake" firing rate was linearly related to that in the alarmed state. This was found to hold for monkey motor cortex and cat auditory, visual, and parietal cortices. An interpretation was advanced that when the animal falls asleep there is a fixed increase in excitation coupled with an increase in inhibition since neither a change in excitation or inhibition alone could explain these variations.

2.6 Synthesis of Spontaneous Activity Phenomenology

From this survey we can deduce a general schematic description for the spontaneous activity in the sensory nervous system (Fig. 2.4). I propose two different types of spontaneous activity generators: peripheral ones whose outputs are not affected by the state of the animal, and central cortical ones that are affected by the state of the animal. In addition, both peripheral and cortical noise sources are based upon Poisson-like processes. In the periphery because the stochastic, usually Poisson-type, release of transmitter substance in the receptor-first neuron synapses, centrally because the neurons receive thousands of other independent cortical inputs. At intermediate levels in the brainstem, midbrain, and thalamus, the most complex interval histograms show up. This may arise because of the action of the local circuits that modify the incoming peripherally generated spontaneous activity, and by the centrifugal effects that signal the state of the animal.

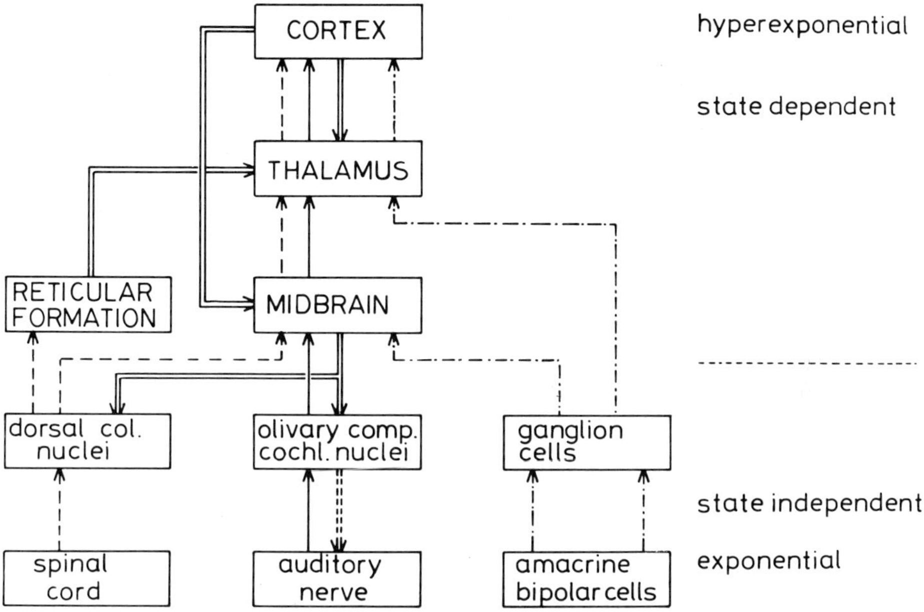

Fig. 2.4. Schematic overview of the interaction of two proposed spontaneous activity generators. Most data can be understood if one postulates a state-independent spontaneous activity generator in the peripheral receptors and a state-dependent one in the neocortex. *Dashed lines* indicate activity flow in the somatosensory system, *full lines* represent the auditory system, and *dot-dash lines* the visual system all from the periphery upwards. *Double solid lines* represent descending pathways from the cortex, and activity generated in the reticular formation. The state-independent generators are postulated to produce exponential interval distributions, the state-dependent ones are characterized by hyperexponential distributions

At the most peripheral level, the auditory nerve, as well as at the cortical level, a general distribution that fits most of the interval histograms and the occurrence of clusters is the hyper exponential distribution:

$$I_h(t) = p_1\lambda_1 \exp(-\lambda_1 t) + (1-p_1)\lambda_2 \exp(-\lambda_2 t) \;, \tag{2.9}$$

where $(1-p_1)$ is the probability of occurrence of spike clusters (pairs, triplets, etc.). For the auditory nerve this probability is very low, and for all practical purposes p_1 equals 1. To include effects of absolute refractory mechanisms t can be replaced by $(t-R)$. For the cortex p_1 depends on the state of the animal. This p_1 is high during arousal and becomes smaller during sleep and especially light anesthesia when cluster formation is very high. During deep anesthesia both λ_1 and λ_2 tend to zero. There are some additional findings that suggest p_1 higher for simple visual cells than for complex cell. At intermediate levels unimodal interval distributions can uniformly be fit by a variant of the inverse Gaussian distribution [Eq. (B.6)] which includes the variance of the membrane noise of the spike generating cell (Pernier and Gerin 1971). To include bimodal and multimodal interval histograms again hyper distributions can be used:

$$Y_h(t) = p_1 Y_1(t) + p_2 Y_2(t) + p_3 Y_3(t) + \ldots , \tag{2.10}$$

where $\Sigma_i p_i = 1$ and $Y_i(t)$ are inverse Gaussian distributions. For the brainstem level, including the cochlear nuclei, the olivary complex, the retinal ganglion cells (because they are third-order neurons) and the dorsal column nuclei in most cases p_3 and higher are close to zero: most interval histograms are uni- or bimodal. There is not much known about a possible effect of the animal's state on p_1 and p_2. At the level of the thalamus unimodal distributions are found in the auditory, ventrobasal, and nonspecific thalamus, while most activity in the LGN appears bi- or multimodal. In the LGN burst activity ($p_2 \neq 0$) seemed to be associated with Y-cells, unimodal histograms with X-cells. However, the effect of the state of the animal is fairly large: light anesthesia causes bursting activity. At the intermediate levels interruption of peripheral input is known to abolish spontaneous activity in the ventral cochlear nucleus – it did not recover within a period of over 2 months – and to markedly reduce spontaneous activity in about 60% of the cells in the LGN. The dorsal cochlear nucleus which also gets its peripheral input directly from the auditory nerve did not suffer from interruption of peripheral activity. This suggests either independent generation of spontaneous activity (DCN) or input from the more central parts of the nervous system (possibly also VCN).

The description of the spontaneous activity given here does not really pertain to the mechanisms of generation, although it assumes parallel, more or less independent inputs to explain the observed histograms. Many other and more elaborate models are advanced and we will deal with some of them, especially those to explain the findings in the thalamus, in Chapter 4.

It must be noted that some of the results obtained must be ascribed to the level of anesthesia. This is known to induce burst-like activity and hence a tendency to bimodal and multimodal interval histograms. For the most peripheral levels of the central nervous system anesthesia has no effect. e.g., on the activity of the auditory nerve. In general, recordings from the cortex require only light anesthesia or are performed in unanesthetized, restrained animals. At both levels the spontaneous activity mostly has monomodal interval distributions. Recordings from brainstem and midbrain structures always require rather deep anesthesia and one wonders if this could not account for the majority of complex findings in these structures.

3 Receptive Fields

3.1 Receptotopic and Nonreceptotopic Receptive Fields

Originally (Hartline 1938) the receptive field (RF) was defined as "the region of the retina which must be illuminated to obtain a response of a given fiber. The location of the receptive field is fixed, its extent however depends upon the intensity and size of the spot of light used to explore it, and upon the condition of adaptation; these factors must, therefore, be specified in identifying it". A somewhat similar concept was defined earlier (Adrian et al. 1931) but named the receptive area, comprising "the cutaneous area from which a single mechanoreceptor can be stimulated". Both these receptive field concepts actually describe what is now known as the excitatory RF center. Mechanoreceptive units have only excitatory RF's, sometimes consisting of multiple excitatory zones distributed on circular or oval areas with well-defined boundaries as in the rapidly adapting and type-I slowly adapting units. More gradually changing sensitivity is found for the RF's of the Pacinian corpuscles and the type-II slowly adapting mechanoreceptors (e.g., Johansson 1978; Ray and Krüger 1983). Visual RF's measured for retinal ganglion cells are more complex, they combine an excitatory or inhibitory center with an antagonistic surround. The RF's are concentric. To encompass this, Kuffler (1953) extended the receptive field concept to "all areas of the receptor surface in functional connection with a ganglion cell". A still more general definition was given by Sejnowski (1976a) as "the receptive field of a neuron is the set of sensory receptors which significantly affects its rate of firing".

Both in the visual and somatosensory system the receptive fields are of spatial dimensions. Cutaneous receptive fields are areas on the skin in one-to-one relationship to the outside world that affects it. Visual receptive fields are areas on the retina arising from projection of the outside world: scenes that are seen under a constant angle activate the same receptive fields. Receptive fields therefore are defined upon the receptor surface. Using the definition of Sejnowski, the receptive field concept can be extended to the auditory system. In this case the receptors are the haircells and the receptor surface is the cochlear partition of the inner ear in case of mammals. The cochlear partition codes sound frequency in a tonotopic way: high frequencies near the stapes and lower frequencies progressively away from the stapes. The receptive field is one-dimensional: distance from the stapes on which sound frequency is mapped in a logarithmic way. Auditory receptotopic receptive fields therefore are defined in the frequency domain, and although they are basically one-dimensional, in

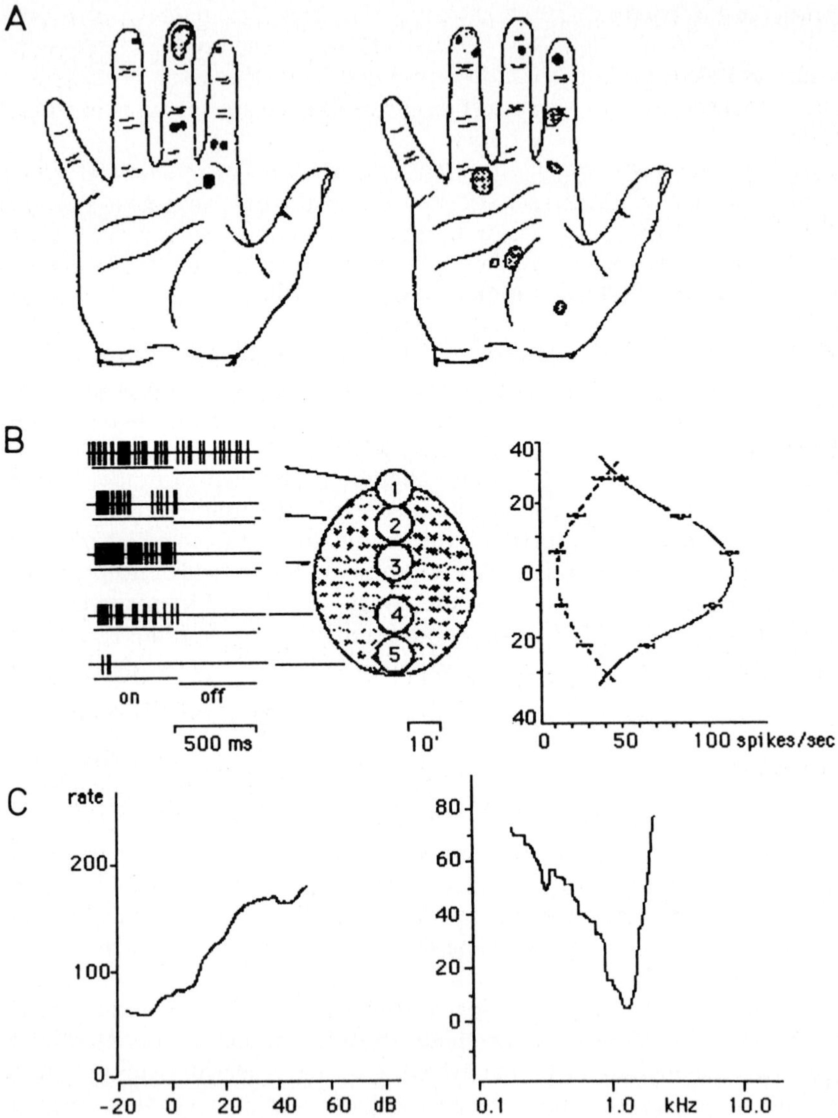

Fig. 3.1 A–C. Receptive fields for the somatosensory (**A**), the visual (**B**), and the auditory system (**C**). In the somatosensory system receptive fields of cutaneous neurons corespond to a small part of, e.g., skin surface on the palm of the hand (after Johansson 1978). In the visual system (**B**) small areas on the retina constitute the result of a projection of the receptive field. In the auditory system (**C**) it is the extent of the activation pattern along the cochlea that determines the receptive field. The *left hand side* describes the rate-intensity function of the nerve fiber of which the tuning curve is shown on the *right*. (Liberman and Kiang 1978)

accordance with Hartline's definition their extent depends on sound intensity, their borders have been named tuning curves (Kiang et al. 1965) or frequency-threshold curves (Evans 1975). The area enclosed by the tuning curve is generally called the response area. For auditory nerve fibers response areas to single frequency stimuli are always excitatory.

Thus in the somatosensory, visual, and auditory sensory systems RF's are based on the extent to which the receptor surface influences activity in sensory nerve fibers. The RF may represent spatial dimensions which are either coincident with the action of the environmental stimulus such as in cutaneous receptors, or are based on a transformation or image of a scene in the outside world such as in the retina. The RF may also represent spectral dimensions as for the cochlea. Receptotopic projections of these receptor surfaces upon higher centers in the nervous system have been described extensively (e.g., Woolsey 1982). These projections are termed somatotopic, retinotopic, and cochleotopic (or tonotopic). An overview of receptive fields is shown in Fig. 3.1

Box C. Conformal Mapping

The mappings we are concerned with transform a two-dimensional surface $\{z\}$ into a two dimensional surface $\{w\}$, where the surface can be described using complex variables $z = x+iy$ and $w = u + iv$. A mapping is conformal in a point if it preserves the angle between two differentiable arcs passing through that point.

At a point in which a function $w = u+iv$ is analytic, its real part $u = u(x, y)$ and imaginary part $v = v(x, y)$ are connected by the Cauchy-Riemann differential equations:

$$\frac{\partial u}{\partial x} = \frac{\partial v}{\partial y}, \quad \frac{\partial u}{\partial y} = -\frac{\partial v}{\partial x}. \tag{C.1}$$

The mapping from (x, y) to (u, v) is conformal, in fact any analytic function of a complex variable represents a conformal mapping with the exception of points where $\frac{dw}{dz} = 0$. The magnification factor of the mapping is given by $\left|\frac{dw}{dz}\right|$.

We will consider two basic types of mapping: the linear mapping and the exponential mapping.

A *linear* fractional mapping $w = \frac{az+b}{cz+d}$ with $ad-bc \neq 0$ transforms all circles in the z-plane into circles in the w-plane. When $c = 0$ a simple linear transformation is obtained. The inverse of a linear transformation is also a linear transformation, as is the succession of two (or more) of them.

Box C (continued)

An *exponential* mapping $w = e^z$ has the property that $u^2 + v^2 = e^x$ (equation of a circle) and that $\frac{v}{u} = \tan y$ (equation of lines through the origin). It transforms lines that are parallel to the imaginary axis in the z-plane into circles around the origin, and lines that are parallel to the real axis in the z-plane to rays emanating from the origin in the w-plane. All inclined straight lines in the z-plane are transformed into logarithmic spirals in the w-plane. The inverse mapping is called the *logarithmic* mapping; here we have to be aware of certain difficulties regarding single valuedness (Riemann surface). The scale factor of the exponential mapping is given by e^x, and that of the logarithmic mapping by $|z^{-1}|$.

Some examples and illustrations of logarithmic mappings are shown in Fig. C.1. For detailed information consult Paliouras (1975).

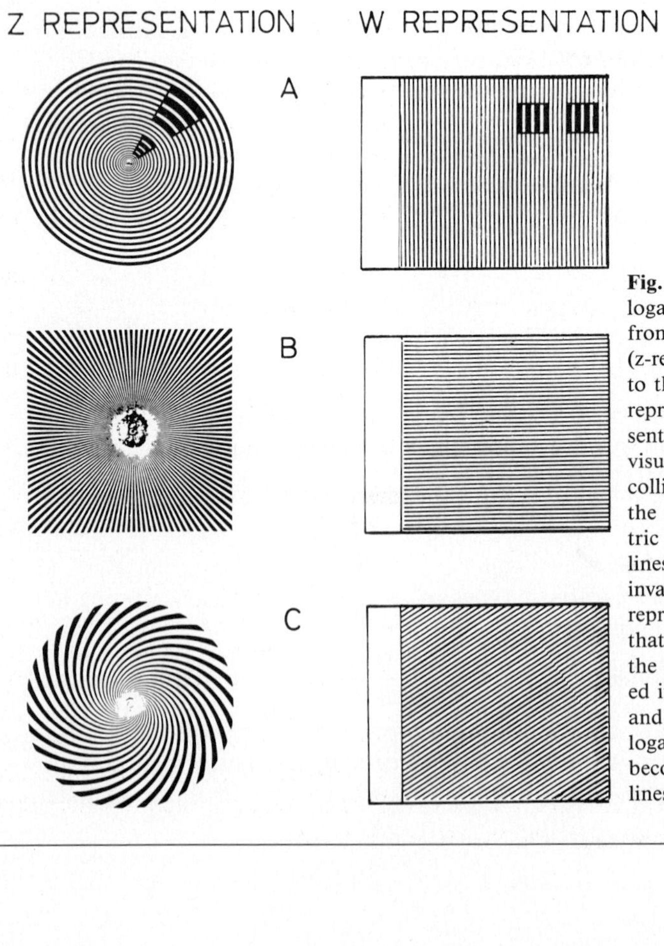

Fig. C.1 A–C. Complex logarithmic mapping from one representation (z-representation, related to the retina) to another representation (w-representation, related to visual cortex or superior colliculus). **A** represents the mapping of concentric circles into vertical lines and illustrates size invariancy of the W-representation, **B** shows that rays emanating from the center are transformed into horizontal lines, and **C** illustrates that logarithmic spirals become inclined straight lines. (Schwartz 1980)

The cochleotopic mapping can be considered as a point-to-sheet representation: isofrequency planes are found at all levels of the auditory nervous system (e.g., Aitkin et al. 1984). Results for the inferior colliculus indicate that the cochlear place is nearly linearly related to depth of penetration in the colliculus for the apical half of the cat cochlea, the more basal cochlea has a significantly larger representation (Merzenich and Reid 1974). The same slightly overrepresented basal cochlea representation is found for the auditory cortex (Merzenich et al. 1975).

Spatial receptive fields like the retinotopic ones undergo a characteristic distortion when mapped upon, e.g., the superior colliculus and the striate cortex. Schwartz (1977a, b, 1980, 1984) has described the mapping on the striate cortex as a logarithmic conformal mapping:

$$w = \ln(z+a) \ , \tag{3.1}$$

which maps foveal parts of the receptive field overrepresented at the expense of more peripheral receptive fields. One of the properties of the logarithmic conformal mapping is that a multiplication of size in the image plane – the retina – results in a translation in the projection plane – the cortex. This mapping therefore is size invariant (Fig. C.1). Such a mapping has also been shown to apply to the superior colliculus in rhesus monkey (Ottes et al. 1986). It has been implied (Schwartz 1977a) that straight lines in the somatosensory cortex are mapped onto logarithmic spirals in the cutaneous periphery. Thus the inverse mapping from receptor surface to cortex in this case is also logarithmic. The earliest suggestion for complex logarithmic functions to describe the retinocortical mapping came from Fischer (1973a) on the basis of accurate determination of receptive field sizes in the cortex as a function of retinal excentricity.

In a first approximation the cochleotopic map on the inferior colliculus and the auditory cortex can be considered as a linear mapping of length along the cochlea (measured from the stapes) upon the radial dimension in the colliculus and the tangential dimension on the cortex. Since innervation density is nearly constant along the cochlear partition, there is obviously no need for magnification of a "foveal" part as in the visual system, except for possible signs of an "acoustic fovea", e.g., in bats.

While spatial dimensions are directly coded via cutaneous receptive fields and through the retinal image via visual receptive fields, coding of auditory space requires information from both ears. Through the studies of Knudsen and colleges (e.g., Knudsen and Konishi 1978), it has become clear that in the auditory midbrain of the barn owl an orderly, representation of auditory space is found. The MLD of the barn owl consists of a tonotopic region in which for each frequency binaural information is compared (in all probability interaural time and spectral differences). This information is combined in the space mapped region of the MLD – which does not have a tonotopic organization – to form an orderly arrangement of spatial receptive fields. In addition these spatial receptive fields are transferred to the optic tectum where they are brought in register with the visual space maps.

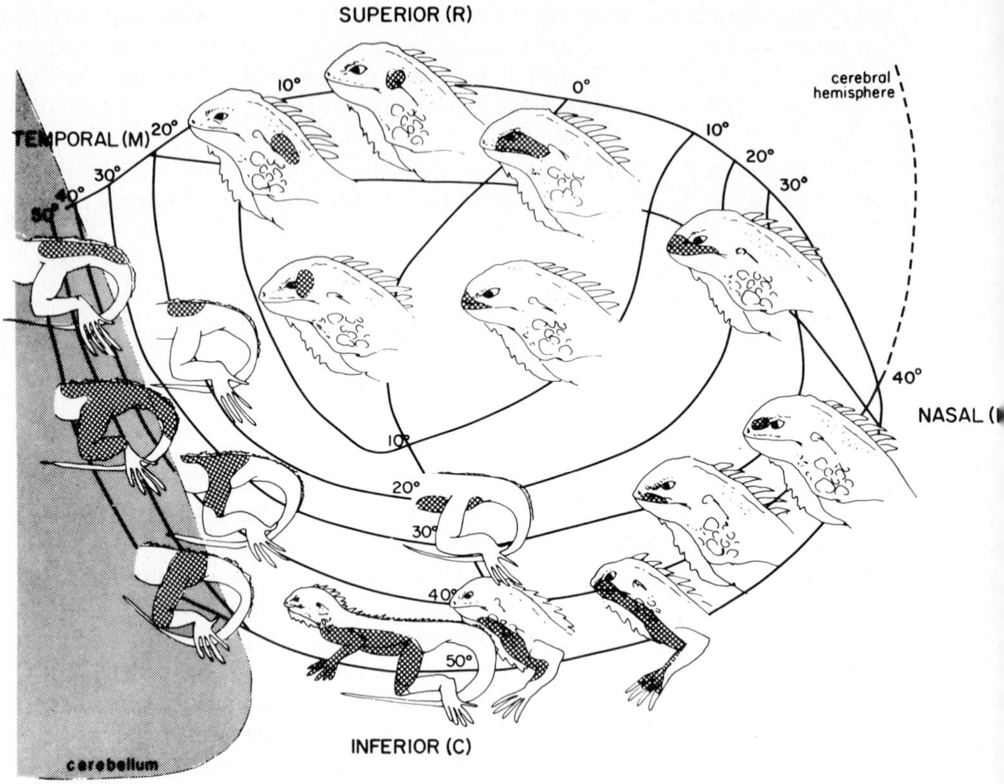

Fig. 3.2. Visual and cutaneous maps in the superior colliculus are in register. This figure shows the size and location of the cutaneous receptive fields mapped onto the visual hemi-field. (Stein and Gaither 1981)

This combination of various sensory maps of space in the optic tectum or superior colliculus is an important characteristic. The maps exist independently but are in register (e.g., Stein et al. 1976; Stein and Gaither 1981). This holds especially for the visual and somatosensory system (Fig. 3.2) and also for the visual and the auditory system in animals where the eyes do not move with respect to the head, as in the barn owl. In monkeys, which perform saccadic eye movements, the auditory and visual receptive fields stay in register regardless of the eye position (Jay and Sparks 1984), a situation which exists in the cat only when the eyes are in the central position (Harris et al. 1980).

This being in register of visual and auditory spatial receptive fields requires that auditory space is also mapped in a way comparable to the mapping of the visual field, i.e., in a conformal logarithmic way. It essentially requires that the azimuth is logarithmically transformed or mapped onto the superior colliculus surface. Azimuth θ is basically coded through interaural time differences which for a spherical head are:

$$\Delta t = \frac{r}{c} (\theta + \sin \theta) , \tag{3.2}$$

were r is head radius and c is the sound velocity. One would therefore expect to find an overrepresentation of neurons that are sensitive to small interaural time differences in those nuclei where this feature is represented. In space-coded nuclei one should expect that the frontal directions are overrepresented and that these RF's are smaller than those for more lateral directions.

What is common in these concepts of "receptive fields" is that they all are essentially static (algebraic) of nature. It is only recently that the dynamic aspects of the receptive field gained specific attention, I will return to this in Section 3.4.

3.2 Center-Surround Structure

A characteristic feature of receptive fields is that they generally have a center-surround structure which is antagonistic, i.e., an excitatory center is accompanied by an inhibitory surround and vice versa. The best-known examples are the concentric center-surround RF's of retinal ganglion cells. It is assumed that the formation of a center-surround structure is based upon the convergence of excitatory and inhibitory inputs on the retinal ganglion cells.

Cutaneous mechanoreceptors have purely excitatory receptive fields. However, at the level of the dorsal column nuclei and the lemnisco-thalamic system these excitatory receptive field centers are surrounded by inhibitory areas (Gordon 1973). This is thought to be caused by the action of inhibitory interneurons (Schmidt 1973).

Interestingly, primary auditory nerve fibers demonstrate something as a silent inhibitory surround. When an auditory nerve fiber is stimulated with a tone at the characteristic frequency, and simultaneously a second tone of sufficient intensity is present in certain frequency regions below and above CF and adjacent to the tuning curve, it is found that the second tone inhibits the activity produced by the first. The second tone does not produce or reduce (spontaneous) activity in this auditory nerve fiber when presented alone. This phenomenon is called two-tone suppression, can also be demonstrated in intracellular recordings from inner haircells, and does not require an intact nerve supply. This center-surround phenomenon is therefore not due to neuronal interaction or convergence (e.g., Kiang 1984). True neuronal interaction type center-surround receptive fields in the auditory system have been observed in the dorsal cochlear nucleus (Evans and Nelson 1973; Voigt and Young 1980) and higher auditory centers. Also the spatial nonreceptotopic RF's in the barn-owl's auditory midbrain show a center-surround structure (Knudsen and Konishi 1978).

Taking into account that the retinal ganglion cells are third-order neurons, we tentatively conclude that neural interaction type center-surround receptive fields only arise at a sufficiently high level in the CNS as a consequence to the convergence of excitatory and inhibitory interneurons.

The center surround structure can formally be described (Rodieck 1965) as resulting from the summation of a narrow excitatory Gaussian and a wider inhibitory Gaussian:

$$\mathrm{RF}\ (\mathrm{r}) = \mathrm{g_e}\ \frac{1}{2\pi\sigma_e^2}\ \mathrm{e}^{-\mathrm{r}^2/2\sigma_e^2} - \mathrm{g_i}\ \frac{1}{2\pi\sigma_i^2}\ \mathrm{e}^{-\mathrm{r}^2/2\sigma_i^2}\ , \tag{3.3}$$

where σ_e^2 and σ_i^2 are the variances of the excitatory and inhibitory curves and g_e and g_i are adjustable weighting factors (Fig. 3.3d).

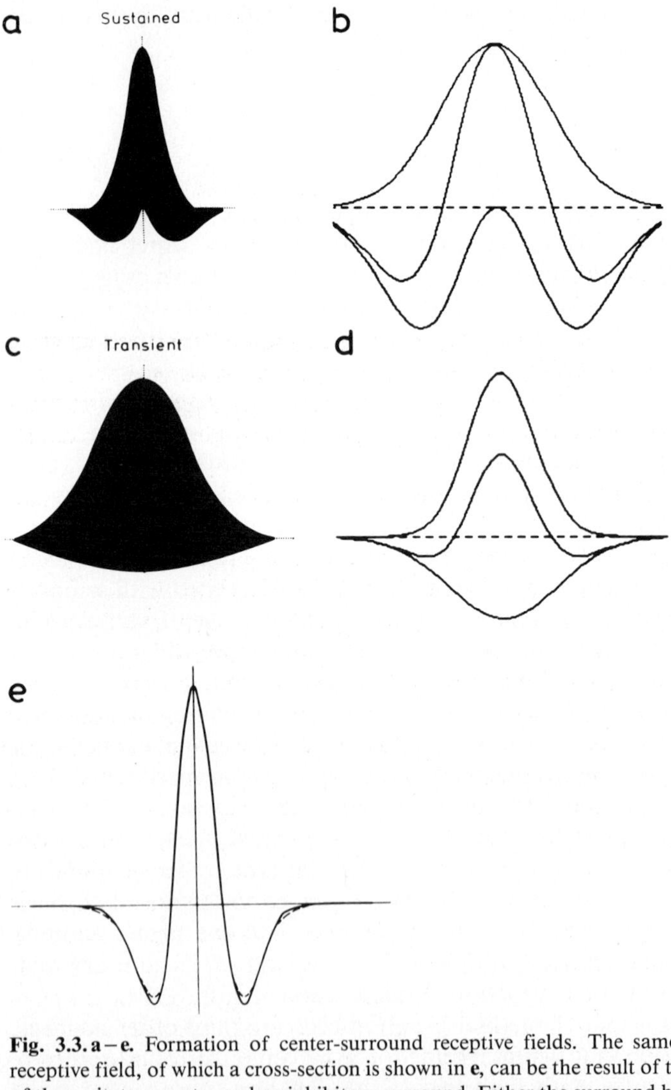

Fig. 3.3. a–e. Formation of center-surround receptive fields. The same "Mexican hat"-shaped receptive field, of which a cross-section is shown in **e**, can be the result of two different interactions of the excitatory center and an inhibitory surround. Either the surround has a "bagel-shape" (**a, b**) or is contiguous across the center (**c, d**). (After Hickey et al. 1973; Hammond 1975; Marr 1981)

An alternative description is to consider the center-surround structure as arising from taking the second spatial derivative or Laplacian of a Gaussian shaped input (e.g., Marr 1981).

$$\Delta G(r) = \frac{1}{\pi\sigma^4}\left(1 - \frac{r^2}{2\sigma^2}\right) e^{-r^2/2\sigma^2} , \tag{3.4}$$

where

$$\Delta = \frac{\partial^2}{\partial x^2} + \frac{\partial^2}{\partial y^2} , \qquad r^2 = x^2 + y^2$$

and

$$G(r) = \frac{1}{2\pi\sigma^2} e^{-r^2/2\sigma^2} .$$

Equation (3.4) may also be written as

$$\Delta G(r) = \frac{2}{\sigma^2} G(r) - \frac{r^2}{\sigma^4} G(r) , \tag{3.5}$$

which combines a Gaussian-shaped excitatory input and an annulus-shaped inhibitory effect of about the same extension (Fig. 3.3b). In both descriptions it has been tacitly assumed that excitation and inhibition have the same dynamic properties. Both descriptions can give nearly the same center-surround structure (Fig. 3.3e) as shown by Marr.

Taking the sum of a Gaussian-shaped excitatory center and a Gaussian-shaped inhibitory surround as in Eq. (3.3) assumed that the inhibitory effect extends across the entire excitatory center. The spatial derivative model [Eq. (3.5)] has the consequence that inhibition is nearly absent in the center (see Fig. 3.1). The difference between the two models can be made very small for specific selections of σ_i/σ_e (cf. Marr 1981). Evidence for the validity of both models has been given by Hickey et al. (1973) and Hammond (1975). They both assigned the sum of Gaussians model – or better difference of Gaussians – to transient cells presumably Y-type ganglion cells (Fig. 3.3c), and the spatial derivative model to the sustained firing X-type ganglion cells (Fig. 3.3a). In the latter case, Hickey et al. (1973) suggested that there was a clear absence of inhibition in the excitatory center, while Hammond (1975) merely suggested a decrease of the inhibitory effect in the center. It was suggested that the excitatory and inhibitory Gaussians had the same extension in transient cells, but that the inhibitory annulus in case of the sustained cells was more extensive than the central mechanism.

In the case of the two-tone suppression phenomenon in the cochlea basically the last model [Eq. (3.5)] or a generalization thereof is applicable when rate suppression is considered. The so-called synchrony suppression extends into the excitatory area but can be regarded as a rather trivial competitive effect between the two stimuli.

3.3 Receptive Field Changes in Central Neurons: Feature Detectors?

Peripheral auditory neurons are sharply tuned, i.e., their receptive field is small. From the level of the superior olivary complex on, broadly tuned auditory units have been found. These seem to coexist with sharply tuned cells. This situation continues up to the level of the primary auditory cortex. Nerve cells more sharply tuned than in the auditory periphery are not found in more central structures (e.g., Aitkin et al. 1984). In the frequency domain one does not observe a clear specialization for auditory neurons when going from the periphery to the auditory cortex (Fig. 3.4). A notable exception has been reported by Suga (1973), who described neurons in the auditory cortex of the bat that are "specialized" to detect the presence of very specific frequencies, i.e., those reflected by the target. Such neurons have a very narrow excitatory area sandwiched between large inhibitory areas, typically the width of the excitatory area is only 5% of the characteristic frequency even at levels 60 dB above threshold.

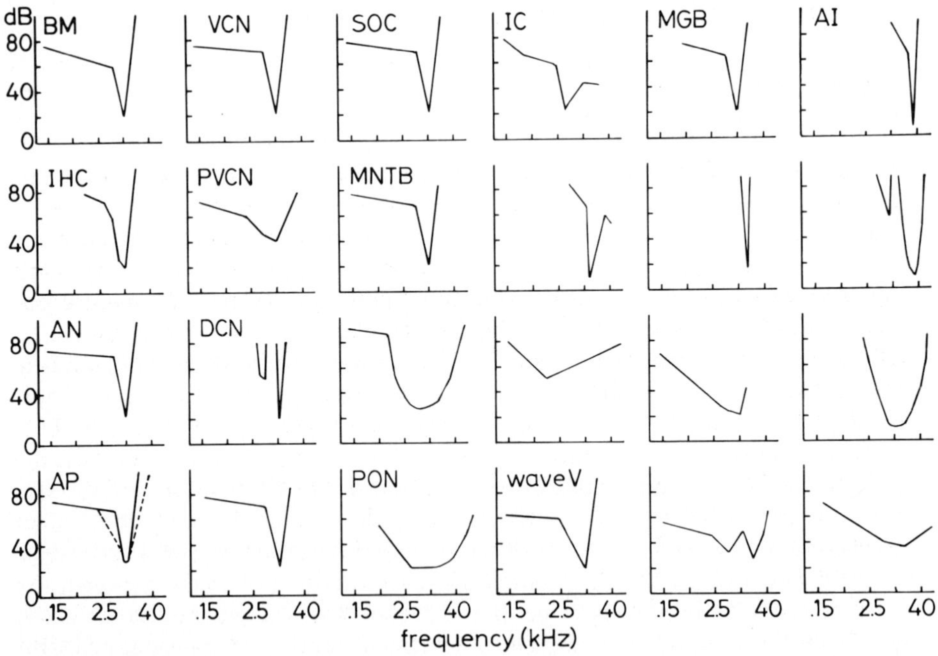

Fig. 3.4. Survey of frequency-tuning curves in the auditory nervous system. Abbreviations: first column: *BM* basilar membrane; *IHC* inner hair cell; *AN* auditory nerve fibers; *AP* compound action potential of the auditory nerve. Second column: *VCN* ventral cochlear nucleus; *PVCN* posterior VCN; *DCN* dorsal cochlear nucleus. Third column: *SOC* superior olivary complex; *MNTB* medial nucleus of the trapezoid body; *PON* pre-olivary nuclei. Fourth column: *IC* inferior colliculus; *wave V* a component of the auditory brainstem response originating mainly from the lateral lemniscus (Moller and Janetta 1981). The *two final columns* show examples from single-unit tuning curves from *MGB* medial geniculate body, and *AI* primary auditory cortex

For the visual system the concentric center-surround structure found for retinal ganglion cells is also observed in the next station, the lateral geniculate body. The size and shape of the RF remains largely the same, but a suppressive surround is added beyond the normal surround structure (e.g., Hammond 1972). Various models have been forwarded to account for this additional surround, but the most likely explanation is that it is caused by the inhibitory effect of interneurons within the lateral geniculate nucleus (e.g., Stevens and Gerstein 1976). In the striate cortex one observes elongated or "linear" receptive fields, i.e., a linear organization of the ON and OFF regions. By simple convergence of a number of LGN cells, these "simple" oriented receptive fields can be formed.

Different combinations of LGN cells results in different preferred orientations for the cortical simple cells (Hubel and Wiesel 1962). Other suggestions are that the excitatory input to a cortical cell is circularly symmetric and that intracortical lateral inhibition results in the elongated shape of the receptive fields (Creutzfeldt et al. 1974; Schiller et al. 1976). This has been formalized by Schwartz (1977a, b) by describing the required type of inhibition to occur through a unidirectional lateral inhibition, a modification of the omnidirectional lateral inhibition operator given in Eq. (3.4):

$$\Delta_\theta = \cos^2\theta \frac{\partial^2}{\partial x^2} + \sin^2\theta \frac{\partial^2}{\partial x^2} \; . \tag{3.6}$$

Besides these simple units, more complex units can be distinguished which usually have larger receptive fields and in the earlier models were thought to arise by convergence of simple cells, although evidence has been presented that complex cells receive direct input from LGN neurons. Such complex cells respond to oriented lines or edges anywhere in the visual field. One of the main problems is whether this specialization is due to strict serial processing or can to a large extent also be explained on the basis of parallel processing (e.g., Stone and Dreher 1982). In Chapter 8 we will discuss the evidence that has been obtained from simultaneous recording of several neural units and the implications this has for models that account for the increasing specialization of cells.

As a result of investigating retinal ganglion cells in anuran amphibians, Lettvin et al. (1961) caused a major change in the interpretation of cell properties in visual neurophysiology. This was a shift in attention from the image space (the neural response) to the object space (the objects causing the response). In a way the investigators started to look through the eyes of the animal. This can be called the retrospective approach (Segundo 1970), the reverse approach (e.g., Eggermont et al. 1983b), or in a more restricted sense, the *feature detection* approach. The feature detection approach is guided by considering only those parameters of stimuli that are throught to be significant for the animal in question. Thus Lettvin et al. described a certain class of retinal ganglion cells as "bug" detectors. This is basically the approach that has been advanced by Hubel and Wiesel by naming cells and cortical areas after a pronounced trigger feature such as an oriented line or edge. Although most visual cortical cells respond to a rather wide range of orientations, taking only the preferred orientation simplified the classification. In the same way auditory neurons at elevated intensities are responsive to a rather wide range of frequencies but classifying the neurons according to their characteristic frequency makes an ordering possible and meaningful. Thus the feature detector concept is useful above all as a convenient qualitative classifier, it does not necessarily imply that the brain uses it in making decisions.

In the auditory cortex of bats, Suga et al. (1981) observed neurons tuned to the distance of the reflecting target (for the emitted sonar beam), other neurons were found to be tuned to the velocity of the target relative to the animal. These trigger features represent special stimuli or stimulus properties that are relevant to the animal and evoke a large gradient in neural activity for small variations in that stimulus. In other words, the response of a feature detecting neuron is sensitive to some stimulus parameters which produce a significant change in firing rate when varied around the trigger feature, while other parameters leave the response more or less invariant.

Suga (1973) elucidated the two extremes that can form the neural basis of pattern recognition. One extreme is that the recognition of a particular sound is determined by the spatiotemporal pattern of discharges of auditory neurons at the higher levels of the central nervous system. The other extreme is that there are higher-order neurons that are specialized to respond to particular acoustic signals and that these are essential for the recognition of sound. Suga considers that speech and most animal sounds have three information-bearing elements: a constant frequency component, a noise component, and an FM component, and proceeds to show that neurons might be specialized to process these information-bearing elements. In the same view one might consider the oriented lines and edges to be information bearing elements of visual scenes. Schwartz (1977b, 1980) considers the presence of an orderly representation of such sensitive elements a consequence of the complex logarithmic mapping from retina to cortex that repeats itself on a local level. This suggests that the mapping actually limits or determines what features will be represented.

It may be more convenient for later discussions to talk about feature-extracting neurons. Feature extraction refers basically to the representation of a complex sensory stimulus by a set of information-bearing elements (features) which retain the salient information, but which discard redundant or irrelevant aspects of the stimulus. In the remainder of this book, I hope to make it clear that the unambiguous identification of a stimulus must in general be a *collective property* of the firing of many neurons, not even strictly confined to one specific brain area. This does not exclude that neurons may well be quite specialized. In the superior temporal sulcus of the macaque monkey Bruce et al. (1981), among 383 neurons tested for auditory visual and somato sensory stimuli, found a few neurons that had relatively small receptive fields and responded best to specific stimuli such as faces. A similar finding was reported by Perrett et al. (1982), who found in monkey's temporal cortex about 10% of nearly 500 single units studied that had a selective response to faces. The authors concluded from the properties of those neurons that they were part of a system specialized to code for faces or features present in faces. Damage to the equivalent region in human temporal cortex seems to produce prosopagnoisia, or difficulty in face recognition (Luria 1973; Geschwind 1979).

3.4 Temporal Properties of the Receptive Field

Auditory, somatosensory, and visual receptive field sizes depend on stimulus velocity. The speed of a stimulus passing over the retina, the velocity of a skin stroke or the rate of frequency modulation (movement of the activity pattern over the cochlear partition) influence the extent and character of the receptive field. In this respect we must differentiate between movement-sensitive neurons where moving stimuli elicit a significantly stronger response than stationary ones, and movement-specific neurons which respond only to moving stimuli and not to stationary ones. The receptive field in the latter situation can be determined only which moving stimuli. In the first case the RF size generally increases for larger stimulus velocities.

For the amphibian retinal ganglion cells the firing rate is a slowly saturating function of stimulus velocity (Grüsser and Grüsser-Cornehls 1973); a power function with an exponent between 0.5 and 0.95 generally describes the data well. The RF size seems to be roughly proportional to the firing rate elicited by the (moving) stimulus. In the cat, movement-sensitive neurons in A 19 show a preferred range of velocities, usually between 40 and 90 degrees per second. Above and below that range the firing rate of the cell decreases. In the middle suprasylvian cortex of the cat, neurons' firing rate increases for velocities up to 100 degrees per second, then stays constant, and for velocities above 500 degrees per second abruptly decreases.

In the somatosensory system the mechanoreceptors of the "hair" and field type (Burgess and Perl 1973) are already sensitive to stimulus velocity. In the auditory system FM-sensitive neurons are found in the cochlear nucleus and progressively more in higher central nuclei. FM-specific neurons, however, are first encountered at the level of the inferior colliculus and then constitute only 2% of the population. At the auditory cortex of bats, this percentage is increased up to 14% (Suga 1973). The temporal aspects can be combined with the spatial aspects of the receptive field into spatiotemporal RF's which describe the dynamic sensitivity of, e.g., visual neurons (e.g., Stevens and Gerstein 1976).

In most sensory neurons one may distinguish between transient and sustained responses to an abrupt commencing and then continuing stimulus. The transient may appear at the onset but also at the end of the stimulus. The transient may be an increase in the neuronal firing rate or a decrease, the same distinction can be made for the sustained part of the response. In the retina we distinguish as major neuron classes the sustained cells (X-cells) and the transient cells (Y-cells) which have separate projections up to the visual cortex. In the visual system most responses to temporally structured stimuli are very complex due to the interaction of the excitatory and inhibitory parts of the receptive field with their respective temporal properties (Fig. 3.5 C).

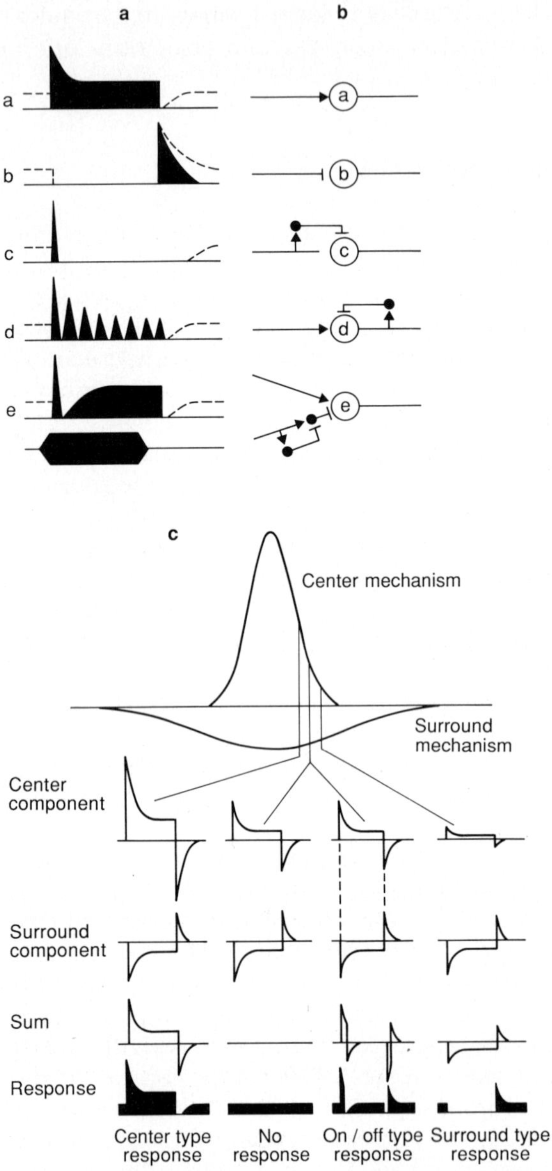

Fig. 3.5a–c. Temporal aspects of auditory **(a)** and visual **(b)** receptive fields with suggested mechanisms **(b, c)**. In the auditory system the explanations are based on temporal interactions between spike trains, in the visual system spatial properties are taken into account as well. (Suga 1973; Rodieck 1968)

In the auditory system a quite extensive range of temporal patterns are found (Fig. 3.5 A, B). The response of mammalian auditory nerve fibers is sustained but with an initial transient (part A), the sustained value is reached approximately 30–50 ms after toneburst onset. This phenomenon is probably due to both refractory properties of the auditory nerve fibers and to a much larger extent to the transmitter-receptor interaction in the synapse. After cessation of the stimulus there is a depression of the spontaneous firing rate. A comprehensive model-oriented review has been published recently (Eggermont 1985a) that relates this phenomenon to the synapse kinetics. The auditory nerve fibers innervate the cochlear nucleus complex where, besides the primary-like responses as found in the auditory nerve, four new temporal response patterns emerge (e.g., Aitkin et al. 1984). The primary-like pattern is found mainly in the anteroventral part of the cochlear nucleus and exhibited by those bushy cells that are innervated by one large end bulb. The schematic wiring diagram in Fig. 3.5 Ba is therefore appropriate. In the posterioventral part (PVCN) one observes pure onset transients (Fig. 3.5 Ac), the cells responsible for it are the octopus cells and the excitatory response is terminated by delayed inhibition from thin collaterals. In the model explanation an interneuron is interposed to provide the same effect; however, in the PVCN no interneurons exhibiting this effect are known. In the interstitial nucleus the sharp onset transient is followed by a slowly adapting sustained part. A so-called chopper-type (Fig. 3.5 Ad) is also observed in PVCN and is produced by stellate cells innervated

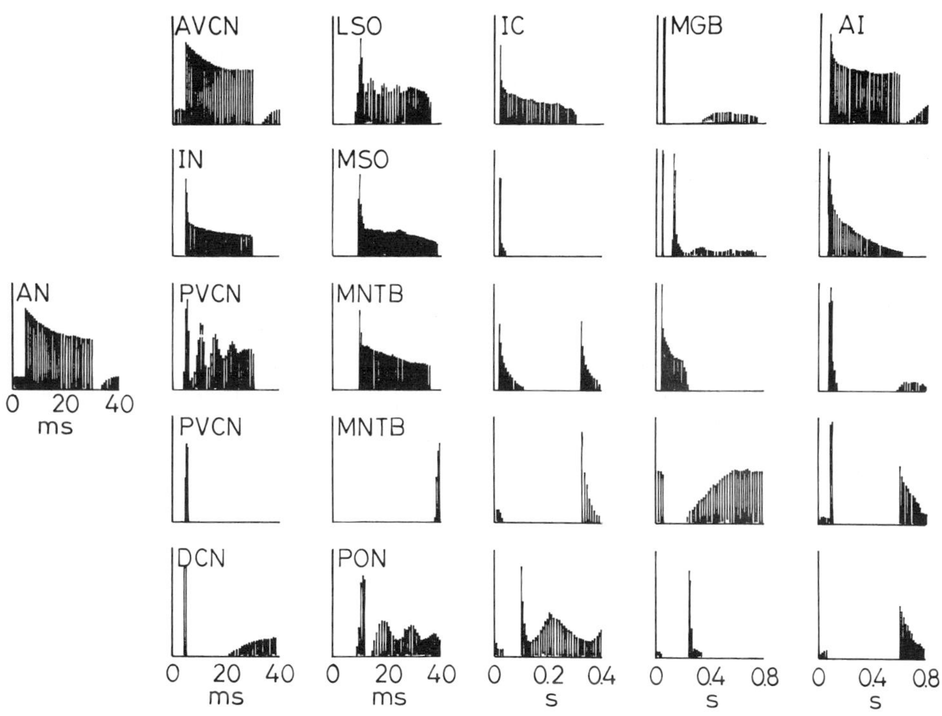

Fig. 3.6. Overview of temporal response properties in the auditory nervous system. Shown are PST histograms to tone bursts at the characteristic frequency for the auditory nerve (*AN*), the cochlear nuclei (*AVCN, PVCN, DCN, IN* interstitial nucleus), the brainstem level (*LSO* lateral superior olive; *MSO* medial superior olive; *MNTB* medial nucleus of the trapezoid body, and *PON* preolivary nuclei), the midbrain (*IC* inferior colliculus), the thalamus (*MGB* medial geniculate body), and the auditory cortex (*AI*). Note the change in time scale from 40 ms on the left to 0.8 seconds on the right

by primary fibers through many small synaptic terminals. The firing pattern shows as an oscillation superimposed on a primary-like pattern, the period of oscillation is determined mainly by refractory mechanisms, and bears no relation to the characteristic frequency of the neuron. The explaining model which proposes recurrent inhibition therefore cannot be the correct one for the cochlear nucleus. Finally, in the dorsal cochlear nucleus so-called pauser neurons are found, which after an optional initial transient, slowly build up a sustained response (Fig. 3.5 Ae). The responsible model is probably as suggested in the figure. It must be noted that there are generally many models that explain the same phenomenon. Only by taking into account other properties of the neurons such as receptive field and spontaneous and driven activity can one (sometimes) favor one model above the other.

The next level in the auditory nervous system features one new pattern of response: the off transient (Fig. 3.5 Ab). Such neurons are found in the medial nucleus of the trapezoid body. In other parts of the olivary complex various modifications of previously found patterns occur. The inferior colliculus which – in contrast to the superior colliculus for the visual system – is a mandatory station for all nervous activity going to higher parts of the brain, shows also on-off cells and most activity is strongly phasic in character. The same can be said for the medial geniculate body, the auditory nucleus in the thalamus. Surprisingly, the primary response patterns, albeit on a slower time scale, return in the auditory cortex. An overview is given in Fig. 3.6 and references can be found in the review by Aitkin et al. (1984). When considering the changes in temporal pattern of responses that occur from auditory periphery and cortex, one must realize that anesthesia has a very pronounced effect on most response patterns: in deep anesthesia the only remaining patterns are the on- and off-transient ones. Most investigations in midbrain and thalamus require relatively deep anesthesia, while recordings from the auditory cortex can be performed from awake restrained cats. The lack of primary-like response types in brainstem and midbrain nuclei may well be due to the anesthesia used.

In general, the temporal response properties are shown independently of the RF. This assumes these properties to be separable, which for a significant fraction of central auditory neurons does not hold (e.g., Eggermont et al. 1981). It is rather straightforward to combine the response properties in both the spectral and temporal domains: spectrotemporal receptive fields (Aertsen and Johannesma 1981 a, b). The spectrotemporal receptive field (STRF) describes the average spectrogram of a sound that elicits responses in an auditory neu-

Fig. 3.7 a–i. Various types of spectro-temporal sensitivities (RFs) to tonal stimuli for neurons in the Torus Semicircularis (the auditory midbrain) of the grassfrog. The *horizontal axis* represents time *before* an action potential. The *vertical axes*, representing frequency, have a logarithmic scale. Intensity of the average pre-action potential stimulus is coded in gray. Activation is represented by a darker than background gray level, inhibition or absence of activity by a lighter than background gray level. **a** Monomodal RF. **b–d** Bimodal RF's. **e** Broadly tuned RF. **f** Narrowly tuned inhibitory RF. **g** Broad inhibitory RF with a postinhibitory activation. **h** Example of excitatory RF with inhibitory side band at higher frequencies. **i** Example of postactivation inhibition. (Epping and Eggermont 1985)

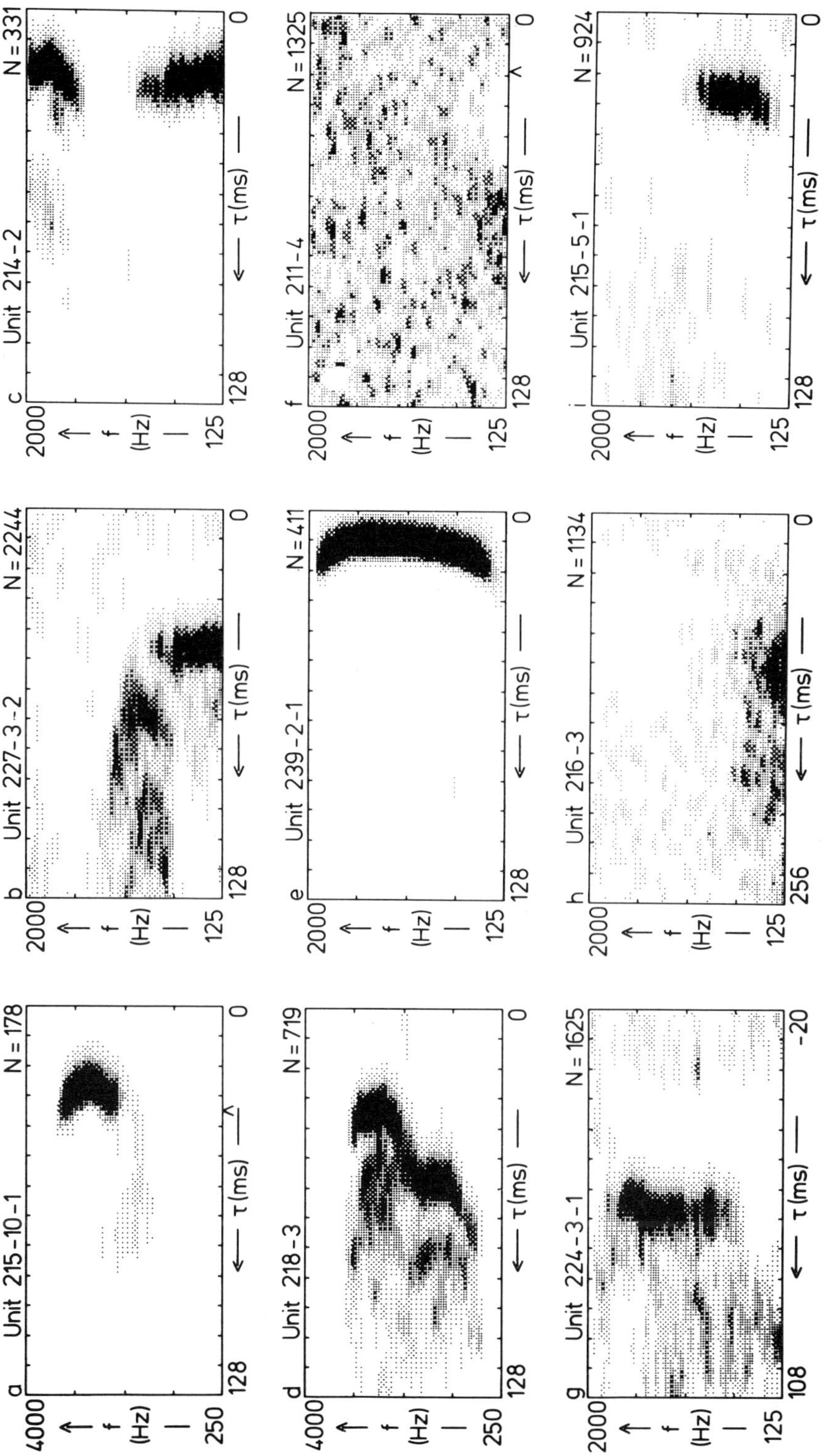
a Unit 215-10-1 N=178
b Unit 227-3-2 N=2244
c Unit 214-2 N=331
d Unit 218-3 N=719
e Unit 239-2-1 N=41
f Unit 211-4 N=1325
g Unit 224-3-1 N=1625
h Unit 216-3 N=1134
i Unit 215-5-1 N=924
f (Hz)
τ (ms)

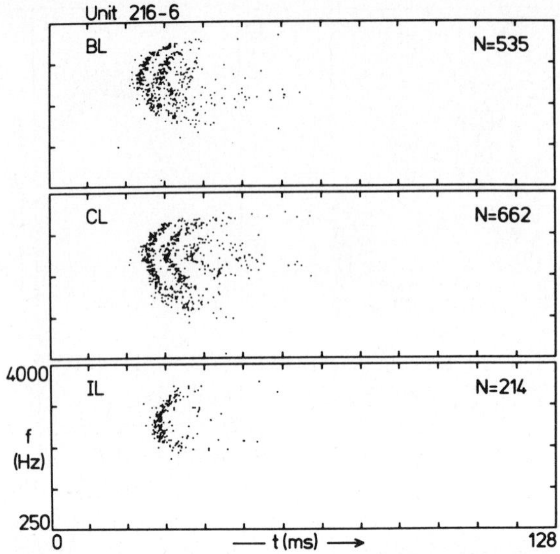

Fig. 3.8. Dot display of neural activity as a function of time after the onset of a tonepip, and as a function of tonepip frequency for bilateral (*BL*), contralateral (*CL*) and ipsilateral (*IL*) stimulation. The *horizontal axis* represents time *after* tonepip onset, the *vertical axis* represents frequency on a logarithmic scale. One observes a chopper-type pattern of response for the BL and CL conditions, and an onset type of response for the IL condition. (Epping and Eggermont 1986)

ron. One may alternatively say that the STRF represents the sonogram of the preferred sound for that neuron (Fig. 3.7). It is noted that the term spectrotemporal as used here does not necessarily require temporal changes in the stimulus, i.e., a stimulus moving over the receptor sheet. The STRF basically gives a combined representation of the spectral sensitivity of the neuron and its latency or response pattern as a function of the frequency. In the visual system, spatiotemporal receptive fields indicate the sensitivity for direction and/or velocity of moving stimuli (e.g., Orban 1984). The equivalent in the auditory system would be frequency-modulated tones, specifying up- or downward modulation and the rate thereof. When a Gaussian white noise is used as a stimulus, rapid changes in activity of various parts of the receptor surface take place. The resulting STRF (Eggermont et al. 1983c) may indicate prefered combinations and sometimes even prefered changes in the stimulus (Epping and Eggermont 1985). An equivalent representation is to show the dot display of the neural activity both as a function of stimulus frequency and time after stimulus onset when this can be defined. In Fig. 3.8 one observes that the extent and clearness of a chopper pattern depends on stimulus frequency as well on certain spatial aspects of the sound.

Is has been found (e.g., Johannesma and Eggermont 1983) that STRF's of auditory neurons in the midbrain of the grassfrog depend to a large extent on what type of stimulus is used. Mapping the STRF with pure tones or toneburst gives different results than exploring it with continuous noise. This forms a serious restriction of the STRF as a single-unit concept and can only be understood if the neuron in question takes a different place within a neuron ensemble for different stimulus situations. To explore this, simultaneous recordings of a number of units should be performed. This will be considered in later chapters.

4 Single-Neuron Models

An important aspect of the modeling of neural networks is the choice of the individual neuron model. When the neural network is very large the details of the modeling of the single neurons become less important. Even the fact that the outputs of the individual neurons are highly irregular can be neglected and some models consider only the average firing rate as the determining input or output value. For very large numbers of neurons, the firing activity thus becomes a rather deterministic process and the neural net shows a behavior comparable to the macroscopic parameters of a gas consisting of a large number of individual molecules. In this case temperature and pressure are the externally measurable parameters that reflect the intermolecular interactions. Although on the microscopic level everything behaves stochastically, at the macroscopic level the process can be described deterministically. Most neural network models have been extensively reviewed by Levine (1983). We will deal with such models in the next chapter.

When we are dealing with few-neuron or local-circuit network models the number of inputs to each neuron, the strength of the individual synaptic connections, the time course of the postsynaptic potentials as well as the time dependence of the neural threshold becomes important. Such models are already simplifications that do not take into account the way the membrane potential depends on ion transport across the cell membrane or the way postsynaptic potentials are attenuated and filtered in the dendritic tree branches. Single-neuron models have been designed basically for two reasons: the explanation of spontaneous activity, and the explanation of changes in receptive fields with respect to size, shape, spatial and temporal parameters. The first type of model seeks to explain the stochastic activity of the neuron as measured in its spontaneous action potential production, the second type is deterministic and relates to wiring diagrams. While the second type of model primarily relates to neural structure, the first type relates to neuronal processes.

Neural process models again can be divided into two classes. Models that describe the time course of the membrane potential and relate this to the action potential output form one kind. The other kind of model views the neuron as a black box that transforms incoming point processes into an output process. Ideally, models should contain aspects of both mechanisms, and recently such models have been formulated by Johannesma and van den Boogaard (1985).

I will not deal exhaustively with single-neuron models since rather extensive treatments can be found in Holden (1976) and MacGregor and Lewis (1977). In the following survey I will very selectively deal with stochastic models for single-neuron firing and will treat the subject in a way comparable to that of Fienberg (1974).

4.1 Diffusion Models

Chapter 2 emphasized the role of the transmitter release in the synapse as one of the sources of the stochastic spontaneous activity of neurons. We measure the spontaneous activity at the output of the neuron. For central neurons this is a result of input from spontaneously active peripheral neurons. The neuron transforms one point process into another. On the basis of a comparison of input and output spike trains we might be able to infer some neuron properties. This is the system-theoretical approach and I will deal with that in Chapter 9. Starting with the transmitter release we may study the fluctuations in postsynaptic membrane permeability. Modeling such a process will describe the membrane potential in terms of the stochastic opening and closing of ion channels, which has been described more deterministically by the Hodgkin-Huxley equations. Such an approach would be too detailed for our purposes. I choose the way in between (Fig. 4.1): the membrane potential is described deterministically and its time course will result in the stochastic generation of action potential spike trains.

An assumption at the basis of most models is that the process to be modeled is stationary. This means that the parameters that describe the process are the same for every time interval under consideration. This allows the experimenter to calculate averages such as mean firing rate, number of intervals with a certain length or distributions. For a stationary process it makes sense to estimate

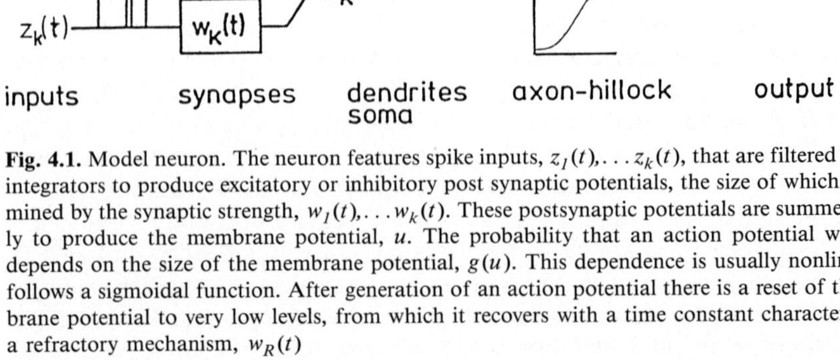

Fig. 4.1. Model neuron. The neuron features spike inputs, $z_1(t), \ldots z_k(t)$, that are filtered by leaky integrators to produce excitatory or inhibitory post synaptic potentials, the size of which is determined by the synaptic strength, $w_1(t), \ldots w_k(t)$. These postsynaptic potentials are summed linearly to produce the membrane potential, u. The probability that an action potential will occur depends on the size of the membrane potential, $g(u)$. This dependence is usually nonlinear and follows a sigmoidal function. After generation of an action potential there is a reset of the membrane potential to very low levels, from which it recovers with a time constant characteristic for a refractory mechanism, $w_R(t)$

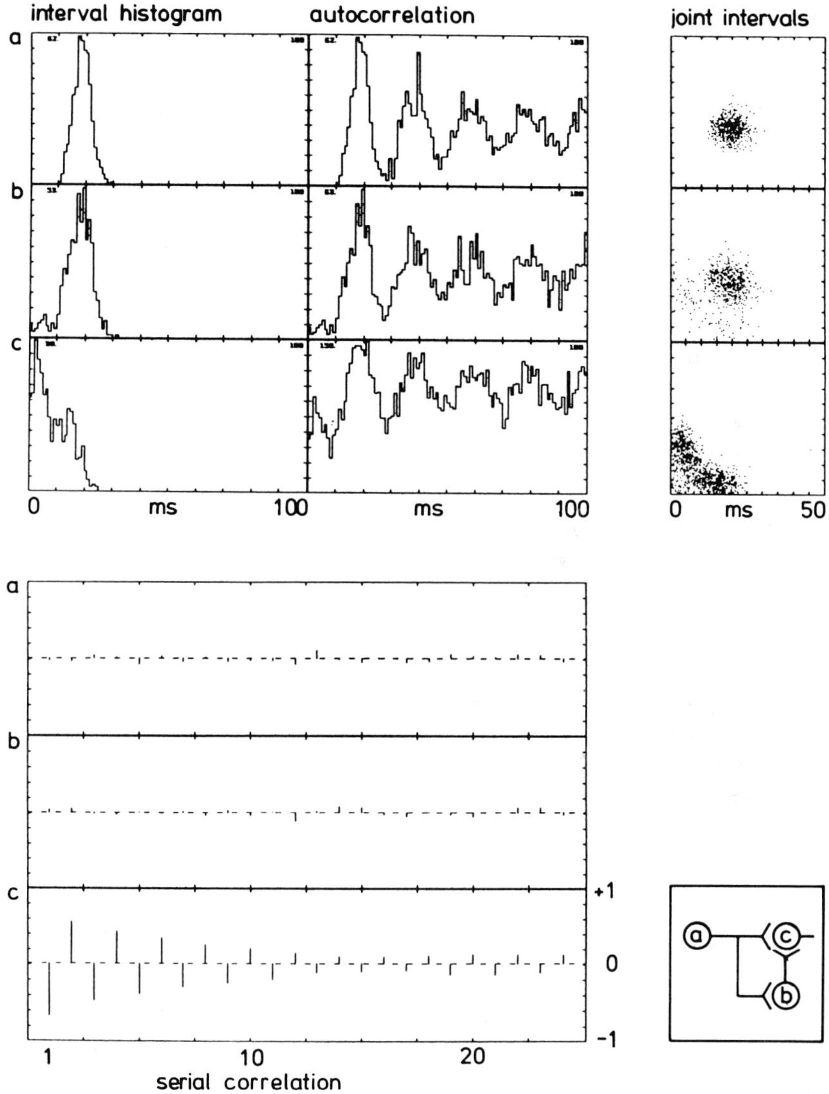

Fig. 4.2 a – c. Characterization of action potential sequences by interval histograms, autocorrelation functions, joint-interval scattergrams and serial correlation coefficients. For a simple three-neuron model, shown as an *insert* in the *lower right corner*, these functions are calculated and shown for each of the three neurons. The neuron indicated with **a** is a spontaneously active driver neuron with a more or less periodic spike pattern as shown in the various plots labeled with **a**; the serial correlation is, however, very small, indicating that the successive spike intervals are independent. Neuron **b** which receives excitatory input from **a** and is also spontaneous (Poisson distributed) shows only a slight modification in its firing pattern corresponding to **a**. Neuron **c** receives excitatory input from both **a** and **b** and shows this periodic component although not as clear as for each of the driver neurons; its successive intervals are no longer independent but show an alteration of long and short intervals as becomes clear from both the joint-interval scattergram and the plot of the first 25 serial correlation coefficients

the interspike interval distribution by constructing the corresponding histogram. When the stochastic process that generates the action potentials is a renewal process (see Box D) the interval distribution is sufficient to characterize the process. In other situations, the correlations between contiguous and noncontiguous intervals need to be determined (see Fig. 4.2). Instead of the interval statistic in which the length of the interval is the stochastic variable, one may also use a counting statistic in which the cumulative number of events in a certain interval is considered as a discretely changing stochastic variable. In principle both statistics should give the same results, especially for the more simple processes, but we have seen that counting statistics may suggest models that differ considerably from those resulting from interval statistics (Chap. 2, Sect. 2.3):

Box D. Renewal Theory

Consider non-negative stochastic variables $X_1, X_2, \ldots X_N$ which are identically distributed. Such variables are independent if their N-dimensional joint probability density is equal to the product of the N marginal densities:

$$f(X_1, X_2, \ldots, X_N) = \prod_{i=1}^{N} f(X_i) \ . \tag{D.1}$$

The values of the stochastic variables will be interpreted as the intervals between contiguous events in a point process: X_1 is the interval from the origin to the first point of the process and X_i $(i = 2, 3, \ldots)$ is the interval between the $(i-1)$st and i-th points following the origin (Cox 1962; Cox and Miller 1970; Cox and Isham 1980; most of what follows is based on these references).

When all X_i $(i = 1, 2, 3, \ldots)$ have the same probability density function (p.d.f.) the process is called an *ordinary renewal process*. Effectively then the process is defined as if there is a point occurring at the origin. If the first interval has a different pdf from all other X_i which still have the same pdf, then the process is called a *modified renewal* process: the process does not start from a point at the origin but rather at some position in an ongoing interval.

A special case of a modified renewal process arises when $f_1(X)$ has the following form:

$$f_1(X) = \frac{1-F(X)}{\mu} \ , \tag{D.2}$$

where $\mu = E[X_i]$ $(i = 2, 3, \ldots)$ and $F(X)$ is the distribution function corresponding to $f(X)$. This process is called an equilibrium or *stationary renewal process.* It can be shown (see references above) that the ordinary

Box D (continued)

renewal process is a stationary renewal process if and only if it is a Poisson process, i.e., has an exponential pdf.

For large time t, the forward recurrence time V_t in an ordinary renewal process has the pdf

$$f(V_t) = \frac{\mathscr{F}(V_t)}{\mu} = \frac{1-F(X)}{\mu} \ , \tag{D.3}$$

where $\mathscr{F}(V_t)$ is called the *survivor function* and μ is the mean interval as before (also called mean failure time). Comparing Eqs. (D.2) and (D.3) it follows that an equilibrium renewal process can be regarded as an ordinary renewal process in which the system has been running a long time before the first observation has been made.

Define $S_n = X_1 + \ldots + X_n$ as the time up the nth event in a renewal process and denote by $g_n(X)$ the pdf associated with S_n. Since a renewal process is defined in terms of a sequence of independent intervals, properties based on intervals are particularly simple. Thus we obtain for $g_n(X)$:

$$g_n(X) = f^{(n)}(X) \quad \text{for an ordinary renewal process} \ , \tag{D.4}$$

and

$$g_n(X) = f_1(X) * f^{(n-1)}(X) \quad \text{for a modified renewal process} \ , \tag{D.4b}$$

where $*$ denotes convolution and $f^{(n)}(X)$ denotes the result of an n-fold convolution of $f(X)$.

One may also be interested in counting events up to a certain time: $N(t)$ is number of events in $(0,t)$. If we denote the distribution function of g_n by G_n, the counting properties can be obtained from the interval properties by using

$$N(t) < k \quad \text{if and only if} \quad S_k > t \ ,$$

which results in

$$\mathrm{Prob}\{N(t) = k\} = G_k(t) - G_{k+1}(t) \ . \tag{D.5}$$

The expected value of $N(t)$ is called the *renewal function* $H(t)$:

$$H(t) = E[N(t)] = \sum_{n=1}^{\infty} n\{G_n(t) - G_{n+1}(t)\} = \sum_{n=1}^{\infty} G_n(t) \ . \tag{D.6}$$

The *renewal density* is defined as the derivative of $H(t)$:

$$h(t) = \frac{dH(t)}{dt} = \sum_{n=1}^{\infty} g_n(t) \ . \tag{D.7}$$

For the ordinary renewal process $h(t)$ can be interpreted as

$$h(t) = \lim_{\delta \downarrow 0} \frac{1}{\delta} \mathrm{Prob}\{N(t:t+\delta) > 0\} \ .$$

Box D (continued)

For the special case of an equilibrium renewal process it holds that

$$H(t) = \frac{t}{\mu} , \tag{D.8}$$

and thus

$$h(t) = \frac{1}{\mu} . \tag{D.9}$$

The so-called integral equation of renewal theory, or simply the renewal equation, is given by

$$h_m(t) = f_1(t) + \int_0^t h_m(t-u) f(u) du \tag{D.10a}$$

for the modified renewal process (h_m), and for the ordinary renewal process (h_0) by

$$h_0(t) = f(t) + \int_0^t h_0(t-u) f(u) du . \tag{D.10b}$$

The asymptotic behavior is given as

$$\lim_{t \to \infty} h(t) = \frac{1}{\mu} . \tag{D.11}$$

A special type of renewal process is the Poisson process with rate λ for which the intervals are independently and exponentially distributed. It is an ordinary renewal process since

$$\begin{aligned} F_1(t) = F(t) &= 1 - e^{-\lambda t};\ t > 0 \\ &= 0 \qquad , t \leqslant 0 \end{aligned} \tag{D.12}$$

with $H(t) = \lambda t = t/\mu$ (D.13)

and $\mathrm{Prob}\,\{N_t = 0\} = e^{-\lambda t},\ t > 0$ (D.14)

$$\mathrm{Prob}\,\{N_t = n\} = \frac{(\lambda t)^n}{n!} e^{-\lambda t},\ n = 0, 1, 2, \ldots;\ t > 0 \tag{D.15}$$

$$\mathrm{Prob}\,\{X_n < t\} = F^{(n)}(t) = \int_0^t \frac{\lambda^n \tau^{n-1}}{(n-1)!} e^{\lambda \tau} d\tau . \tag{D.16}$$

For inhomogeneous Poisson processes $\lambda = \lambda(t)$, and for inhomogeneous renewal processes $\mu = \mu(t)$. In these cases $\mu(t)$ is sometimes called an age-specific failure time. This can be extended to a situation where μ also depends on the state variable X and then $\mu(X, t)$ is an age-wear-specific failure time.

Now let us consider a neuron with membrane potential u, resting membrane potential u_r, and threshold level S. The simplest model assumes that excitatory stimuli arrive according to a Poisson process with intensity λ_e. The neuron has a threshold equal to k stimulus sizes, and the neuron fires after k stimuli have arrived. The waiting time to firing is then given by a gamma distribution (see Box B). This very simple model can be extended with inhibitory inputs that arrive with rate λ_i. Assume the size of excitatory and inhibitory inputs the same and the u is reset to u_r after u reaches S. In case one also assumes that $\lambda_e = \lambda_i$, the process can be viewed as a simple random walk process with absorbing barier S. The distribution of first passage times (or interspike intervals) is given as (Gerstein and Mandelbrot 1964):

$$f(t) = \frac{1}{\sqrt{4\pi}} S\, t^{-3/2} \exp\left(-\frac{S^2}{4t}\right), \quad t \geqslant 0 . \tag{4.1}$$

When the intensities of the excitatory and inhibitory processes are different, one may define a so-called drift velocity, $c = \lambda_e - \lambda_i$, and the interval distribution becomes:

$$f_c(t) = \frac{1}{\sqrt{4\pi}} S\, t^{-3/2} \exp\left(-\frac{(S-ct)^2}{4t}\right) . \tag{4.2}$$

These distributions are of the inverse Gaussian type (see Box B). One may consider such models as linear integration models without decay: incoming inputs either augment or decrease the current membrane potential value. In addition, the model is, except for the stochastic input, quite deterministic. Incorporation of a stochastic variation in the membrane potential with variance σ^2 leads to (Pernier and Gerin 1971):

$$f_\sigma(t) = \frac{1}{\sigma\sqrt{4\pi}} S\, t^{-3/2} \exp\left(-\frac{(S-ct)^2}{2\sigma^2 t}\right) , \tag{4.3}$$

which seems to provide a better fit to interval histograms than Eq. (4.2), not unexpectedly, since there is one more free parameter that can be used in the fitting procedure. The most serious objection against the model is that in a real neuron there is a decay of the membrane potential towards the resting value whenever its value due to inputs differs from u_r. Such models are linear integration models with exponential decay as studied by, e.g., Stein (1965) and Johannesma (1968), and we will come back to these models somewhat later in this chapter.

A special class of the renewal models is formed for situations in which a neuron receives input from a relatively large number of independent neurons and in which each input is rather small relative to the threshold S. In this case the random walk model can be studied in the diffusion approximation.

Box E. Master Equation

Consider a neuron with membrane potential u(t) and assume that u(t) changes in discrete and very small unit steps (relative to threshold S) that can either increase or decrease u. Indicate u by the number of unit steps $0, 1, \ldots, m, \ldots, S$. If we start from a certain membrane potential at time $t = 0$, we may ask what the probability is that after $(n+1)$ jumps the membrane potential has a value equal to m. We indicate this probability by $F(m; n+1)$. Before the last jump the value of u was either $(m-1)$, m, or $(m+1)$.

Define Λ_e as the probability of a positive increase and Λ_i as the probability of a negative increase then:

$$F(m; n+1) = \Lambda_e F(m-1; n) + (1 - \Lambda_e - \Lambda_i) F(m; n) + \Lambda_i F(m+1; n) \ . \tag{E.1}$$

Now, consider time discretized in steps Δt and define $\lambda_e = \frac{\Lambda_e}{\Delta t}$ and $\lambda_i = \frac{\Lambda_i}{\Delta t}$ as the transition rates. Take Δt so small that at most one unit change in u can take place. Then Eq. (E.1) can be written as

$$\frac{F(m; t+\Delta t) - F(m; t)}{\Delta t} = \lambda_e F(m-1; t) - (\lambda_e + \lambda_i) F(m; t) + \lambda_i F(m+1; t) \ . \tag{E.2}$$

In the limit $\Delta t \to 0$ a so-called *master equation* results:

$$\frac{dF(m; t)}{dt} = \lambda_e F(m-1; t) - (\lambda_e + \lambda_i) F(m; t) + \lambda_i F(m+1; t) \ . \tag{E.3}$$

In stationary situations the principle of detailed balance holds:

$$\lambda_e F(m-1; t) = \lambda_i F(m+1; t) \ . \tag{E.4}$$

This can be used to go from discrete changes in u(t) toward continuous changes in u(t). Let the momentary value of u(t) change in small steps $\varepsilon \ll 1$. In that case because of the detailed balance condition $\lambda_e \simeq \lambda_i$; call both equal to λ:

$$\frac{dF(m; t)}{dt} = \lambda \{F(m - \varepsilon; t) - 2\, F(m; t) + F(m + \varepsilon; t)\} \ . \tag{E.5}$$

By expanding F in a Taylor series up to second order in ε we arrive at

$$\frac{dF(m, t)}{dt} \simeq \lambda \left\{ F(m; t) - \frac{dF(m, t)}{dm} \cdot \varepsilon + \frac{1}{2} \frac{d^2F(m, t)}{dm} \varepsilon^2 - 2\, F(m; t) + F(m; t) + \frac{dF(m, t)}{dm} \varepsilon + \frac{1}{2} \frac{d^2F(m, t)}{dm} \varepsilon^2 \right\} , \tag{E.6}$$

Box E (continued)

resulting in

$$\frac{dF(m;t)}{dt} \simeq \lambda \cdot \frac{d^2F(m;t)}{dm^2} \cdot \varepsilon^2 ,$$

a diffusion equation or Fokker-Planck equation. (E.7)

By solving the master equation, either directly or through conversion into equations for the moments of F, one arrives at the properties of the distributions F(m; t). [For more details see van Kampen (1981) and Gardiner (1985).]

In the following diffusion model we consider the excitatory postsynaptic potentials (epsp's) to be of size ε and to arrive with rate λ_e, the inhibitory postsynaptic potentials (ipsp's) are of size i and arrive with rate λ_i. The membrane potential is assumed to decay to its resting value in an exponential way with time constant τ. Let F(u; t) be the cumulative distribution for the membrane potential to have the value m at time t provided its value was u_0 at time zero. Let f(m, t) be the corresponding probability density function. Following Fienberg (1974) the evolution equation for the cumulative distribution reads:

$$\begin{aligned} F(m;t+\Delta t) = {} & F\{(u-u_r)e^{\Delta t/\tau}+u_r;t\}\cdot\{1-(\lambda_e+\lambda_i)\Delta t\} \\ & +F\{(u-u_r-\varepsilon)e^{\Delta t/\tau}+u_r;t\}\cdot\lambda_e\Delta t \\ & +F\{(u-u_r+i)e^{\Delta t/\tau}+u_r;t\}\cdot\lambda_i\Delta t , \end{aligned} \tag{4.4}$$

where the first term on the right hand side of the equation describes the probability that the membrane potential has the value u and no epsp's or ipsp's occur in (t, t+Δt). The second term describes the probability that the membrane potential at t equals (u$-\varepsilon$) and that in (t, t+Δt) one epsp arrives, the third term then gives the probability that the membrane potential is (u+i) at t and that in the next Δt one ipsp arrives. The simultaneous occurrence of one epsp and one ipsp is considered negligible. In this case the diffusion or Fokker-Planck equation, which can be derived as explained in Box E, reads:

$$\begin{aligned} \frac{\partial f(u,t)}{\partial t} = {} & -\frac{\partial}{\partial u}\left\{\left(\varepsilon\lambda_e - i\lambda_i - \frac{u-u_r}{\tau}\right) f(u,t)\right\} \\ & +\frac{1}{2}(\varepsilon^2\lambda_e+i^2\lambda_i)\frac{\partial^2 f(u;t)}{\partial u^2} . \end{aligned} \tag{4.5}$$

In general, the Fokker-Planck equation is the equation that describes the transition probability density, f, of a process X that satisfies

$$dX_t = \mu(X_t)\,dt + \sigma(X_t)\,dW_t , \tag{4.6}$$

where $W = \{W_t, t \geqslant 0\}$ is a standard Wiener process and $\mu(X_t)$ and $\sigma(X_t)$ are the mean and standard deviation of the stochastic variable X_t. In this case the Fokker-Planck equation is

$$\frac{\partial f(x,t)}{\partial t} = -\frac{\partial}{\partial x}\{\mu(x) f\} + \frac{1}{2}\frac{\partial^2}{\partial x^2}\{\sigma^2(x) f\} . \tag{4.7}$$

A comparison between Eqs. (4.5) and (4.7) shows that we may interpret $\mu(x) = -\frac{u-u_r}{\tau} + (\varepsilon\lambda_e - i\lambda_i)$ as the mean value of the membrane potential, and $\sigma^2 = (\varepsilon^2\lambda + i^2\lambda_i)$ as its variance. If we take for simplicity $u_r = 0$, then the solution of Eq. (4.5) – which also can be considered as the forward equation of the Ornstein-Uhlenbeck process – reads:

$$f(u;t) = \frac{1}{\sqrt{\pi\tau(1-e^{-2t/\tau})}\cdot\sigma} \exp\left\{-\frac{[(u-\mu\tau)-(u_0-\mu\tau)e^{-t/\tau}]^2}{\tau(1-e^{-2t/\tau})\cdot\sigma^2}\right\} . \tag{4.8}$$

In the special case that the threshold of firing value $S = \mu\tau$, an expression for the interspike interval distribution can be obtained in closed form

$$f(t) = \frac{2(S-u_0)}{\sqrt{\mu}\cdot\sigma}\cdot\tau^{-3/2}\cdot(e^{2t/\tau}-1)^{-3/2}\cdot\exp\left\{-\frac{[(S-u_0)^2+2t/\tau]^2}{\tau(e^{2t/\tau}-1)\cdot\sigma^2}\right\} . \tag{4.9}$$

In other situations the solution is available only as the Laplace transform of f(t), which is so complicated that no inversion is known (Lansky 1983). For the general situation with any S, Ricciardi and Sacerdote (1979) derived expressions for the mean M and variance V of the interspike interval distribution. In a first approximation they consider $\mu \simeq 0$ and $\sigma^2 \gg \frac{S^2}{\tau}$, i.e., no excess of epsp's with respect to ipsp's and a very noisy membrane potential, and take $\mu_0 = 0$. Thus they arrived at

$$M \sim \sqrt{\pi\tau}\cdot\frac{S}{\sigma} \tag{4.10}$$

and

$$V \sim \sqrt{\pi\tau}\cdot\frac{S}{\sigma}\cdot\tau\ln 4 , \tag{4.11}$$

suggesting that M and V are linearly related and that the output process will be approximately Poisson.

From the above results, the pdf for the intervals is readily obtained for $\tau = \infty$, i.e., nondecaying increments and decrements as in a perfect integrator. Lansky (1983) gives:

$$f(t) = \frac{S-u_0}{\sqrt{2\pi}\cdot\sigma}\, t^{-3/2} \exp\left\{-\frac{(S-u_0-\mu t)^2}{2\sigma^2 t}\right\} . \tag{4.12}$$

This distribution is again the inverse Gaussian distribution, and is also in slightly different form, the solution of the random walk model [see Eqs. (4.1) to (4.3)]. The moments of this distribution are (e.g., Yang and Chen 1978):

$$M = \frac{S - u_0}{\mu} \quad \text{and} \quad V = \frac{1}{2} M \frac{\sigma^2}{\mu^2} . \tag{4.13}$$

We note that the interval distributions from Eqs. (4.9) and (4.12) are unimodal distributions. In our survey of spontaneous activity in Chapter 3 we have come across bimodal and multimodal distributions. Suggestions have been forwarded to approximate these distributions as a linear combination of these inverse Gaussians (Pernier and Gerin 1975) in the way we introduced the hyperexponential distribution (Box B). Pernier and Gerin applied this hyperinverse Gaussian model to interval histograms of 71 neurons recorded in the rabbit's neocortex and reported that all distributions could be well approximated. Another approach was followed by Giglmayr (1979), who used a diffusion model with positive drift and introduced into this model an age-wear-specific failure rate to take into account the relative refractory period as a randomly distributed time of reduced sensitivity of the nerve cell following each spike generation. The absolute refractory period was introduced as a translation in this failure rate. With this model it was possible to obtain multimodal pdf's for the interspike intervals. Relative refractory periods have been introduced in diffusion models also in more deterministic ways. Several formulations from literature include Hagiware (1954): $S(t) = S_0 \exp\left(\frac{\theta}{t}\right)$, $\theta > 0$, Geisler and Goldberg (1966): $S(t) = S_0 + \frac{\exp(-t/\theta)}{1 - \exp(-t/\theta)}$, $\theta > 0$, and Clay and Goel (1973): $S(t) = S_0 + \delta \exp(-t/\psi)$ with $\delta / S_0 \gg 1$. Another possibility is related to that of Giglmayer (1979) and keeps the threshold fixed but reduces the input with a factor $S_0/S(t)$, where $S(t)$ can be any of the above functions (Johannesma 1968).

4.2 Discrete Input Models

As we have seen in the previous paragraph, the neuron's membrane potential is often modeled analogously to a diffusion process. Such models require a white noise input into a leaky integrator with constant threshold followed by instant reset of the membrane potential when the threshold value is reached. The white noise input was assumed to be formed from the summated activity of a large number of neurons independently contributing a small postsynaptic potential. Such an assumption may be valid for neurons in the cortex, but generally not for, e.g., sensory neurons in the thalamus. Single neurons subjected to a rather deterministic input cannot be modeled by a diffusion process, although a random walk approach may still be valid.

One of the most simple discrete input models considers a neuron that receives Poisson input (parameter μ). The neuron gives a response to the k-th input that falls within a certain integration period T, provided the inputs do not occur during the refractory period. It is assumed that the integration period is exponentially distributed with mean T. It is then possible to set up a Master equation and to calculate the distribution of the intervals in case $0 \leqslant t \leqslant T$ (Ten Hoopen and Reuver 1965):

$$F_k(t) = 1 - \sum_{i=0}^{k-1} \frac{(\mu T)^i}{i!} \exp(-\mu T) , \tag{4.14}$$

where the second term on the right hand side gives the probability of all input situations that do not lead to k inputs in interval T.

Such a model is still unrealistic, since in real neurons membrane potentials decay toward the resting value. An alternative to this has been proposed by Ten Hoopen (1966c) by taking exponentially distributed durations of epsp's and ipsp's that change the value of u by one unit. For this renewal model a master equation has been derived and compared to the case of exponentially decaying epsp's. For both cases the same interval mean is obtained:

$$\mu(t) = \lambda_e \tau \{1 - \exp(-t/\tau)\} , \tag{4.15}$$

where λ_e is the arrival rate of the epsp's and τ is the decay time constant or the mean of the exponentially distributed epsp duration. For the variance one finds for the exponential decay model:

$$\sigma^2_{\exp}(t) = \lambda_e \tau \{1 - \exp(-2t/\tau)\}/2 , \tag{4.16a}$$

but for the exponentially distributed case

$$\sigma^2_{\mathrm{e.d.}}(t) = \frac{2}{1 + \exp(-t/\tau)} \cdot \sigma^2_{\exp}(t) = \mu(t) , \tag{4.16b}$$

which suggests a Poisson process. Only in case that $\tau \to \infty$, i.e., not decay, does the exponential distribution model approach the exponential decay model in both mean and variance.

A more general discrete input model was formulated by van den Boogaard and Johannesma (1984). They assumed that action potentials (modeled as δ-functions) are subjected to a first-order linear integration. The membrane potential u(t) is modeled by a random walk, which means that u(t) is a state variable subject to decay (parameter b), deterministic stimulus input (optional, v) and stochastic input from other neurons (Z with weighting w):

$$\frac{d}{dt} U = -bu + v + wZ . \tag{4.17}$$

U and Z are stochastic variables, u is the current value of U. This model can be illustrated by the block diagram in Fig. 4.1.

The state space consists of three regions (Fig. 4.3): the active region where U may generate impulses, e.g., according to a nonlinear pulse generator g(u) which equals zero when $u < -d$ and $g(u) = \nu$ if $u \geqslant -d$. Thereafter there is an immediate reset of the membrane potential with constant amount w. The

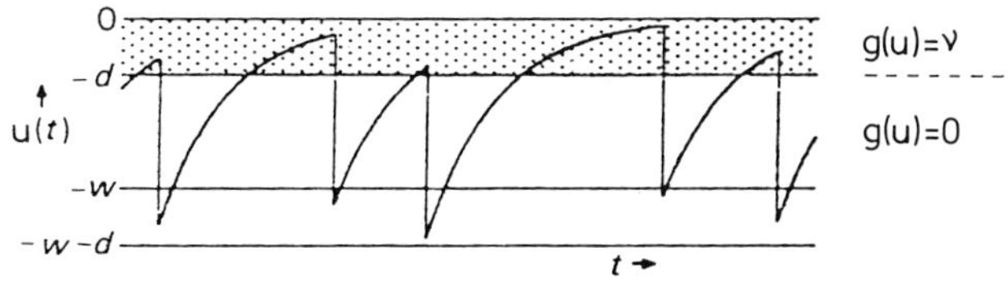

Fig. 4.3. State space diagram of the neuron model from Fig. 4.1; for explanation see text. (After van den Boogaard and Johannesma 1984)

waiting time for a next event T is now considered to be the sum of two components: a dead period $T_A(U)$, the time required to enter the active region, and T_S the survival time in the active region. T_s is considered exponentially distributed:

$$f_S(t) = \nu \exp(-\nu t) \tag{4.18}$$

and the probability density function of the dead period is given by

$$f_A(t) = (\nu \exp bt)\left(-\frac{w}{d} + \exp bt\right)^{(\nu/b)-1} . \tag{4.19}$$

Furthermore, $T_A(U)$ and T_S are assumed independent, thus the distribution of the waiting time f_T is given by the convolution of f_A and f_S. For the scaled waiting time $\hat{T} = bT$ the pdf is given by:

$$f_{\hat{T}}(\tau) = \begin{cases} 0 & , \tau < \ln 1/\delta \\ \varrho^2 \exp(-\varrho t) \int\limits_0^{\exp(\tau - 1/\delta)} \left(\xi + \frac{1}{\delta}\right)^{\varrho} \xi^{\varrho-1} d\xi & , \ln\frac{1}{\delta} < \tau \leqslant \ln\left(1 + \frac{1}{\delta}\right) \\ \varrho^2 \exp(-\varrho t) \int\limits_0^1 \left(\xi + \frac{1}{\delta}\right)^{\varrho} \xi^{\varrho-1} d\xi & , \tau > \ln\left(1 + \frac{1}{\delta}\right) , \end{cases} \tag{4.20}$$

where $\delta = \frac{d}{w}$ and $\varrho = \frac{\nu}{b}$ determine completely the scaled distribution function. It is easily seen that the absolute refractory period is determined only by δ and that the distribution has an exponential tail given by ϱ. In addition $f_{\hat{T}}$ does not completely characterize the point process since this is not an ordinary renewal process. Figure 4.4 gives interval distributions for $\delta = 0.6$ and various values for ϱ.

The point process generated by this model turns out to be a self exciting point process. Figure 4.1 in essence described the model. The input Poisson process N is filtered with a linear filter W_N and results in shot noise. There is a filtered feedback to simulate the refractory mechanism. A comparable model has been described by Teich and Saleh (1981) for a shot-noise-driven self-exciting point process in photon detection. It was intended to model the spontaneous discharge of retinal ganglion cells. Other models that incorporate refractory mechanisms use either a recovery function or a stochastically vary-

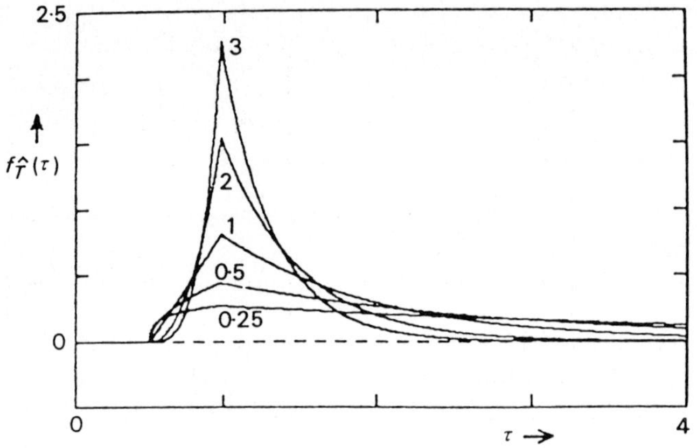

Fig. 4.4. Interval distributions for the neuron model, for explanation see text. (After van den Boogaard and Johannesma 1984)

ing absolute dead time. Teich and Diament (1980) use a recovery function of the form $\frac{x^2}{1+x^2}$ which for a recovery time constant τ takes the form $\frac{t^2}{t^2+\tau^2}$. For a mean output rate λ_0 for a certain stimulus intensity the scaled interval distribution reads

$$f(t/\tau) = \lambda_0 \exp(-\lambda_0 t) \cdot \frac{t^2}{t^2+\tau^2} \cdot \exp\left\{\lambda_0 \tau \tan^{-1}\left(\frac{t}{\tau}\right)\right\} . \tag{4.21}$$

This unimodal distribution has a mode $t_m = \left(\frac{2}{\lambda_0}\right)^{1/3} \cdot \tau^{2/3}$. The unimodal interval histograms that we have considered so far are mostly found in peripheral sensory and motor systems; more central neurons such as in the thalamus generally have multimodal interval histograms. More advanced models have been proposed but will not be discussed here, the reader is referred to Holden (1976) for further details.

4.3 Neural Interaction – Black Box – Models

Especially in the thalamus, bimodal and multimodal interval histograms have been found. Bishop et al. (1964) suggested explaining these results by the interaction of an excitatory and inhibitory input to a LGN cell. This model has been elaborated upon by Ten Hoopen (1966a, b) and Ten Hoopen and Reuver (1968). In this set of papers they consider basically three classes of models (Fig. 4.5): the independent model (a), the feedforward model (b), and the feedback model (c). The last two models have dependent excitatory and inhibitory inputs $\phi(t)$ and $\psi(t)$ respectively. We will discuss these three models in some

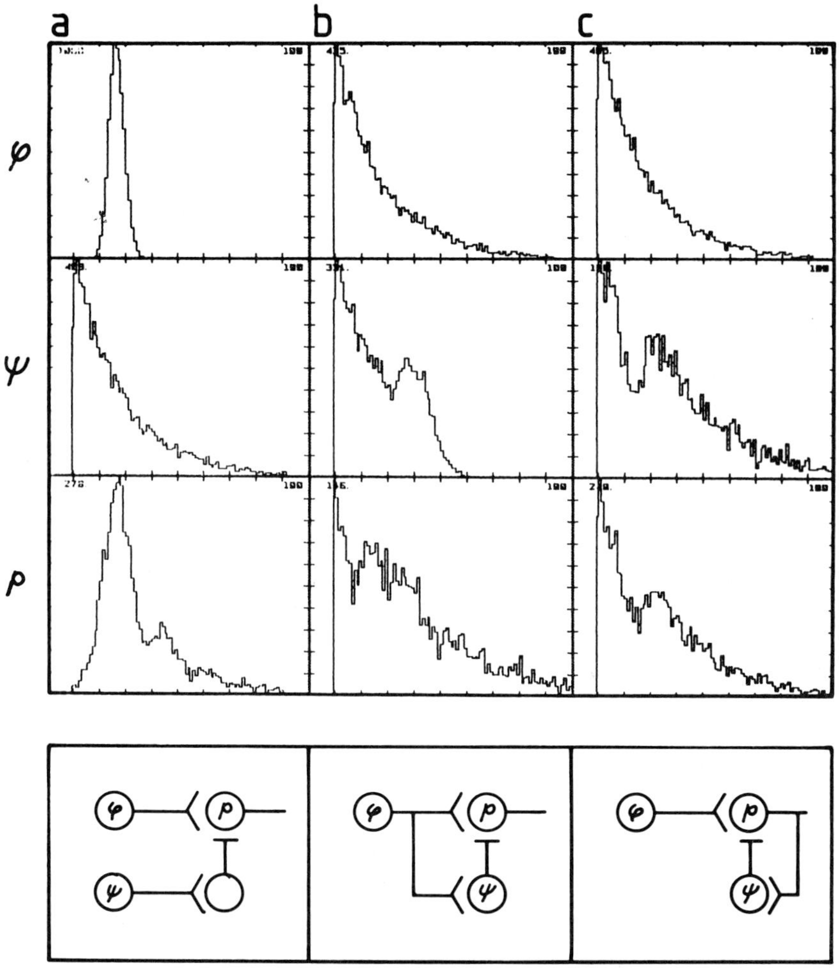

Fig. 4.5 a–c. Model simulations of the neural networks proposed for the LGN. The columns indicate the three models: independent (**a**), feed-forward (**b**), and feed-back (**c**). The rows represent the interval histograms for the φ, ψ and p process under the three model conditions. Bimodality for model **a** can only be produced when the driver neurons, or at least one of them are quasi-periodic. The other models function as well with Poisson input

detail. The independent model features $\phi(t)$ as a monomodal distribution and $\psi(t)$ as a Poisson process with intensity μ. The resulting output process $p(t)$ is assumed to result from the deletion of the next excitatory pulse by any inhibitory pulse and that any undeleted excitatory pulse provokes a response in the thalamic cell.

Box F. Laplace Transforms

Let $s = \sigma + i\omega$ be a complex frequency then $X(s) = \int_{-\infty}^{\infty} x(t)e^{-st}\,dt$ is called the Laplace transform of x(t), indicated by $L\{x(t)\}$. There is also an inverse transformation:

$$x(t) = L^{-1}\{X(s)\} = \frac{1}{2\pi i}\int_{\sigma - i\infty}^{\sigma + i\infty} X(s)e^{st}\,ds \ .$$

Functions x(t) that are $\neq 0$ for $t<0$ and $t>0$ have a double sided Laplace transform. A faster convergence is obtained for functions that are identical to zero for $t<0$. In the remainder we will assume that all functions x(t) are multiplied by the unit step function U(t) which takes the values 0 for $t<0$; 1/2 for $t=0$, and 1 for $t>0$.

Some properties of the Laplace transform (Spiegel 1965):

1. Linearity $\quad L\{c_1x_1(t)+c_2x_2(t)\} = c_1X_1(s)+c_2X_2(s)$.

2. Translation $\quad L\{x(t-a)\} = e^{-as}X(s)$
   $L\{e^{at}x(t)\} = X(s-a)$.

3. Scaling $\quad L\{x(at)\} = \frac{1}{a}X\left(\frac{s}{a}\right)$.

4. Differentiation $\quad L\{x^{(n)}(t)\} = s^nX(s) - s^{n-1}x^{(1)}(0) - s^{n-2}x^{(2)}(0) - \ldots - x^{(n-1)}(0)$.

5. Integration $\quad L\left\{\int_0^t x(u)\,du\right\} = \frac{1}{s}X(s)$.

6. Multiplication $\quad L\{t^nx(t)\} = (-1)^n\frac{d^n}{ds^n}X(s)$.

7. Division $\quad L\left\{\frac{1}{t}x(t)\right\} = \int_s^{\infty} X(u)\,du$.

8. For periodic functions $x(t+T) = x(t)$ with $T>0$:

$$L\{x(t)\} = \frac{\int_0^T e^{st}x(t)\,dt}{1-e^{-sT}} \ .$$

9. Convolution $\quad L\left\{\int_0^t x(u)y(t-u)\,du)\right\} = X(s)Y(s)$.

Box F (continued)

Some simple Laplace transforms

x(t)	*X(s)*
$\delta(t-t_0)$	e^{-st_0}
$\delta(t)$	1
$U(t-t_0)$	$\frac{1}{s}e^{-st_0}$
$U(t)$	$\frac{1}{s}$
$tU(t)$	$\frac{1}{s^2}$
$t^aU(t)$	$\frac{\Gamma(a+1)}{s^{a+1}}$
$e^{-bt}U(t)$	$\frac{1}{s+b}$
$te^{-bt}U(t)$	$\frac{1}{(s+b)^2}$
$t^ae^{-bt}U(t)$	$\frac{\Gamma(a+1)}{(s+b)^{a+1}}$
sin at	$\frac{a}{s^2+a^2}$
cos at	$\frac{s}{s^2+a^2}$

For the independent model with a Poisson inhibitory process the Laplace transform of the output distribution reads

$$P(s) = \frac{\Phi(s+\mu)}{1+\Phi(s+\mu)-\Phi(s)}, \tag{4.22}$$

where $\Phi(s) = L\{\phi(t)\}$. For the thalamus and especially the LGN a reasonable choice for $\phi(t)$ is the γ-distribution. This means that $\Phi(s) = \frac{\lambda^\alpha}{(s+\lambda)^\alpha}$ resulting in

$$P_\gamma(s) = \frac{\lambda^\alpha (s+\lambda)^\alpha}{(s+\lambda+\mu)^\alpha (s+\lambda)^\alpha + \lambda^\alpha (s+\lambda)^\alpha - \lambda^\alpha (s+\lambda+\mu)^\alpha}$$
$$= \left\{ 1 + \left[\frac{s(s+\lambda+\mu)}{\lambda(s+\lambda)} \right]^\alpha \right\}^{-1} . \tag{4.23}$$

It is claimed (Bishop et al. 1964) that for $\alpha > 20$ a bimodal output interspike interval distribution may result.

Actual calculations based on Fourier transformation of Eq. (4.23) show that multimodal interval histograms do occur for $\alpha > 10$ and μ, λ about $10\,s^{-1}$. The interval distribution calculated on basis on Eq. (4.23) is shown in Fig. 4.6. A simulation using model neurons is shown in the first column of Fig. 4.5 (for details see legend of Fig. 4.5).

The feedforward inhibition model was originally (Ten Hoopen and Reuver 1968) formulated as "every ϕ-event deleted or not triggers a sequence of ψ-events". This can best be envisioned by assuming that the ϕ-events excite an inhibitory interneuron. The assumption made was that $\phi(t)$ is Poisson with parameter μ and $\psi(t)$ is arbitrary monomodal. The Laplace transform of the output distribution reads

$$P(s) = \frac{\mu[1 - \Psi(s+\mu)]}{s + \mu - \mu\Psi(s+\mu)} . \tag{4.24}$$

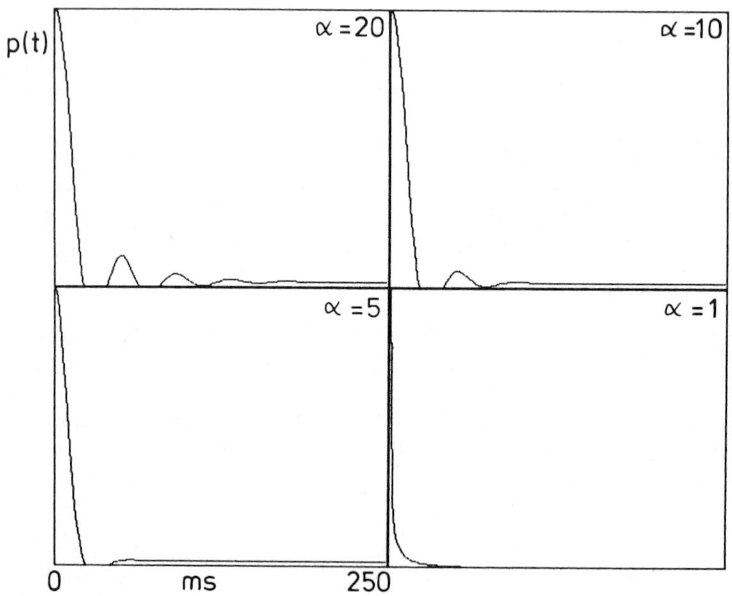

Fig. 4.6. Result of evaluating the inverse Laplace transform of Eq. (4.23) for various parameter values. We used $\lambda = \mu = 10\,s^{-1}$, s = 0, 1280 Hz, α = 1, 5, 10, and 20. One observes that for a = 20 the interval histogram starts to look like the one in Fig. 2.3; note that in the simulation the vertical axis is linear!

In the situation that $\Psi(t)$ is a gamma process, $\Psi(s) = \left(\frac{\lambda}{s+\lambda}\right)^{\alpha}$ and

$$P(s) = \frac{\mu\left[1-\left(\frac{\lambda}{s+\mu+\lambda}\right)^{\alpha}\right]}{s+\mu\left[1-\left(\frac{\lambda}{s+\mu+\lambda}\right)^{\alpha}\right]} \; . \tag{4.25}$$

As can be seen from this expression, the tail of the distribution will be Poisson like: $\lim\limits_{s\to\infty} P(s) = \frac{\mu}{s+\mu}$. If any, there is only a weak suggestion of bimodality. Simulations are shown in the second column of Fig. 4.5.

The feedback inhibition model assumed in the original description (Ten Hoopen and Reuver 1968) that "only undeleted ϕ event (p-events) have a trigger effect on $\psi(t)$". For an excitatory process $\phi(t)$ that is Poisson with parameter μ it was derived that:

$$P(s) = \frac{\mu}{s+\mu} + \frac{\mu}{(s+\mu)^2} \cdot \frac{1-\Psi(s+\mu)}{1-\Psi(s+\mu)+\mu\Psi(s+\mu)} \cdot [-\Psi(s+\mu)] \; . \tag{4.26}$$

A simulation is shown in Fig. 4.5, last column.

Most of these expressions are quite untractable analytically, but some insight into the signal processing capacities of such "local circuits" can be obtained by simulation and by numerically solving the equations. Sanderson et al. (1973) also considered the independent deletion model for which they defined an inhibition strength ϱ as the ratio of the mean rates of the ψ and ϕ processes: $\varrho = \mu_\psi/\mu_\phi$, and a mean rate transfer function $T(\varrho) = \mu_p/\mu_\phi$. As interacting processes they considered gamma distributions of order n ranging from $n = 1$ (Poisson) to $n = \infty$ (regular process) and could show that $T(\varrho)$ depended explicitly on ϱ and n (the distribution parameter). If we indicate by $T(\varrho\,|\,n_1, n_2)$ the mean rate transfer function obtained for a gamma distribution or order n_1 inhibiting a gamma distribution of order n_2 than the following results were reported:

$$T(\varrho\,|\,1,\infty) = e^{-\varrho} \; ; \quad T(\varrho\,|\,1,n) = \left(1+\frac{\varrho}{n}\right)^{-n} \; ; \quad T(\varrho\,|\,1,1) = (1+\varrho)^{-1}$$

$$T(\varrho\,|\,n,1) = 1-\varrho\left\{1-\left(1+\frac{1}{n\varrho}\right)^{-n}\right\} \; , \quad T(\varrho\,|\,\infty,1) = 1-\varrho(1-e^{-1/\varrho}) \; ;$$

$$T(\varrho\,|\,\infty,\infty) = 1-\varrho \; .$$

It is not surprising that in the limit of small ϱ all mean rate transfer functions converge to 1, suggesting that there is no difference in this respect between regular spike trains or Poisson spike trains.

The question now is whether these stochastic models fit in with the wiring diagrams for the thalamus. For this purpose we will give a short review of some properties of LGN circuitry. One to six retinal ganglion cells converge on each

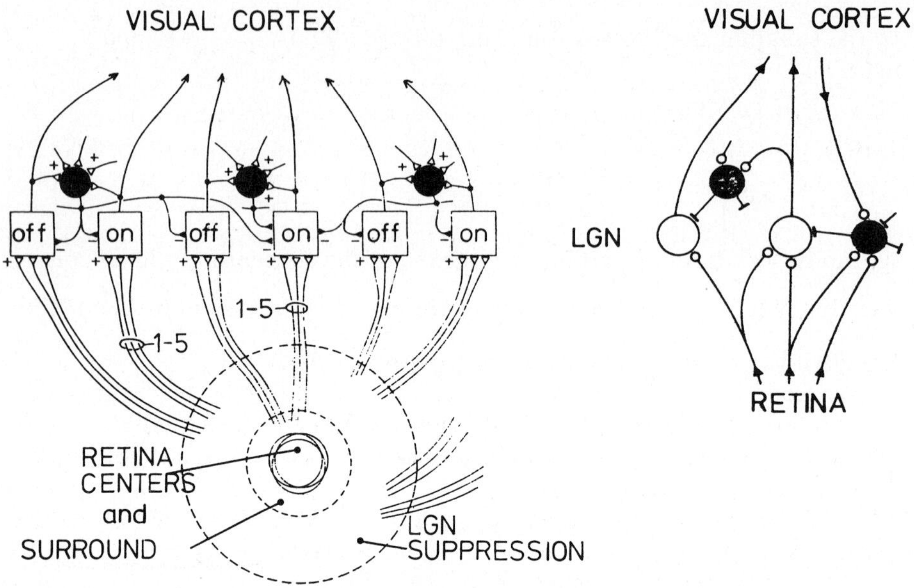

Fig. 4.7. Overview of thalamic neural circuitry. The two types of inter neurons and their effect, together with the interaction of the on- and off-cells are outlined. (After Dubin and Cleland 1977; Levick et al. 1972)

principal cell in the LGN (Dubin and Cleland 1977). There is a class of intrageniculate interneurons that receive direct input from retinal ganglion cells and from the visual cortex. These interneurons exert an inhibitory action on the principal cells. The triade: optic nerve terminal, interneuron dendrite, and principal cell dendrite synapse has been described by Shepherd (1979) as a general property of the synaptic organization of thalamic nuclei. Stevens and Gerstein (1976) came to the same wiring diagram on the basis of simultaneous recording from LGN cells. Previously Levick et al. (1972) had advanced the idea that interneurons are activated by collaterals from the principal cells that recurrently inhibit the same or neighboring principal cells. Dubin and Cleland (1977) showed that these interneurons indeed exist, but are located in a layer just above the classic LGN boundaries. They termed these neurons "perigeniculate" and made plausible that these interneurons do not receive cortical input. In the LGN therefore both feedforward (via the intrageniculate interneurons) and feedback (via the perigeniculate interneurons) inhibition is found. In addition one might consider the cortical inhibitory action via the intrageniculate interneurons as independent from the retinogeniculate excitatory input. The problem is whether this can exist independently from the feedforward inhibition. (An overview of thalamic circuitry is given in Fig. 4.7.)

The stochastic models that assume independent inputs appear to be less realistic than those assuming dependent input channels, since inhibitory action exerted by the intrageniculate neurons depends on the excitation by the retinal ganglion cells that also terminate upon the principal cell. The feedback inhibition produced by the perigeniculate cells is clearly dependent on the principal cell output which in turn depends partially on ganglion cell input. This situation corresponds to the model where only undeleted ϕ-events (thus p-events) produce ψ-events. Since neither the $\phi(t)$ nor $\psi(t)$ distribution will be Poisson an even more general equation than (4.26) has to be used. The inhibitory effect due to intrageniculate interneurons which receive retinal as well as cortical inputs must be described by the model where all ϕ events produce ψ events. In practice, both types of dependent inhibitory interactions are present, therefore some compound mechanism governs the output statistic of the LGN neurons. The intrinsic variability of thalamic neurons and their dependency on the (sleep) state of the brain is probably mediated by the intrageniculate interneurons as well.

How does this relate to the *receptive field* properties of the LGN neurons? RF's of LGN cells are of the center-surround type as found in the retinal ganglion cells but in addition have a so-called suppressive surround. Stimulation of this surround only suppresses activity produced by stimulating either the center or the antagonistic surround. In some models this extra inhibitory surround arises purely through convergence of retinal ganglion cells, in others the collateral recurrent inhibition through the perigeniculate interneurons is responsible (Fig. 4.7). It thus seems that for the LGN the wiring diagrams and some of the stochastic models are clearly compatible.

5 Neural Network Models

Models of large groups of neurons can be very diverse. In some models individual parameters of real life neurons can still be distinguished. In the models of Little (1974) and Shaw and Vasudevan (1974) one can still speak about a membrane potential and the individual action potentials. Simplifications and generalizations occur on both sides. One extreme is to consider only the average firing rate of a neuron as conveying information (e.g., Sejnowski 1976a), another extreme is to consider neurons as logical switching elements (e.g., McCulloch and Pitts 1943). It is interesting to note, however, that in the limit of large numbers of neurons both extremes are capable of representing the same neural net. We will discuss a limited sample of neural network models at this point and will start with the logical neuron models, then advance with more realistic Little-type neurons and finally treat some examples of continuous modeling of neural nets. In later chapters we will discuss more stereotyped network models which deal either exclusively with a particular brain structure or with particular properties such as plasticity, learning and associative memory.

5.1 Neurons as Logical Switching Elements

In 1943 McCulloch and Pitts published a paper under the title *A Logical Calculus of the Ideas Immanent in Nervous Activity.* The paper is not easy to read because of the use of logical operators but contains a wealth of ideas. The authors assume that the activity of neurons is all-or-none and that a certain number of synapses must be excited within a certain integration time to fire the neuron. The only delay in the neural net is the synaptic delay, τ, which is subsequently taken as the unit clock pulse for the discrete time model. It was assumed that any inhibition prevents excitation of the neuron at the time of its action. In addition the structure of the net was considered fixed in time. Any neuron was assumed to have N inputs $\xi_1, \ldots, \xi_N$ with weighting factors $w_1, \ldots, w_N$, a threshold θ, and an output η. The neural firing logic is described by

$$\eta(t+\tau) = \sigma\left\{\sum_i w_i\, \xi_i(t) - \theta(t)\right\}, \tag{5.1}$$

with $\sigma\{x\} = 1$ for $x > 0$ and otherwise 0.

Such logical neurons were not considered to represent the action of a single reallife neuron, rather it was assumed that with a small number of such logical neurons a real neuron can be modeled. As a result of the operations that can be performed by large numbers of logical neurons, it was concluded that every net with a memory, an adequate afferent input, and motor output behaves like a Turing machine, i.e., can perform any thinkable calculation. In principle it was proved that a large net of logical neurons can act as a brain.

Equation (5.1) was also used by Caianiello (1961) in slightly different form and then called the neuronic equation:

$$\eta_j(t) = \sigma \left\{ \sum_{i,n} w_{ij}\, \xi_i(t - n\tau) - \theta_j(t) \right\} .$$

Here it is assumed that a neuron i can influence the firing of neuron j up to $n_0\tau$ time steps in the future, thereby leaving the realm of Markov processes. An equation as the one given above represents a sequential machine, i.e., a machine in which the values of variables at one interval are determined by inputs on that interval and values of those variables at previous intervals.

McCulloch and Pitts (1943) were actually among the first to state the later famous Hebb learning principle (e.g., Hebb 1949) in the following words: "the phenomena of learning, which are of a character persisting over most physiological changes in nervous activity, seem to require the possibility of permanent alterations in the structure of nets. The simplest such alteration is the formation of new synapses or equivalent local depression of threshold. We assume that some axonal termination cannot at first excite the succeeding neuron; but if at any time the neuron fires, and the axonal terminations are simultaneously excited, they become synapses of the ordinary kind, henceforth capable of exciting the neuron. The loss of an inhibitory synapse gives an entirely equivalent result".

Caianiello introduced this learning element in his so-called mnemonic equations.

$$\frac{dw_{ji}(t)}{dt} = [\alpha\, \xi_i(t - n\tau)\, \eta_j(t) - \beta \cdot \sigma\{w_{ij}(t) - w_{ij}(0)\}] \cdot w_{ij}(t)\, \sigma\, \{W_{ij} - w_{ij}(t)\} ,$$

where α, β and W_{ij} are constants and parameters. The mnemonic equations reflect the assumption made above that learning coresponds to changes in efficacy of synaptic connections. It is important to note that the changes in $w_{ij}(t)$ occur rather slowly so that in Eq. (5.1) or the neuronic equivalent the w_{ij} may be treated as constant.

The assumption that introduced the logic into the switching elements was that a neuron fires if and only if the algebraic sum of all potentials impinging on its axons exceeds a certain threshold. By assuming output firings separated by a minimum of one delay, τ, the potentials are provided by all the connected neurons which fired in the previous cycle. By introducing a more general firing rule the neuron may become more realistic as in the so-called Little-neurons (Little 1979).

5.2 Little-Neurons as Network Elements: Introduction of Probability

Assume time discretized into instants τ that are of the order of one synaptic delay or one refractory period. In such an instant a neuron can fire at most only once. The state of the brain or the neural net is assumed to be given by a vector $\Xi(t) = |\xi_1, \ldots, \xi_N\rangle$ (see Box H in Chap. 6 for notation) where $\xi_i = +1$ represents firing of neuron i and $\xi_i = -1$ represents absence of firing. At time t the net can therefore be in 2^N states. A neuron i undergoes a change in membrane potential w_{ij} as a consequence of the arrival of a spike from neuron j ($\xi_j = +1$). The contribution of all neurons to the change in membrane potential of neuron i is therefore given by

$$\sum_j \left(\frac{\xi_j + 1}{2}\right) \cdot w_{ij} \ .$$

The probability that the ith neuron will fire is introduced analogous to the probability of spin flips in the Ising model (see Box G):

$$p(\eta_i = +1) = \left\{ \exp - \beta \left[\sum_j \left(\frac{\xi_j + 1}{2}\right) \cdot w_{ij} - \theta_i \right] + 1 \right\}^{-1} \ . \tag{5.2}$$

Since $p(\eta_i = -1) = 1 - p(\eta_i = +1)$, we can combine the probabilities of firing and nonfiring into one equation

$$p(\eta_i) = \left\{ \exp - \beta \eta_i \left[\sum_j \left(\frac{\xi_j + 1}{2}\right) \cdot w_{ij} - \theta_i \right] + 1 \right\}^{-1} \ . \tag{5.3}$$

In general, the threshold θ and the factor β will depend on the neuron i but we consider them presently as the same for all neurons.

It is assumed that the N neurons are independent and the operator P changes the initial state $|\xi_1, \ldots, \xi_N\rangle$ into state $|\xi'_1, \ldots, \xi'_N\rangle$. Then

$$\langle \xi'_1, \ldots, \xi'_N | P | \xi_1, \ldots, \xi_N \rangle = \prod_{i=1}^{N} p(\eta_i) \tag{5.4}$$

defines a $2^N \times 2^N$ matrix whose elements give the probability that after one cycle (time step τ) a particular state Ξ is transformed into state Ξ'.

This can also conveniently be written in the following form (Little and Shaw 1978)

$$P_{\eta\xi} = \frac{\exp \beta M_{\eta\xi}}{\sum_\zeta \exp \beta M_{\zeta\xi}} \ , \tag{5.5}$$

where $P_{\eta\xi}$ is the probability of going from firing pattern $|\xi_1, \ldots, \xi_N\rangle$ to firing pattern $|\eta_1, \ldots, \eta_N\rangle$, and

$$M_{\eta\xi} = \sum_{i=1}^{N} \eta_i \left[\sum_{j=1}^{N} \left(\frac{\xi_j + 1}{2}\right) w_{ij} - \theta_i \right] \ . \tag{5.6}$$

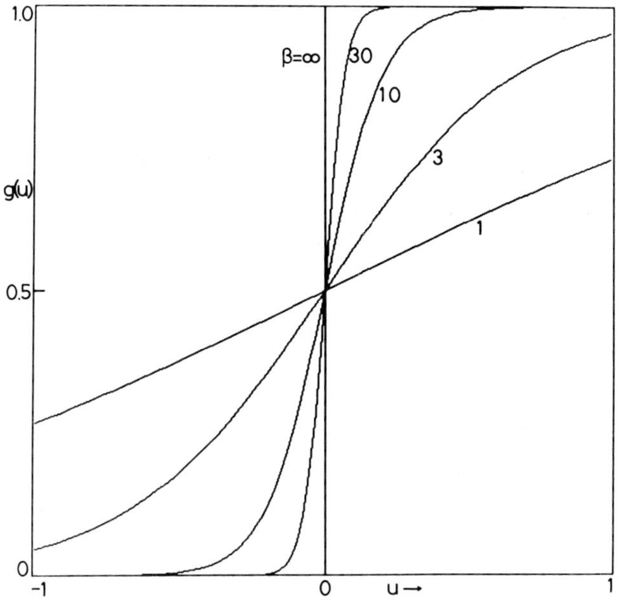

Fig. 5.1. Probability of output spikes, $g(u)$, as a function of the membrane potential (u) for various membrane noise (β) conditions. For very large β (low temperature limit) the neuron behaves as a logical switching element, for $\beta = 1$ the neuron's output is linearly related to the membrane potential (high temperature limit). Observe that the "Little neuron" in many ways resembles the neuron model shown in Fig. 4.1

The Eq. (5.5) shows some resemblance to the expression for the probability density function for the phase space [cf. Eq. (G.3)].

An interesting derivation of the dependence of the parameter β on the synaptic noise in the neurons has been given by Shaw and Vasudevan (1974):

$$\beta_i = 2\sqrt{\frac{2}{\pi}} \cdot \delta_i^{-1} , \tag{5.7}$$

with δ_i^2 the variance of the summated input to neuron i. Therefore the noise factor β, which was shown to be inversely proportional to absolute temperature in the Ising model, for the neural model is inversely proportional to the variance of the pooled synaptic input acting upon a neuron. If $\beta \rightarrow \infty$ (equivalent to $T = 0$, i.e., no thermal noise) then the neuron becomes a deterministic switching device. If $\beta \rightarrow 0$ (equivalent to the high-temperature limit) then a linearization of Eq. (5.5) is possible (Fig. 5.1):

$$P_{\eta\xi} = \frac{1 + \beta M_{\eta\xi}}{\sum\limits_{\zeta} (1 + \beta M_{\zeta\xi})} . \tag{5.8}$$

Because of the large noise every state becomes equally probable; for N neurons there are 2^N states and thus

$$P_{\eta\xi} = \frac{1+\beta M_{\eta\xi}}{2^N} \ . \tag{5.9}$$

It must be remarked that the Little model neuron can account for nearly all unimodal interval histograms encountered in Chapter 2, but because of the underlying Markov assumption cannot account for any type of multimodal interval histogram (Thompson and Gibson 1981 a, b).

5.3 Statistical Theories of Memory

Little (1974) assumed that the state of the brain at a particular instant of time corresponds to the configuration of spins in one row of the lattice in a crystal. The existence of a correlation between two states of the brain which are separated by a long period of time is directly analogous to occurrence of long-range order in a ferromagnetic crystal. This makes the formal treatment in a way comparable to the two-dimensional Ising model. There is, however, a problem since in the Ising model P is a symmetric operator: spins affect rows on both sides, and in the neuron model P is not symmetric due to causality. Nevertheless P is considered diagonalizable.

Matrix P with elements $p_{\eta\xi}$ is a strictly positive, nonsymmetric matrix with elements $\leqslant 1$ (because the elements describe transition probabilities). It has a largest eigenvalue equal to 1 and all other eigenvalues are smaller than 1.

Assume that after m steps state $|\xi_1, \ldots, \xi_N\rangle$ is transformed into $|\eta_1, \ldots, \eta_N\rangle$ then

$$\langle \Xi_\eta | P^m | \Xi_\xi \rangle = P^m_{\eta\xi} \ . \tag{5.10}$$

If P has eigenvectors ϕ_r with components $\phi_r(\xi)$: $\Xi_\xi = \sum_r \phi_r(\xi)$ and because $\phi_r(\xi)$ is normalized to unity:

$$\sum_\xi \phi_r(\xi)\phi_s(\xi) = \delta_{rs} \ ,$$

we obtain

$$\langle \Xi_\eta | P^m | \Xi_\xi \rangle = \sum_r \lambda_r^m \phi_r(\eta)\phi_r(\xi) \ . \tag{5.11}$$

If m is sufficiently large only one λ_r (equal to 1 or very close to 1) survives. The probability of obtaining Ξ_η after m cylces is $\Gamma(\eta)$. Let the initial state be obtained again after M (very large) cycles [use Eq. (5.11) and the orthonormality relation]:

$$\Gamma(\eta) = \frac{\sum_\xi \langle \Xi_\xi | P^{M-m} | \Xi_\eta \rangle \langle \Xi_\eta | P^m | \Xi_\xi \rangle}{\sum_\xi \langle \Xi_\xi | P^M | \Xi_\xi \rangle} = \frac{\sum \lambda_r^M \phi_r^2(\eta)}{\sum_r \lambda_r^M} = \phi^2_{max}(\eta) \ . \tag{5.12}$$

In general, the joint probability of reaching state Ξ_ζ after one cycle provided it was Ξ_η after m cycles (1 > m)

$$\Gamma(\zeta,\eta) = \sum_{\xi} \langle \Xi_\xi | P^{M-1} | \Xi_\zeta \rangle \langle \Xi_\zeta | P^{1-m} | \Xi_\eta \rangle \langle \Xi_\eta | P^m | \Xi_\zeta \rangle / \sum_{\xi} \langle \Xi_\xi | P^M | \Xi_\xi \rangle$$

$$= \frac{\sum_r \lambda_r^M \phi_r^2(\zeta)\phi_r^2(\eta)}{\sum_r \lambda_r^M} = \phi_{max}^2(\zeta)\phi_{max}^2(\eta) \; . \qquad (5.13)$$

Therefore $\Gamma(\zeta,\eta) = \Gamma(\zeta)\cdot\Gamma(\eta)$, thus states Ξ_η and Ξ_ζ are independent and there are no persistent states.

Only in case that λ_{max} is degenerate then there are at least two eigenfunctions ϕ_1 and ϕ_2 belonging to λ_{max} and

$$\Gamma(\zeta,\eta) \neq \Gamma(\zeta)\cdot\Gamma(\eta)$$

thus a persistent state may occur, i.e., the neural net is endowed with memory.

Memory is formed generally by learning. In the present model it is assumed (Little and Shaw 1978) that w_{ij} is determined by the time average $\overline{C\xi_i\eta_j}$ of the past correlated firing history of the neurons i and j: C is a function of the long-term firing history of the neurons i and j.

We have seen that memory states of the brain can be represented by eigenvectors. Such eigenvectors are linear combinations of the 2^N states of a neural network of N neurons. These linear combinations are considered not to be physiologically interpretable. Shaw (1978) and Roney and Shaw (1980) have adopted the assembly definition of John (1972): the response of a single stimulus presentation averaged across many neurons of an assembly would be the same as the PSTH or averaged response of a single neuron in the assembly to many presentations of stimuli to the animal. Assume n assemblies in the population of N neurons which are characterized by their average firing rate which was considered either larger (+), equal (0), or smaller (−) than the spontaneous activity. A pattern χ is introduced that describes the total firing for each assembly in a given time step. If $m = N/n$ is sufficiently large so that a given χ^A dominates in a linear combination of the 2^N states and in such χ^A only one assembly has + while all others are 0. Thus the linear combination now becomes equivalent to χ^A and this is considered physiologically interpretable. It is further assumed that χ^A represents a set of equivalent brain states. Such a network can learn n^2 transitions $\chi^A \rightarrow \chi^B$. Roney and Shaw (1980) introduced a modified Hebb hypothesis: the individual w_{ij} are assumed dependent on the correlation of firings between the assembly in which neuron i is a member and the assembly in which j is a number.

The physiological realisation could be a cortical hypercolumn with $N \approx 10^3$, each assembly could be an orientation column ($n \approx 30$) and thus $m \approx 35$. This means that a 30×30 matrix describes all transitions between the 3^{30} neural network states but only the 30 assembly states are of importance and physiologically interpretable (e.g., Shaw et al. 1982).

In the concepts of Roney et al. (1979) and Shaw et al. (1982) assemblies are tightly coupled neurons featuring so-called dendritic bundles which even might show electrotonic coupling. Thus basically all neurons in an assembly behave in the same way. If then an assembly represents a given state of the neural net, then each neuron out of that assembly represents that state. This is easily extrapolated to the notion that "whenever two stimuli can be distinguished, in normal life or in a psychophysical experiment, then proper analysis of the impulses occurring in a single neuron would enable them to be distinguished with equal or greater reliability" (Barlow 1985). Thus an assembly in the sense of John (1972) and Shaw is actually a very reliable single neuron – because average rate is used – that is protected against faulty responses by sheer redundancy. This would indicate that the large number of nerve cells in the brain mostly serve to introduce reliability and that the actual coding and information transfer requires only the square root of the number of nerve cells, if at least we may extrapolate the ratio of number of assemblies to the number of cells in a cortical column to hold for the entire brain.

A slightly different model is advanced by Hopfield (1982): the neurons act asynchronously and are two-state McCulloch-Pitts threshold devices involving sudden $0 \rightarrow 1$ or $1 \rightarrow 0$ changes at random times. It appears that content-addressable memory is one of the simplest collective properties of a system of such highly interconnected neurons. The connection matrix $W = \{w_{ij}\}$ contains many memories simultaneously and these can be reconstructed individually from partial information in an initial state, i.e., by association. It was also noted that this model was isomorphic with the Ising model. But because W is a symmetric but random matrix its physical counterpart was supposed to be a spin glass model with many local stable states. In a subsequent paper, Hopfield (1984) demonstrated that a continuous model with high gain had the same stable states as the stochastic switching model.

Let C_i and R_i be the membrane capacitance and resistance of neuron i with membrane potential u_i and short term average firing rate V_i, then

$$C_i \frac{du_i}{dt} = \sum_j w_{ij} V_j - \frac{u_i}{R_i} + I_i \tag{5.14}$$

$u_i = g_i^{-1}(V_i)$ is the "inverse" output-input relation , (5.15)

$w_{ij} V_j$ represents the electric current input into cell i produced by cell j, I_i is any other fixed current to neuron i.

This set of coupled linear-nonlinear equations is thus capable of simulating the same network as the McCulloch-Pitts neurons. Hopfield showed this with action potentials of "size" $V_0 \delta(t-t_i)$ with firing probability $F g(u)$ where F is maximum firing rate. When V_0 is small the evolution of the state of the network in the form of a Master equation can be written as

$$\frac{\partial P}{\partial t} = \sum_i \frac{1}{RC} \frac{\partial u_i P}{\partial u_i} - \sum_j \frac{\partial P}{\partial u_j} \cdot \frac{V_0 F}{C} \sum_i w_{ij} g(u_j) + \frac{V_0^2 F}{2C^2} \sum_{i,j,k} g(u_k) w_{ik} w_{jk} \frac{\partial^2 P}{\partial u_i \partial u_j} , \tag{5.16}$$

where $P(u_1, \ldots u_i, \ldots u_N, t)\, du_1 \ldots du_i \ldots du_N$ is the probability that at input 1 has value $u_1, \ldots$, input i has value $u_i, \ldots$ etc.

The master equation for this system does not have a detailed balance, indicating that the quantal noise is not characterized by a temperature.

Box G. Statistical Mechanics and the Ising Model

We consider systems comprising N identical units $i = 1, \ldots, N$. To each units is assigned a set of three coordinates q_1, q_2, q_3, and a set of three momenta p_1, p_2, p_3. (If the units have mass m then p_i is defined as $p_i = m \frac{\partial q_i}{\partial t} = m\dot{q}_i$.) N units have 3N coordinates and 3N momenta.

One may describe the state of the system as a point in a 6N-dimensional phase space. The time evolution of the system can then be represented by a trajectory in this phase space. Its movement through phase space is governed by a Hamiltonian $H(p, q)$ through the equations of motion:

$$\frac{\partial q_i}{\partial t} = \frac{\partial H}{\partial p_i} \; ; \quad \frac{\partial p_i}{\partial t} = -\frac{\partial H}{\partial q_i} \; . \tag{G.1}$$

A common way to describe the dynamics of the system is by the evolution of a probability density function $\varrho(p, q; t)$. By definition,

$$\varrho\, d\Omega = \text{Prob \{system at t in volume element at } d\Omega \text{ around } (p, q) \text{ in phase space\}} \; .$$

For the systems considered, $\varrho\, d\Omega$ is the fractional number of points in volume element $d\Omega$ around (p, q). Because N is conserved, Eq. (G.1) leads to Liouville's theorem:

$$\frac{\partial \varrho}{\partial t} + \sum_{i=1}^{3N} \left(\frac{\partial \varrho}{\partial q_i} \frac{\partial H}{\partial p_i} - \frac{\partial H}{\partial q_i} \frac{\partial \varrho}{\partial p_i} \right) = 0 \; . \tag{G.2}$$

In stationary situations $\frac{\partial \varrho}{\partial t} = 0$, and

$$\varrho(p, q) = \frac{\exp[-\beta H(p, q)]}{\sum\limits_{(p, q)} \exp[-\beta H(p, q)]} \; . \tag{G.3}$$

The numerator $Z = \sum\limits_{(p,q)} \exp[-\beta H(p, q)]$ is called the partition function, and is a normalization constant to ensure $\int\limits_{\Omega} \varrho\, d\Omega = 1$, $\beta = (k_B T)^{-1}$ with k_B Boltzmann's constant and T is the absolute temperature.

Box G (continued)

In classical systems $H(p, q)$ is an expression of the energy of a system in terms of coordinates and momenta. Sometimes such a Hamiltonian cannot be defined in this sense, but generally it can be in terms of extensive quantities. These are quantities that are proportional to N, such that the extensive quantities of subsystems add, but their probabilities multiply.

As an example consider the *Ising spin model* describing a ferromagnetic lattice where magnetic spins can take two positions $\xi_i = \pm 1$. At the site of each spin there is assumed a molecular magnetic field due to the surrounding spins:

$$h_i = \sum_j w_{ij} \xi_j \,, \tag{G.4}$$

where w_{ij} is the exchange interaction. In addition, an external field h_i^0 may be present. The effect of the field $(h_i + h_i^0)$ is such that spins tend to align, this is expressed through the correlation function $h_i \xi_i > 0$.

It is assumed that the interactions are only local and a Hamiltonian is defined according to

$$H\{\Xi\} = -\sum_{\langle i,j \rangle} w_{ij} \xi_i \xi_j - \sum_i h_i^0 \xi_i \,, \tag{G.5}$$

where $\langle i, j \rangle$ indicates a nearby spin pair, Ξ is a particular spin configuration $|\xi_1, \ldots, \xi_N\rangle$, and

$$\varrho\{\Xi\} = \frac{\exp(-\beta H\{\Xi\})}{\sum_{\Xi} \exp(-\beta H\{\Xi\})} \,. \tag{G.6}$$

Now consider a two-dimensional lattice with N rows and N columns. Let Ξ_α $(\alpha = 1, \ldots, N)$ be all spin coordinates of row α, with boundary condition $\Xi_{N+1} = \Xi_1$. A lattice configuration is described by $(\Xi_1, \ldots, \Xi_N)$. Define the *exchange energy within a row* as

$$E(\Xi) = -w \sum_{k=1}^{N} \xi_k \xi_{k+1} - h^0 \sum_{k=1}^{N} \xi_k \,, \tag{G.7}$$

where the exchange and the effect of the external field are considered isotropic. Define the *exchange energy between rows*

$$E(\Xi, \Xi') = -w \sum_k \xi_k \xi_k' \,, \tag{G.8}$$

with $\Xi = |\xi_1, \ldots, \xi_N\rangle$ and $\Xi' = |\xi_1', \ldots, \xi_N'\rangle$.

Box G (continued)

Define a matrix with elements

$$\langle \Xi | P | \Xi' \rangle = \exp(-\beta\{E(\Xi, \Xi') + E(\Xi)\}) \ , \tag{G.9}$$

which governs the transition probability between states Ξ and Ξ'. Using the bra(C)ket notation: $\langle \Xi |$ represents a row vector and $| \Xi \rangle$ a column vector, the partition function is defined as:

$$Z = \mathrm{Tr}\, P^N = \sum_{\Xi_1} \ldots \sum_{\Xi_N} \langle \Xi_1 | P | \Xi_2 \rangle \langle \Xi_2 \rangle \langle \Xi_2 | P | \Xi_3 \rangle \ldots \langle \Xi_N | P | \Xi_1 \rangle \ . \tag{G.10}$$

Since P is a symmetric matrix, it can be diagonalized. The diagonal then represents the eigenvalues of P: $\lambda_1, \ldots, \lambda_N$. Then P^N is also diagonal with eigenvalues $(\lambda_1)^N, \ldots, (\lambda_N)^N$, and $Z = \sum_\alpha (\lambda_\alpha)^N$.

Let $\varrho(\Xi; t+\Delta t)$ be the probability of finding the system in state Ξ at $t+\Delta t$, then

$$\varrho(\Xi; t+\Delta t) = \sum_{\Xi'} \langle \Xi | P | \Xi' \rangle \cdot \varrho(\Xi'; t) \cdot \Delta t \ , \tag{G.11}$$

which leads to a master equation

$$\frac{\partial \varrho(\Xi;t)}{\partial t} = \sum_{\Xi \neq \Xi'} \left\langle \Xi \left| \frac{\partial P}{\partial t} \right| \Xi' \right\rangle \cdot \varrho(\Xi', t) - \sum_{\Xi'' \neq \Xi} \left\langle \Xi'' \left| \frac{\partial P}{\partial t} \right| \Xi \right\rangle \cdot \varrho(\Xi; t) \ . \tag{G.12}$$

If the principle of detailed balance holds $\left(\text{write } \dot{P} = \frac{\partial P}{\partial t}\right)$:

$$\langle \Xi' | \dot{P} | \Xi \rangle \exp(-\beta H(\Xi)) = \langle \Xi | \dot{P} \rangle \Xi' \rangle \exp[-\beta H(\Xi')] \ . \tag{G.13}$$

According to the Perron-Frobenius theorem, there is only one eigenvalue of P^N that is equal to 1: $\lambda_1 = 1$, all other λ_r are smaller than 1, provided P^N is irreducible. Then Eq. (G.11) becomes for $t \to \infty$

$$\varrho(\Xi; t=\infty) = \lim_{N\to\infty} \sum_{\Xi'} \langle \Xi | P^N | \Xi' \rangle \cdot \varrho(\Xi'; 0) \ . \tag{G.14}$$

With $\langle \xi_1 |$ and $| \xi_1' \rangle$ the left respectively right eigenvectors belonging to the largest eigenvalue λ_1 this leads to

$$\varrho(\Xi; t=\infty) = \sum_{\Xi'} \langle \xi_1 | \xi_1' \rangle \cdot \varrho(\Xi'; 0) \ . \tag{G.15}$$

The state Ξ has therefore a finite asymptotic probability if the eigenvectors $\langle \xi_1 |$ and $| \xi_1' \rangle$ are nonorthogonal. One may speak of a persistent state if both eigenvectors belong to λ_1, i.e., the state is degenerate. A general reference is Huang (1963).

5.4 Statistical Neuromechanics

Statistical mechanics deals with properties of ensembles of identical and interacting units (see Box G). The ensemble can be in a continuum of states or in a number of discrete states. A noise source (e.g., temperature) introduces transitions between possible states of the ensemble. One generally distinguishes between the stationary properties of the ensemble and its dynamics. The stationary properties are completely determined by the extensive quantities of the system, i.e., those quantities of which the value is proportional to the number of units in the ensemble. Hamiltonians are extensive quantities and one has developed the habit of calling all quantities that have this extensiveness property Hamiltonians. The dynamics of the ensemble are governed by the master equation; this generally assumes a Markov property; the next state of the system is determined only by the present state and the time until the next state. Only in case the Markov assumption holds, i.e., when the principle of detailed balance holds, can the ensemble be described by a Hamiltonian.

The principles of statistical mechanics can be applied to neural nets consisting of identical units. To apply this to the cortex will require that one only considers, e.g., the ensemble of pyramidal cells as in Braitenberg's (1974) skeleton cortex. The state of the cortex must then be equated with the state of the pyramidal cells. The state of a neural net can be represented either as a binary code as in Little's (1974) neurons, as the probability of firing of a neuron, as its average firing frequency (short-term average), as the average firing rate of its assemblies (Roney and Shaw 1978) etc.

One may now proceed in two ways. The classical way has been described by Cowan (1968), who formulates the equations of motion that represent the changes in nervous activity that result from interactions within and between assemblies in nervous nets. In this case the statistics are related to ensemble averages. Let P_r be the probability of firing of the rth neuron, then

$$\frac{dP_r}{dT} = (\gamma_r + \alpha_r^{-1} \sum_r w_{sr} P_s)\, P_r (1-P_r) \qquad r = 1, \ldots N;\ s = 1, \ldots N\ . \tag{5.16}$$

γ_r and α_r are growth coefficients, w_{sr} is a coupling coefficient. T is dimensionless time in units of the mean interneuronal delay. For a neural net with reciprocal connections the equations transform to

$$\frac{dv_r}{dT} = \sum_s \lambda_{sr} \frac{\partial H}{\partial v_r}\ , \tag{5.17}$$

where $v_r = \ln \left\{ \frac{P_r}{P_r^*} \cdot \frac{1}{1-P_r} \right\}$; $\lambda_{sr} = \frac{w_{sr}}{\alpha_s \alpha_r}$ and $H = \sum_r \alpha_r \{(1 + P_r^* \exp v_r) - P_r^* v_r\}$

with P_r^* as the steady-state value of P_r.

The H so defined is a positive constant of motion of a system of neural oscillators, i.e., a Hamiltonian. Then this H is a function of certain logarithms of the probability of firing in the net:

$$H = \sum \alpha_r \, (-P_r^* \ln P_r) - (1-P_r^*) \ln (1-P_r) + P_r^* \ln P_r^* \ . \tag{5.18}$$

Cowan (1968) now argues that if one considers such a single neural oscillator continually bombarded by other neural circuits, then the average behavior of the oscillator can be viewed as if it were a member of a very large ensemble. In such a case there is an invariant measure of the form of a density function:

$$\varrho = \frac{\exp(-\beta H)}{\prod\limits_{\Gamma} \int\limits_{-\infty}^{\infty} \exp(-\beta\, H_r)\, dv_r} \ , \tag{5.19}$$

where β^{-1} is formally analogous to the thermodynamic temperature. It was shown that β is a function of the first and second moment of the distribution of firing probabilities, i.e., a function of the mean and variance of the firing rate.

In this case the Hamiltonian is formulated in the classical sense. Another possibility is to define a system Hamiltonian of an ensemble of the type encountered in the Ising model (See Box G). Hopfield (1982) has proposed such a Hamiltonian

$$H(I) = -\sum_{\langle ij \rangle} J'_{ij}\, \sigma_i \sigma_j - \sum_i h_i^0\, \sigma_i \ , \tag{5.20}$$

in which

$$J'_{ij} = (J_{ij} + J_{ji}); \quad J_{ij} = 1/2\ w_{ij}, \quad \sigma_i = 2S_i - 1; \quad S_i = +1,0; \quad h_i^0 = \sum_j \frac{w_{ij}}{2} - \theta_i ,$$

with w_{ij} the synaptic efficacy; (I is a configuration of S_i's). A Hebb-like learning rule can be incorporated and the dynamics are described by

$$W(I\,|\,J) = \frac{\exp[-\beta H(I)]}{\sum\limits_K \exp[-\beta H(K)]} \ , \tag{5.21}$$

where $W(I\,|\,J)$ is the transition probability between state J at time zero and state I at present. The study of the master equation together with the principle of detailed balance also lead to an Ising-type Hamiltonian. Peretto (1984) derives from Little's results, given

$$H(I\,|\,J) = -\sum_{ij} J_{ij}\, \sigma_i(I)\, \sigma_j(J) - \sum_i h_i^0\, \sigma_i(I) \ , \tag{5.22}$$

that

$$W(I\,|\,J) = \frac{\exp[-\beta H(I\,|\,J)]}{\sum\limits_K \exp[-\beta H(K\,|\,J)]} \ .$$

In the limit of small β, i.e., high spontaneous activity (noise), Peretto shows that

$$H(I) = -\frac{\beta}{4} \sum_{i,j=1}^{N} \left(\sum_{K} J_{ik} J_{kj} \right) \sigma_i(I) \sigma_j(I) \ , \tag{5.23}$$

i.e., the Hamiltonian contains coupling coefficients that are equal to quadratic functions of synaptic efficacies.

The application of statistical mechanical techniques allows a new approach to long-term memory. Both the Hopfield neural net and the Little neural net appear to have the same capacity for long-term memory. The Hopfield model can, because of its Hamiltonian character, give a precise meaning to the idea of associative memory. One may view the phase space as divided into basins of attractions with states of minimal energy; these are the learned states. Noise allows the net to go from one state to another. The dependence of the order parameter of the noise level can be calculated using a mean field approximation:

$$H = -\sum_{ij} J_{ij} \sigma_i \sigma_j \simeq -\sum_{ij} J_{ij} \langle \sigma_j \rangle \sigma_i = \sum_{i} h_i^m \sigma_i \ , \tag{5.24}$$

in which $\langle\,\rangle$ indicates ensemble averaging and h_i^m is the mean field at the site of the ith neuron produced by all other neurons. The number of possible memory states is a logarithmic function of the size of the net and of the noise level (see also Chap. 11):

$$M^* = a \log N + b \ . \tag{5.25}$$

M^* is the number of memory states, a is a constant and b depends on the noise level. In this model stimuli play the role of the external magnetic field in the Ising model. They have a more pronounced effect when the applied stimulus resembles a memorized state of the net because the system settles in a stable state much faster in such a case. When the noise level b exceeds a critical level there is no longer correct associative recall.

The approach of statistical mechanics basically demonstrates the capability of a net composed out of Hopfield or Little neurons to store large numbers of patterns (memory) and the role which is played by the synaptic noise. Learning rules can be incorporated and introduction of dynamics will increase the efficacy of associative recall. However, the statistical mechanics approach which was so powerful in explaining the behavior of macroscopic properties of physical systems in terms of microscopic configurations largely fails when applied to the brain. The most obvious reason is that we do not know meaningful macroscopic observables which can be explained by the collective properties of either spike generation (Little 1974) or state variables (Hopfield 1982). We will later on (Chapters 10 and 13) encounter macroscopic measures as evoked potentials and EEG and the attempts that have been taken to relate them to states of neural nets.

5.5 Neural Field Theories

Field theories introduced by Beurle (1956) relate two quantities which mutually influence each other. These are the field quantities such as the densities of action potentials $\vec{\psi}(r,t)$ which are assumed to satisfy second-order partial differential equations, and the sources or sinks of the field with a strength corresponding to the mean firing rate, $\vec{F}(r,t)$. Both $\vec{\psi}(r,t)$ and $\vec{F}(r,t)$ are continuous functions with existing first- and second-order derivatives.

Griffith (1971) assumes that there exists a linear operator H that relates $\psi(\vec{r},t)$ and $F(\vec{r},t)$. Under the conditions that the connections that a neuron receive from neurons at a distance r fall of as $r^{-1}\exp\left(\frac{1}{2}\beta r\right)$, this relation reads:

$$H\psi \equiv -\left(\frac{1}{v}\frac{\partial}{\partial t}+\frac{1}{2}\beta\right)^2 \psi+\nabla^2\psi = -4\pi f(\psi) \equiv F \ , \tag{5.26}$$

where v is the conduction velocity of the action potentials along the axons and $f(\psi)$ is a sigmoidal response function that relates F to ψ. A modified field equation was introduced by Fischer (1973b):

$$\left(\tau_0 \frac{\partial}{\partial t} - 1\right) H\psi = F \ , \tag{5.27}$$

where τ_0 is a characteristic time for the decay of the action potential activity. The linear operator was considered to contain an interaction-free term

$$H_0 = r_0^2 \nabla^2 \ , \tag{5.28}$$

where r_0 is a characteristic decay distance for action potential activity, and an interaction term H_1 defined by

$$H_1\psi = \int V(\cdot)\, \psi \, d\vec{r}\, dt \ , \tag{5.29}$$

where $V(\cdot)$ is a potential function depending on the distance in time and space of the strength of source or sinks. When $V(\cdot)$ is positive the interaction is excitatory, when $V(\cdot)$ is negative it is inhibitory. Using the Greens function method, special solutions for $\psi(\vec{r},t)$ are then found for various forms of the potential function. In a slightly different approach, Anninos et al. (1970) considered as a variable the fractional number of active neurons in a neural net at time $t = n\tau$, where τ is the unit delay time. Expressions were derived relating the expected fraction of active neurons at $t = (n+1)\,\tau$ to the fraction active at $t = n\tau$. It was found that for certain threshold conditions a small neural net behaves as a highly nonlinear decision element, not unlike a single neuron in the case of the interactions being only excitatory. Introduction of inhibition in the neural net introduces additional stable states besides saturated firing or complete inactiveness (Wilson and Cowan 1972); these were interpreted as memory states or giving rise to bistable perceptive patterns (Wilson and Cowan 1973).

For a spatially homogeneous net, where postsynaptic potentials summate in a linear fashion and where the neurons are densely interconnected Brannan and Boyce (1981) derived a Fokker-Planck equation [cf. Eq. (4.7)] that has the form

$$\frac{\partial p(x,t)}{\partial t} = -\frac{\partial}{\partial x} J(x,t) , \tag{5.30}$$

where p(x, t) is the probability density function of the proportion of neurons with transmembrane potential x. For $x = \theta$ (threshold) $J(\theta, t)$ is the proportion of neurons discharging with time and can be related to the PSTH of a single neuron via the assembly hypothesis of John (1972).

5.6 Interaction in Neural Nets

Interaction can be considered between individual neurons in a neural net. The problem here is not to formulate neural interaction equations but to define an operational variable that mediates the interaction. For large neural nets one might consider the average firing rate of a neuron as such an operational variable. Sejnowski (1976a) considers a filtered version thereof, the average membrane potential, as the more basic variable. The interaction equations proposed read

$$\begin{aligned} \phi_i &= \varepsilon_i + \sum_j w_{ij} r_j \\ r_i &= r_i^* P\left[\frac{\phi_i - \theta_i}{\sigma_i}\right] , \end{aligned} \tag{5.31}$$

where ϕ_i is the average membrane potential of neuron i, ε_i is the average external input to neuron i, r_i is its average firing rat, r_i^* the maximum firing rate, $P[\cdot]$ the probability of firing determined by threshold θ_i and membrane noise σ_i; w_{ij} are the elements of the connection matrix.

After defining an effective connection matrix $K_{ij} = w_{ij} r_j^*$ both equations from (5.31) can be combined into

$$\phi_i = \varepsilon_i + \sum_j K_{ij} P\left[\frac{\phi_j - \theta_j}{\sigma}\right] , \tag{5.32}$$

with σ the value (constant) of the membrane noise standard deviation of each neuron. Assume that there is an equilibrium solution $\phi_i(o)$ and expand ϕ_i in a Taylor series around $\phi_i(o)$. The first-order terms satisfy an equation

$$\phi_i' = \varepsilon_i' + \lambda^{-1} \sum_j K_{ij}' \phi_j' , \tag{5.33}$$

with $K_{ij}' = K_{ij} P'\left[\frac{\phi_j(o) - \theta_j}{\lambda}\right]$ called the *interaction matrix*. Since $P'[\cdot]$ has a significant value only around $\phi_j(o) = \theta_j$, i.e., around threshold, this first-order equation describes the interaction between spontaneously active neurons.

The second-order equation then describes the interaction between neurons that have membrane potentials just below or just above threshold (where $P''[\cdot]$ is large). Sejnowski (1976a) calls them "border neurons". The main issue in this paper is how feature detection arises from neuronal interaction and this is treated as a variational problem. By operationally defining a feature as a stimulus that produces maximum activity in a neuron, features arise when $\phi' = 0$ and $\phi'' \leqslant 0$. This can only occur when the interaction matrix contains a sufficient number of negative elements, i.e., when the net contains inhibitory connections.

When the membrane potential deviates considerably from the resting value, it decays with time constant τ_0 and the interaction equation becomes

$$\tau_0 \phi_i' + \phi_i = \varepsilon_i + \sum_j K_{ij} \cdot \frac{r_j(\phi_j)}{r_j^*} . \tag{5.34}$$

It is now postulated that "information is coded as correlations between spike trains and processed in a collection of neurons by correlation between membrane potentials" (Sejnowski 1976b).

The covariance between the membrane potentials of neurons i and j, $C_{ij}(t, t')$ then satisfies the following equation

$$\left[\tau_0 \frac{d}{dt} + H(t)\right] C_{ij}(t, t') = C_{\varepsilon\phi}(t, t'); \; H(t) = 1 - K_{ij}'(t) , \tag{5.35}$$

where $C_{\varepsilon\phi}(t, t')$ is the covariance between the external input to one neuron i and the membrane potential of the other (j).

From this it becomes clear that the neurons with membrane potentials close to threshold are maximally sensitive to correlated inputs and make the largest contribution to the covariance equation.

The idea of relating information coding to ensemble correlation makes it possible to describe some common neural codes that we have encountered in Chapter 1 in terms of correlations. A special pattern in a single nerve fiber's spike train is then related to the autocorrelation, i.e., to the diagonal terms of the correlation matrix of the neural ensemble. An interval code in a single nerve fiber is then a very simple case of autocorrelation. When all neurons are independent, the ensemble correlation reduces to a single neuron code such as rate code or interval code. Impulse coding in a correlated pair of neurons conveys information described by the cross-correlation, i.e., an off-diagonal term of the correlation matrix. This has been termed the "transmission train" by Frostig et al. (1984), a subset of spikes from both neurons that contribute to the cross-correlation. Sejnowski (1976b) suggests testing the hypothesis that information is processed by correlations between membrane potentials by correlating intracellular recordings. This obviously is very difficult to carry out in the CNS of vertebrates. In Chapter 7 we will show that an estimate of this correlation can be obtained on the basis of correlation of action potential sequences simultaneously recorded from neurons. First of all, however, a survey of experimental techniques that will allow us to gather the data necessary to perform such a calculation will be given in the next chapter.

6 Multiple Separate Recordings from Neural Nets

In order to test the predictions of neural network modeling, it is necessary to observe the activity of neural populations. This requires that the number of simultaneously recorded neural units is considerably increased above what is common in single- or few-unit recordings. The number of units observed is equal to the number of recording devices times the number of units recorded per device. The problem therefore splits in two. On the one hand it comprises efforts to increase the number of recording sites through the design of multielectrode arrays, e.g., using microcircuit fabrication techniques, or through the potentially important optical techniques using photodiode arrays. On the other hand, progress has been made to increase the number of units that can be recorded simultaneously with one electrode through new developments in spike separations techniques.

As far as recording is aimed at testing continuous modeling approaches as discussed in the previous chapter, one might use relatively gross recordings such as intracranially recorded evoked potentials, e.g., subjected to current-source-density analysis (Freeman 1975; Nunez 1981) or use multiunit activity to detect activity in local circuits (e.g., Legatt et al. 1980). At the moment the spatial resolution of the optical recording methods in the vertebrate brain is limited and at each recording site population activity is represented. Thus also this technique when used as a multiunit recording device in vivo currently can only be used to test the continuous modeling results.

Techniques related to discrete neural network models are those using multielectrode arrays which probe the global structure and the single electrode multiunit spike trains separation methods which probe the network at the local circuit level. In this chapter we will survey briefly the multielectrode array and optical recording techniques and deal at length with the multiunit spike train separation techniques.

6.1 Recording from Multiple Sites in the Brain

Various types of multielectrode systems have been designed with the purpose of recording simultaneously from multiple sites in the brain. Often the underlying philosophy is that feature extraction is a property of a neuronal population and by probing such a population at a set of points one might come across "feature-specific spatiotemporal patterns of activity" of the neural population. The recording systems used can be based on mechanically independent electrodes each with their own micromanipulator. The size of the manipulators and electrode attachment systems usually is prohibitive to a large number of electrodes. An example is provided by the system of four individually manipulated microdrives described by Radil-Weiss and Skvaril (1971).

Somewhat less independence of the electrode placements is found in the system designed by Reitböck (1983) featuring 19 independently controllable microelectrodes. The electrodes are fixed in the horizontal plane but can be independently driven vertically. The device, which is rather bulky, consists of a stepping motor which, through a system of brakes and clutches, can be applied to each electrode individually. The interelectrode distance is 200 μm and the step size is 2 μm. A version featuring only seven electrodes was described by Reitböck and Werner (1983). Complete determination of electrode tip distances is obtained with arrays of rigidly coupled microelectrodes. An example is formed by the 5×6 array of 30 microelectrodes with a separation of 160 μm as described by Krüger and Bach (1981), and which was used to record from the monkey visual cortex.

Kuperstein and Whittington (1981) describe a practical 24-channel microelectrode where the recording sites, spaced at 85-μm intervals, are along the edges of a supporting electrode. They applied this to record from the rat trigeminal nerve, the somatosensory cortex of the rat, and the visual cortex of the cat.

Multiwire systems, providing again a higher degree of independence between the electrodes, have been used by Gerstein et al. (1983).

For a more extended overview of the various systems with their manufacturing details the reader is referred to Krüger (1983).

A completely different technique, not yet matching the multielectrode array in its single unit resolution, is provided by the so-called *optical recording technique.* The principle of optically monitoring neuronal activity (Cohen and Salzberg 1978; Grinvald 1984, 1985) is based on the application of voltage-sensitive dyes to the preparation under study. The dye molecules bind to the nerve cell membrane and transform changes in membrane potential into changes in the optical properties of the dye. These changes result in either a change in the absorption of transmitted light or in changes in the level of fluorescence of the stained cells. The signal changes, detected with fast responding photodiodes with a response time of less than 10 μs at temperatures between 13 and 23 °C, are linearly related to changes in membrane potential. Thus when the fluorescence recording is made from the same site as the electrical recording (e.g., in large invertebrate cells), the time course of the two signals is virtually identical. Optical recording thus is a measurement of the intracellular potential. Due to the so far limited resolution of around 50×50 μm in the vertebrate central nervous system the response is inherently a multiunit response. It is appropriate to call the response an "intracellular population response". This may make the interpretation of the optical signals simpler than those of field potentials, which also represent population activity with about the same resolution. The evoked potential signal recorded with large microelectrodes is an extracellularly recorded signal and has complex relationships to the membrane potential changes in the cell population; it is often proportional to a weighted average of first and second derivatives of the membrane potential, depending on recording geometry and distance to the cells. It is assumed that under uniform dye-binding conditions the signal size of the optical measurements is proportional to the membrane area of the cells viewed by the detector. Field potentials generally are also dependent on the cell size, the orientation of the equivalent dipoles and the recording distance. In cases of a so-called "closed field" (Nunez 1981) there will be no – or a very small – net potential at the recording electrode due to cancellation of the individual dipole fields. In these cases optical recordings will still show the activity of the cell cluster. It has been concluded therefore that optical and field potential measurements provide different but complementary information (Orbach et al. 1985).

The transmitted light version of the optical recording technique has been applied to the study of the neurohypophysis in *Xenopus* (Salzberg et al. 1983). It was assumed that synchronized action potential volleys were recorded in the nerve terminals, and it was found that they had relatively long durations (full width at half height approximately 6 ms). This was attributed to temporal dispersion in the timing of the cells in the population, also on basis of the duration of the electrically recorded compound action potential in the neurohypophysis. The recording also revealed very long-lasting (up to 500 ms) after-hyperpolarization.

The fluorescent light method was used to monitor visually evoked activity changes in the optic tectum of the frog (Grinvald et al. 1984). In this case a retinotopic mapping was observed. The optical signals recorded had durations of a few hundred ms, comparable to the durations of electrically evoked surface-recorded evoked potentials. It seems that the optically recorded signal bears some resemblance to the integrated evoked potentials. The number of peaks resolvable with both methods appeared to be the same. It could be demonstrated that the optically recorded response varied with depth – as determined by changing the focal plane of the microscope – indicative of a possible laminar analysis with this technique.

The time resolution of the response appeared to be quite different for the neurohypophysis and the optic tectum. Similar differences in time resolution were found in different areas of the cortex of the anesthetized rat. It proved possible (Orbach et al. 1985) to monitor the activity produced in the somatosensory cortex by the stimulation of the whiskers, and the activity in the visual cortex as produced by a light flash. Characteristically, the spatial and temporal resolution in the somatosensory cortex was far better than in the visual cortex. The activity produced by moving one whisker typically extended about 1000 μm, which is about twice the anatomic dimensions assigned to a "barrel" field; the signal had a duration of about 100 ms. In contrast, a light flash typically stimulated activity in the entire cortical area viewed by the photo diode array (1.8 × 1.8 mm) and the responses were somewhat slower. Preliminary observations on the cat visual cortex revealed an optical response only when bicuculline (blocking intracortical inhibition) was applied.

Although the spatial resolution of the optical methods applied to in vivo preparations is at the moment quite insufficient for multiple separable single neuron analysis, the advantage is that it provides a real-time image of the intracellular population activity of many sites simultaneously. When viewed as a method for functional anatomy, its temporal response gives it an advantage over, e.g., the 2-deoxyglucose methods – even the adaptation of 2-fluoro-2-deoxyglucose in pontron-emission tomography necessarily has to integrate over time – with a time resolution in the minutes to hours range and with at present about the same spatial resolution of about $100 \times 100\ \mu m$. Further technical improvements in microscopy will facilitate the three-dimensional imaging of activity of larger brain structures such as the optic tectum and cortex.

6.2 Separating Multiple-Unit Spike Trains

For a given electrode tip geometry it appears possible to record the activity of only a few neural units; this depends to a large extent on the exposed surface. Metal microelectrodes with an exposed tip of about 10 μm and a 1 kHz impedance in the $1-3\ M\Omega$ range are suited to obtain multiple unit recordings (e.g., Eggermont et al. 1983a). It is known that large cells generally generate larger amplitude spikes (Grover and Buchwald 1970). A population of cells of inhomogeneous size will give rise to spikes of different amplitude when the cell-electrode distance is the same. This occurs because larger cells have a larger dipole moment due to the larger separation of sinks and source, and will present a larger cross-sectional area to the recording electrode. The voltage recorded is expected to decrease with distance between electrode and cell as R^{-1} for a point source (cell soma) or as R^{-2} for a dipole (axon). Schwartz et al. (1976), on the basis of the model results of Rall, estimated that, for distances not too close to the cell, the recorded voltage V depended in a logarithmic way on the distance:

$$V = 390 - 168 \ln R \ , \tag{6.1}$$

where V is in μV and R in cell radii. This means that the voltage gradient (the electric field):

$$\frac{\partial V}{\partial R} = -168\, R^{-1} \tag{6.2}$$

varies inversely with the distance. It is therefore expected that in principle the activity recorded by one and the same electrode from two or more units may be distinguishable on the basis of amplitude and waveform differences, the latter due to volume conduction and filtering in the medium. For few-unit recordings with a single electrode a number of separation schemes have been developed, all using features derived from the spike waveform. It is thereby assumed that the spike waveform and amplitude are characteristic for a particular neuron in a given geometric configuration with respect to the recording electrode.

Classification features are based on either waveform parameters or matched filter approaches. A design of a laboratory for multineuron studies is described by Gerstein et al. (1983): multielectrode recordings are combined with multiunit spike train separation techniques. Schmidt (1984 a, b) extensively describes hardware and computer-oriented approaches to sort multineuron spike trains.

6.2.1 Separation on the Basis of Waveform Parameters

The most simple device to separate multiunit recordings is a peak amplitude discriminator (e.g. Stagg 1973), where the detection of the spike and the feature extraction are nearly at the same point of the spike waveform. This device obviously fails when spike amplitudes are equal but spike waveforms differ. A slightly more elaborate version uses an amplitude window set at some delay after the level-crossing detector (cf. Fig. 6.1). Examples have been presented which act on the electrical signal itself (Calvin 1973) or on its image on an oscilloscope (Hernandez 1980).

When two recording electrodes are positioned on a nerve fiber bundle with some distance between them, the conduction delay offers an additional classification feature (Heetderks and Williams 1975; Looft and Heetderks 1978). The two-electrode configuration offers additional possibilities: one may compare the amplitude of the spike at the two electrodes and use the amplitude ratio together with the vector sum of the amplitudes as the classifier. This has been used for the "stereotrode" approach of McNaughton et al. (1983), who used tightly twisted 25 μm Teflon insulated wires with a tip separation of 6 μm (twice the insulation thickness) to record from the hippocampus in the rat. Alternatively, one may use the conduction delay and some indication of the waveshape (Schmidt and Stromberg 1969) or the conduction delay and the spike amplitudes at both electrodes (Camp and Pinsker 1979).

Variations on the amplitude window discriminators are the two-point-window discriminators first described for computer off-line analysis by Simon (1965) and later built in various on-line hardware configurations (Schmidt 1971; Ramsey 1974; Millar 1983; Lüscher et al. 1983). A claimed optimal version working in real time is based on the comparison of three amplitude values at time points selected prior to an automatic learning procedure. In the actual implementation the distances between the amplitudes of the various spikes are used and subjected to a cluster algorithm (Dinning and Sanderson 1981).

Various three- and more parameter classifiers (cf. Fig. 6.1) comprise that by Andreassen (1979), who used the maximum amplitude, the peak-to-peak amplitude, and the area of the spike, and obtained an error rate in classifying spikes as low as 0.28%. Other classifiers are the ones by Mishelevich (1970), Dill et al. (1970), and Kojima and Bracchi (1980), who all use maximum and minimum amplitudes and some timing information.

Finally, two approaches have been described which use eight parameters. McCann (1973) selects any pair of linear combinations of these parameters – minimum amplitude, time to minimum, time from minimum to next zero

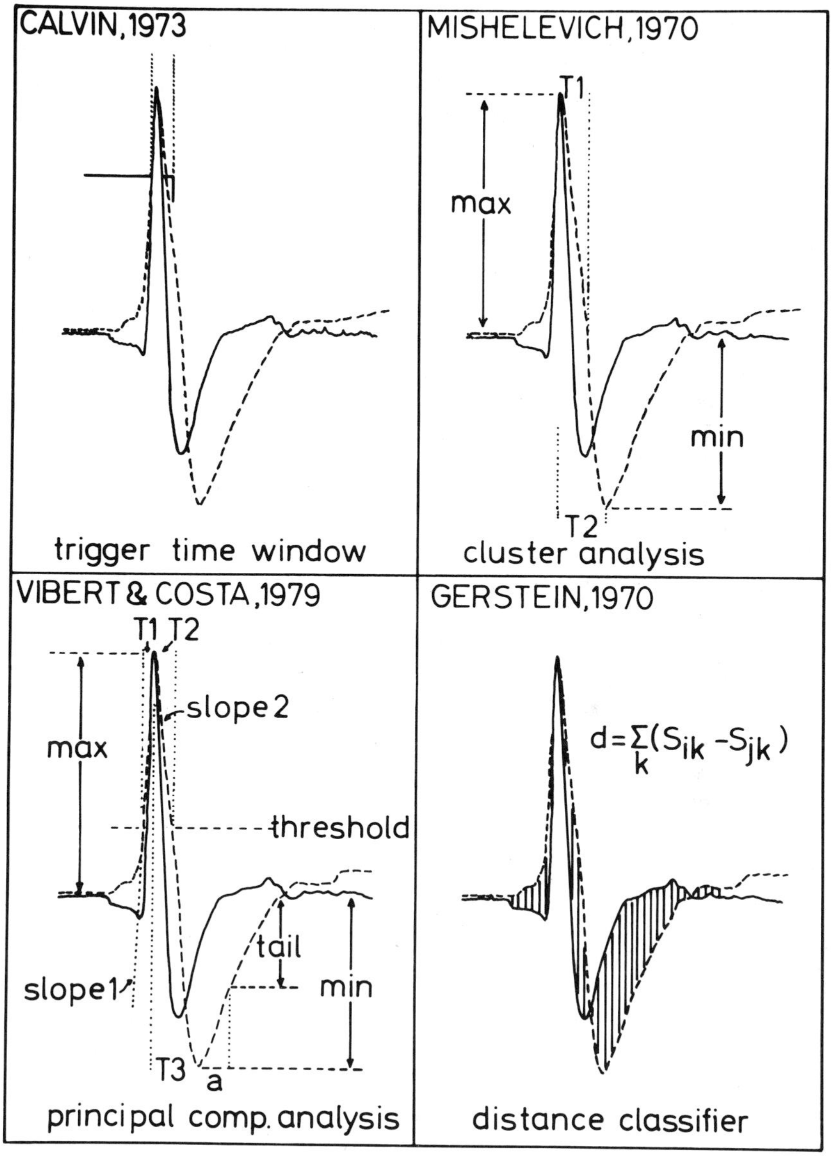

Fig. 6.1. Spike sorting on the basis of waveform parameters. Four techniques are shown: the first is the triggered time window discriminator (Calvin 1973) where the waveform is required to be within a certain amplitude window at a specified time after the trigger to result in an output pulse, the assumption is that this technique can discriminate waveforms with the same peak value but different width. A more elaborate approach (Mishelevich 1970) uses specific points on the waveform such as peak amplitudes, zero crossings, etc. When the number of parameters becomes to large and because of the correlations between their values, principal component analysis may become necessary (Vibert and Costa 1979). It is assumed that spikes generated by different neurons have different waveforms and therefore different extracted parameters. The most complete comparison (Gerstein 1970) uses all points on the waveform to calculate a generalized distance between two (or more) spike waveforms and thus can be considered a template-matching procedure

crossing, area of minimum peak, time from zero crossing to maximum, maximum amplitude, time from maximum to zero crossing, and area of maximum peak – so as to provide the best separation between units. Various pairs might be needed to separate, e.g., a four-unit spike train. Vibert and Costa (1979) subject their eight parameters (cf. Fig. 6.1), which are roughly comparable to those of McCann, to a principal component analysis (see Box H) and then represent the spike waveforms with the two largest principal components. Classification is thereafter done on the basis of cluster analysis.

6.2.2 Spike Separation on the Basis of the Entire Waveform: Template Matching

Template matching procedures are in fact sorting or sieve procedures: representative spike waveforms are either manually selected (e.g., Prochazka and Kornhuber 1973) or selected on the basis of an automated search (Millecchia and McIntyre 1978; D'Hollander and Orban 1979) and may be adopted in the course of the classification procedure by weighted averaging over, e.g., the last 20 spikes of that class (Kent 1971). The comparison with the established or selected template (cf. Fig. 6.1) can be based on a generalized distance criterion involving different weighting factors for different sample points (Gerstein and Clark 1964), on the basis of a best RMS fit between waveform and template in case more templates are present (Keehn 1966), or on basis of the Fourier coefficients of the first eight harmonics of the spike spectrum and that of the template (cf. Appendix of Bessou and Perl 1969).

6.2.3 Optimal Filter Procedures

This procedure is an elaboration on the template matching method: a linear filter is designed that is tuned to respond only to impulses from one unit and to reject background noise as well as other spikes. One may view this as a detection problem of a spike against a background of colored noise. The problem is therefore related to both matched filtering and Wiener filtering. A matched filter is the optimum filter when the signal is corrupted by additive Gaussian white noise. When the noise is not Gaussian, the matched filter is the optimum *linear* filter. In the case of the noise consisting of biological noise with relatively large spikes from other neurons, the noise is neither white nor Gaussian, and matched filter approaches may give very poor results (e.g., Roberts 1979). This is understandable, since spikes with a different "nonmatched" waveform but larger amplitudes might well give similar matched filter outputs.

If we have an electrode signal consisting of the desired spike waveform $f(t)$ and additive, uncorrelated noise $n(t)$ with power N_0, then the frequency response of the optimum linear filter is given by (Blinchikoff and Zverev 1976):

$$\hat{h}_0(i\omega) = G\,\frac{N_0}{2}\,\frac{\hat{f}^*(\omega)}{\hat{n}(\omega)}\,e^{-i\omega t}\;, \tag{6.3}$$

where $\hat{f}(\omega)$ and $\hat{n}(\omega)$ are the Fourier transforms of the spike waveform and the noise, * indicates the complex conjugate of $\hat{f}(\omega)$, the exponential term represents a filter delay t, G is a scaling factor. Matched filters [such as the one in Eq. 6.3)] are independent of the noise level except for a scaling factor, and the maximum signal-to-noise ratio produced by the filter action is independent of signal shape. Therefore it might not be a very useful spike-separating device in cases where the spectra of the spikes tend to differ only marginally. The corresponding a posteriori Wiener filter is given by (DeWeerd and Martens 1978):

$$\hat{h}_\omega(\omega) = \frac{\hat{F}(\omega)}{\hat{F}(\omega)+\hat{N}(\omega)}\;, \tag{6.4}$$

where $\hat{F}(\omega)$ and $\hat{N}(\omega)$ are the power spectra of spike waveform and noise respectively. Note that $\hat{F}(\omega) = \hat{f}(\omega)\,\hat{f}^*(\omega)$ and $\hat{N}(\omega)$ is the Fourier transform of the autocorrelation function of the noise. It is clear from a comparison of Eqs. (6.3) and (6.4) that the Wiener filter disregards phase, while the matched filter specifically makes use of it. One expects therefore a poorer performance from the Wiener filter method. An elaboration of the Wiener filter approach for multiple electrodes, however, has proven successful (Roberts and Hartline 1975). A set of K electrodes was used along a nerve fiber bundle. The template waveform of the ith unit under investigation is then a K-dimensional column vector $\vec{f}_i(t)$, the noise is represented by a K-dim. column vector $\vec{n}(t)$. In fact if the template waveform is discretized into M time bins then we have in fact a $K \times M$ matrix describing it.

Let $\vec{h}_i(-t)$ be the K-dim. column vector representing the impulse response of the filter "matched" to unit i. The filter output to the corresponding template is designed to be equal to one at $t = T$, the epoch of the spike, and to have minimal output when the ongoing noise on the spikes generated by the other observed units are present (Glaser and Ruchkin 1976). This condition reads:

$$\sum_{t=1}^{T} \hat{\vec{h}}_i^a \cdot \hat{\vec{f}}_i = 1\;, \tag{6.5}$$

where $\hat{\vec{h}}_i^a$ is the Fourier transform of the adjoint vector $\vec{h}_i$. Note that an adjoint vector is a transposed, complex conjugated vector, i.e., a column vector $\vec{h}_i$ in its adjoint form is a complex conjugated row vector.

The autocorrelation matrix of the output of the filter to noise is given by the convolution of the noise cross-correlation matrix between the various electrodes, C, and the filter response. The solution giving the optimal linear filter is:

$$\hat{h}_i = b\,\hat{f}_i\left(\sum_{j\neq i} \hat{f}_j\hat{f}_j^a + \hat{C}\right)^{-1}\;, \tag{6.6}$$

which is similar to Eq. (6.4) as far as the denominator is concerned, b being a dimensionless constant. Note that phase information is retained in $\hat{f}_i$. According to Roberts and Hartline, this technique functions optimally if the number of recording channels K is about the same as the number of neurons. Roberts (1979) indicates that the method can also be applied to one unit at a time and then requires less channels. This results in a simpler system but at the expense of repeatedly searching through the multiunit spike file. Roberts and Hartline used their procedure with only two electrodes and had less than 0.5% failures to recognize units. The method was also reported to give a good separation of superimposed waveforms. Elaborate mathematical derivations for a specific recording situation employing also two electrodes are given by Oguztöreli and Stein (1977), Stein et al. (1979) and Andreassen et al. (1979). Their main purpose was to separate motor unit responses from sensory unit responses among others on basis of the difference in direction of movement of the spikes. Their method is also applicable in the case of conduction delay differences for the spikes between the two electrodes. In that case the filter to be used at one electrode is related to the complex conjugate of the filter at the other electrode, the spectral characteristics of the unit potential and its complex conjugate, and the time delay.

6.2.4 Procedures Based on Cluster Analysis in Signal Space

A spike waveform can be described as an ordered array of sample points, i.e., as an N-dimensional vector or as a point in an N-dimensional signal space. Because subsequent sample points of a spike waveform are not independent, the N-dimensional basis of the signal space is not orthogonal. In fact when, e.g., four spike waveform types are present, one might as well reduce the dimensionality of the signal space to 4. The basis then consists of vectors representing the spike waveforms. This is the template matching approach of Prochazka and Kornhuber (1973) which is basically represented in the two upper parts of Fig. 6.2.

For efficient data representation it is desirable to construct an orthogonal basis for the signal space. The coordinates of a point in such a signal space are determined by taking the projection of the spike waveform upon the basis vectors. In practice this may be performed by matched filtering procedures, where the impulse responses of the filters are given by the time-reversed basis vectors (e.g., Glaser and Marks 1968).

The main problem is: which orthogonal basis vectors to use? A heuristic approach is to select such functions whose first two components resemble more or less the spike shapes to be represented. Glaser and Marks (1968) suggest using the Hermite functions:

$$H_n(t) = [2^n n! \sqrt{\pi}]^{1/2} (-1)^n \exp \frac{(t^2)}{2} \frac{d^n}{dt^n} \exp^n(-t^2) \ , \tag{6.7}$$

which are defined over the infinite time interval $(-\infty, \infty)$, and $n = 0, 1, 2, \ldots$. This is a set of orthogonal functions over an infinite time interval and it proved

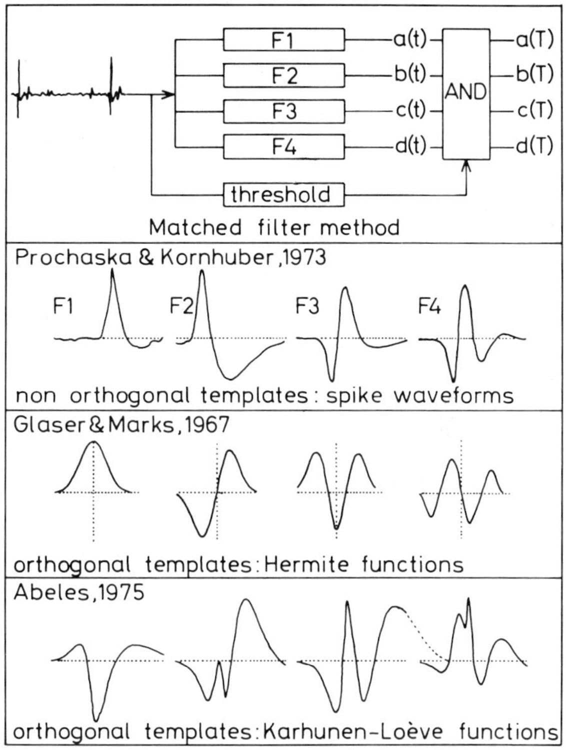

Fig. 6.2. Multidimensional sorting of spike trains. An extension of the template-matching procedure is found in simultaneously matching a given spike wave form with a set of templates ($F_1, \dots F_k$), and selecting that template for which the correlation with the spike waveform is maximal. The templates can be actual spike waveforms in the multiunit record (Prochazka and Kornhuber 1973), general orthogonal templates (e.g., Hermite functions; Glaser and Marks 1967), and orthogonal templates calculated on the basis of the knowledge that only certain spike waveforms in fact do occur (principal components or Karhunen-Loeve functions; Abeles 1975)

possible by using the first two Hermite functions – which basically resemble a Gaussian and its first derivative (cf. Fig. 6.2) – to obtain near orthogonality over a finite interval $-3 \leqq t \leqq +3$ (i.e., a range covering six standard derivations of the first Hermite function).

Because spike waveforms have a rather limited duration, a good alternative would be to look for functions that are orthogonal over a finite interval. Glaser and Marks mention the so-called ultraspherical harmonics which are orthogonal on the interval $(-1, +1)$. Another possibility are the Laguerre functions (Glaser 1971). The main problem with these a priori defined orthogonal functions is that they are too general, i.e., they are in principle able to represent any nonperiodic waveform. However, the convergence is generally very slow and a large number of functions are needed to represent the signal to a certain accuracy. In practice its separation potential, however, appeared to be acceptable (Glaser and Marks 1968).

Box H. Principal Component Analysis

Consider a set of vectors in N-dimensional signal space. These will normally occupy only a limited and spatially concentrated volume in that space. In that case the dimensionality of the space can be reduced with minimal loss in the representation of the set of vectors. Methods to accomplish this – depending on the discipline which uses them – are known as Principal component analysis, Karhunen-Loève expansion, Eigenvector analysis, or Cardinal (characteristic) function analysis.

An example of such a set of vectors is the one representing a collection of spike waveforms. Assume each spike waveform is sampled with 128 points, and thus can be represented as a vector in 128-dimensional space. This vector will be denoted by $|x\rangle$ using the bra(c)ket notation from quantum mechanics (e.g., Dirac 1958). The $|x\rangle$ are column vectors, their transposed $\langle x|$ are called row vectors. The autocorrelation matrix of a spike waveform is given by $|x\rangle\langle x|$. When $|x\rangle$ is a member of a set representing a stochastic process – or a realization of a stochastic process – the expectation $E[|x\rangle\langle x|]$ is the correlation matrix of the process. If in the signal space there is a set of orthonormal basis functions $\{|g_i\rangle\}$, $i = 1,\ldots,k$, then the inner product $\langle g_i|x\rangle$ represents the best fit of $|x\rangle$ by $|g_i\rangle$; i.e., the error $|Er\rangle = |x\rangle - \alpha|g_i\rangle$ is minimal if $\alpha = \langle g_i|x\rangle$. Therefore the best least squares fit of $|x\rangle$ to $\{|g_i\rangle\}$ is of the form:

$$|\tilde{x}\rangle = \alpha_1|g_1\rangle + \alpha_2|g_2\rangle + \ldots + \alpha_k|g_k\rangle \;, \tag{H.1}$$

with $\alpha_i = \langle g_i|x\rangle$.

Let us define the norm of $|x\rangle$ by $\langle x|x\rangle^{1/2}$. Its squared estimate is then equal to:

$$\langle\tilde{x}|\tilde{x}\rangle = \sum_{i=1}^{n} \alpha_i^2 \;. \tag{H.2}$$

The square norm of the error is then given by:

$$\langle Er|Er\rangle = \langle x|x\rangle - \langle\tilde{x}|\tilde{x}\rangle \;. \tag{H.3}$$

The vector space in which the error with respect to $\langle x|x\rangle$ is minimal, is spanned by the eigenvectors (principal components) of the correlation matrix $E[|x\rangle\langle x|]$.

The eigenvalue λ_i associated with each eigenvector $|g_i\rangle$ represents the average amount of energy of the spike that can be approximated by $|g_i\rangle$:

$$E[\langle g_i|x\rangle^2] = \lambda_i \;. \tag{H.4}$$

The basic procedure is to find a basis of reduced dimensionality (as compared to the original N dimensions of the sampled spike waveform) which suffices for representation of the spike waveform (cf. Fig. H.1).

Box H (continued)

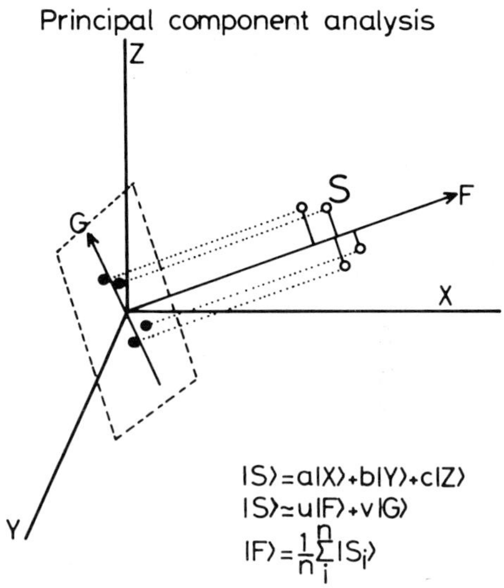

Fig. H.1. Illustration of principal components. Assume a cluster of points in 3D space. This cluster of, say, four points needs normally 12 coordinate values for complete specification. It is possible with relatively small error (when the points are clustered) to represent them with only eight parameters: the projections upon a vector through the origin and the center of gravity of the cluster and upon a vector through the residuals in a plane perpendicular to it. The signal space has thus been reduced to 2D

The following ten steps describe the determination of the eigenvalues and eigenvectors:

1. The spike waveforms of the different units, each one preferably averaged over many realizations to reduce the superimposed noise, are sampled with a frequency at least twice as high as the spectral bandwidth of the spike.
2. Each sampled spike is represented as an N-dimensional vector $|x\rangle$.
3. Many spike waveforms from N different cells ($N \cong 100$) are collected, $\{|x_i\rangle\}$.
4. Each spike vector is normalized:

$$|\xi_i\rangle = \frac{|x_i\rangle}{\langle x_i|x_i\rangle^{1/2}} \; . \tag{H.5}$$

5. The average spike over the cells collected so far is computed:

$$|\bar{\xi}\rangle = \frac{1}{M} \sum_{i=1}^{M} |\xi_i\rangle \; , \; M \leqq N \; . \tag{H.6}$$

6. Each new spike waveform $|\xi_j\rangle$ is shifted in time until $\langle \bar{\xi}|\xi_j\rangle$ is maximal.
7. Then $|\bar{\xi}\rangle$ is updated with $|\xi_j\rangle$ and the new $|\bar{\xi}\rangle$ is used in step 6. This loop 5–7, aimed at lining up the spikes, is repeated until some criterion is met.

Box H (continued)

(The procedure described under (5) to (7) may be replaced by a level detector, the spikes are then automatically lined up. This procedure, however, is not independent of the procedures used for the detection of the spikes in the experimental situation, and will be discussed in the main text.)

8. Compute the correlation matrix, using the shifted waveforms,

$$C = \frac{1}{N} \sum_{i-1}^{N} |\xi_i\rangle\langle\xi_i| \ . \tag{H.7}$$

Note that $C \neq |\bar{\xi}\rangle\langle\bar{\xi}|$, which is the correlation matrix of the average spike waveform over the entire ensemble.

9. Compute the eigenvectors $|g_i\rangle$ and eigenvalues λ_i of matrix C using the characteristic equation:

$$\det |\lambda_i \, I - C| = 0 \ . \tag{H.8}$$

10. The eigenvalues λ_i found are ranked according to decreasing eigenvalue. The first few eigenvectors explain a large percentage of the energy in the spike waveforms, and are selected to represent each new spike in signal space by a set of labels (α_1, α_2, α_3,...). Individual spikes from one neural unit will, because of superimposed noise, appear as clusters in signal space.

Another solution to the problem of spike separation is to design orthogonal functions adapted for the waveform ensemble under consideration. These functions are called eigenfunctions or principal components. Figure 6.3 illustrates the variation in spike waveforms encountered in the frogs midbrain using metal microelectrodes. An a priori classification is given using Roman numerals (e.g., Eggermont et al. 1981). Type I consists of mostly negative going sharp spikes often with a small positivity preceding it, they have been attributed tentatively to arise from axons. Type II spikes are somewhat slower and are always diphasic negative-positive going. The size of type I and type II spikes usually does not exceed 250 μV. Type III spikes are initially positive large amplitude diphasic or triphasic spikes attributed to cell soma activity. Type IV is a variation which has more phases. The large differences in spike amplitude require that the waveforms are normalized to have equal energy (see Box H).

The following assumptions are made: the spike waveform of a given cell is independent of the activity in other cells, the superimposed noise is Gaussian and stationary, and the spikes and noise are additive and statistically independent (Glaser 1971; Abeles 1975; Abeles and Goldstein 1977). The procedure to calculate the eigenvectors is outline in Box H.

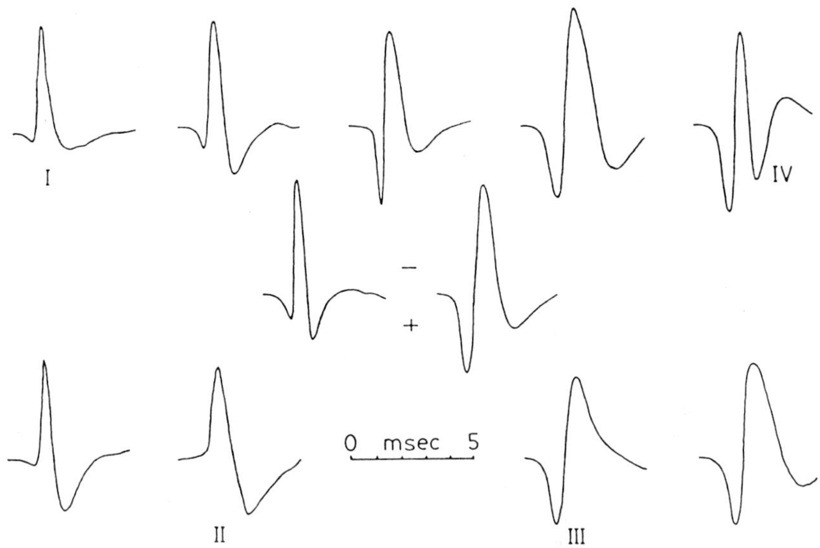

Fig. 6.3. A sample of spike waveforms recorded in the auditory midbrain of the grassfrog. An attempted type indication is given, see also Fig. 6.5

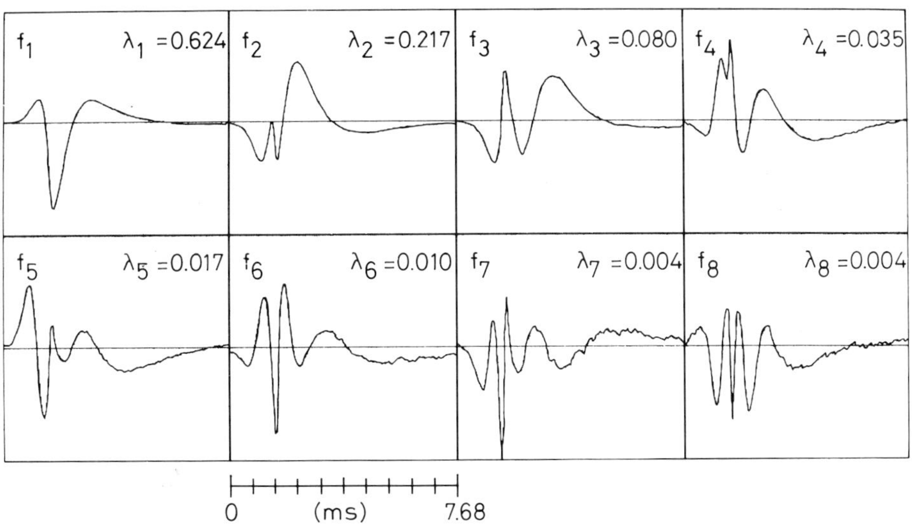

Fig. 6.4. Orthogonal templates for spike waveform decomposition. The indicated eigenvalues (λ_i) represent the fraction of the energy in the population of spike waveforms that is represented by the particular template. One observes the initial rapid decrease and later stabilization in λ_i for increasing i. The *first four templates* represent about 95% of the energy in the spike waveform. (Eggermont et al. 1983a)

The number of eigenvectors required to give a relatively error-free representation of the spike waveform depends on the size of the eigenvalues. Abeles (1975) and Abeles and Goldstein (1977) report for cat auditory cortex spikes values of $\lambda_1 = 0.82$, $\lambda_2 = 0.11$, $\lambda_3 = 0.03$, and use only the two largest principal components, $|g_1\rangle$ and $|g_2\rangle$, to represent a given spike $|x\rangle$

$$|x\rangle = \alpha_1 |g_1\rangle + \alpha_2 |g_2\rangle \tag{6.8}$$

up to about 93% of the spike energy.

Eggermont et al. (1983a) used a similar procedure to compute eigenvectors for the midbrain spikes in the grass frog. They obtained eigenvalues (cf. Fig. 6.4) $\lambda_1 = 0.62$, $\lambda_2 = 0.22$, $\lambda_3 = 0.08$ and $\lambda_4 = 0.03$ together representing about 95% of the energy, and represented spikes by four coordinates:

$$|x\rangle = \alpha_1 |f_1\rangle + \alpha_2 |f_2\rangle + \alpha_3 |f_3\rangle + \alpha_4 |f_4\rangle \ . \tag{6.9}$$

The power of this representation is illustrated in Fig. 6.5 showing averaged spike waveforms of the a priori designed classes I to V with superimposed the reconstruction on basis of Eq. (6.9) and that using the first eight eigenvectors. The representation of the various spike waveforms from the set for which the eigenvectors and eigenvalues were calculated is shown in Fig. 6.6 using the eigenvector pair $|f_2\rangle$, $|f_3\rangle$. It is observed that the a priori designed classes tend to cluster in this representation, which we interpret as an a posteriori justification of the original heuristic classification.

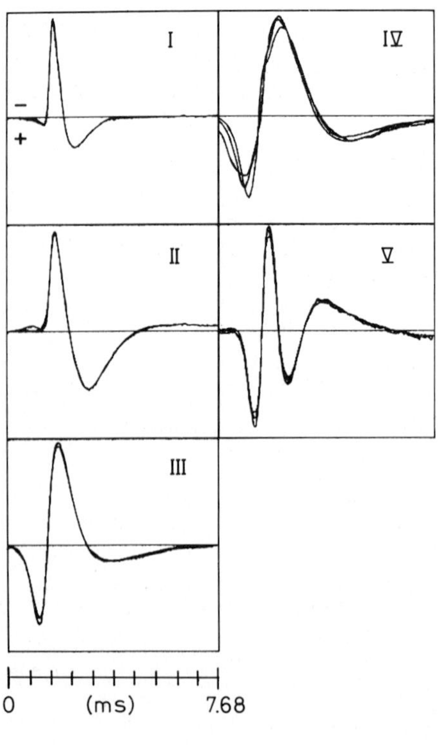

Fig. 6.5. Spike waveform types in the frog's brain. The three main types of spike waveforms recorded with metal microelectrodes are the types *I*, *II*, and *III*. Occasionally the other types are observed. *Type I* is attributed to axons, *type III* usually has the largest amplitude and is thought to originate from the cell body. (Eggermont et al. 1983a)

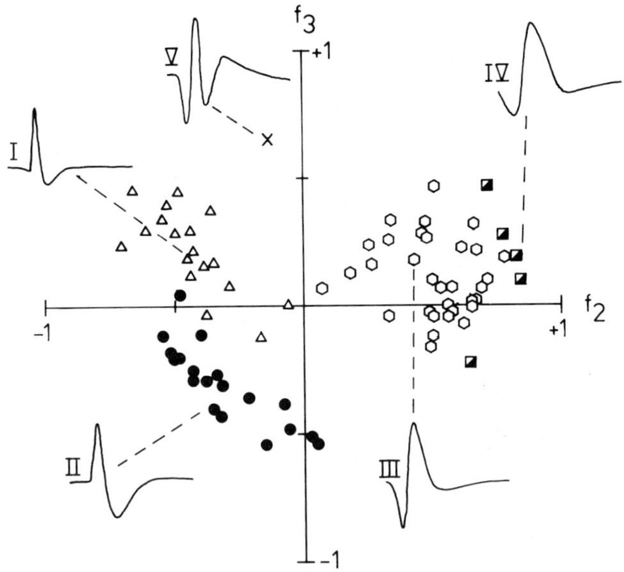

Fig. 6.6. Representation of the learning set of averaged spike waveforms. All 64 waveforms used to construct the orthogonal templates are represented on the plane determined by the second, f_2, and third, f_3, template. (Eggermont et al. 1983a)

It must be noted that by using principal components a considerable reduction in dimensionality of signal space has been achieved. In the case of the cat cortex studies of Abeles (1975), the spikes were originally sampled as 40 data points; the signal space ultimately used was two-dimensional. The frog midbrain study started by sampling the spikes at 128 points while the final space used was four-dimensional. Eggermont et al. (1983a) use the set (α_1, α_2, α_3, α_4) (together with its time of occurrence) as an efficient way to store the entire spike waveform in the computer. This enables reconstruction of the waveforms after separation and allows a visual control of the classification results.

6.2.5 Detection and Classification Procedures

Traditionally, spike detection is performed by a level crossing device which generates an impulse when the electrode signal exceeds a preset value. This value is generally set sufficiently high above the noise level (a common practice is a value of 3σ or more) to avoid artificial triggers by the noise peaks.

When the method of principal components is used the amplitude criterion in the level crossing detectors may be replaced by an energy detection criterion. For instance, assume that the signal size is given by:

$$\langle s|s\rangle \equiv ||s||^2 = \sum_{i=1}^{k} \alpha_i^2(t) , \tag{6.10}$$

where k is generally small. The total error size is then given by:

$$||\mathrm{Er}||^2 = ||\mathrm{x}||^2 - ||\mathrm{s}||^2 \ . \tag{6.11}$$

Abeles and Goldstein (1977) now use three criteria to decide whether there is a spike or not. First of all, $||\mathrm{s}||^2$ has to exceed a predetermined threshold. Secondly, the $||\mathrm{Er}||^2$ must be minimal, at least below a preset value. This is accomplished by constructing a so-called mismatch space $|\mathrm{M}_1\rangle, |\mathrm{M}_2\rangle$, which is orthogonal to $|\mathrm{g}_1\rangle, |\mathrm{g}_2\rangle$. The signal should have minimal energy in the mismatch space:

$$||\mathrm{M}||^2 = \sum_{\mathrm{i}} \beta_{\mathrm{i}}^2 (\mathrm{t}) \ , \tag{6.12}$$

where $\beta_{\mathrm{i}} = \langle \mathrm{x} | \mathrm{M}_{\mathrm{i}} \rangle$.

The total error is considered a good measure for mismatch. If the noise n superimposed on the spike signal s is Gaussian, the N-dimensional noise vector in signal space is uniformly distributed. If a spike is present and can be represented in k dimensions, then on the average per dimension:

$$\sigma_{\mathrm{s}}^2 = \frac{||\mathrm{s}||^2}{\mathrm{k}} \ , \tag{6.13}$$

and because the noise is considered orthogonal to the signal we have on the average per dimension:

$$\sigma_{\mathrm{n}}^2 = \frac{||\mathrm{n}||^2}{\mathrm{N}-\mathrm{k}} \ . \tag{6.14}$$

The variance ratio $\mathrm{F} = \sigma_{\mathrm{s}}^2/\mathrm{o}_{\mathrm{m}}^2$ follows an $\mathrm{F}_{\mathrm{k,N-k}}$ statistic, and criteria for keeping the false positive rate low can be based on this statistic.

Using this procedure, the detection and separation procedure are based on the same method. Wheeler and Heetderks (1982) argue that it can be profitable to study separately the best detection method and the best classification method. Eggermont et al. (1983 a) and Gerstein et al. (1983) use simple Schmitt triggers for spike detection, and principal components for spike classification. Both groups describe a hardware realization of this procedure: Gerstein et al. based on analog techniques, Eggermont et al. on digital techniques. A simple schematic drawing is shown in Fig. 6.7.

When the spike waveforms are represented in a low dimensional signal space, the clusters representing the individual neurons have to be identified. This can be performed using fully automated procedures or by an interactive procedure where the cluster boundaries are indicated by the experimenter. Windows of various shapes are used, such as hexagons (Frostig et al. 1983) or ellipses (Eggermont et al. 1983 a; Gerstein et al. 1983). The procedure used by Eggermont et al. will be illustrated here in Fig. 6.8. This plot basically consists of three parts. Part a in the upper left corner is the display obtained during the experiment and consists of a conventional dot display and the six possible combinations of $(\alpha_1, \alpha_2, \alpha_3, \alpha_4)$ used for judging the separability of the recording.

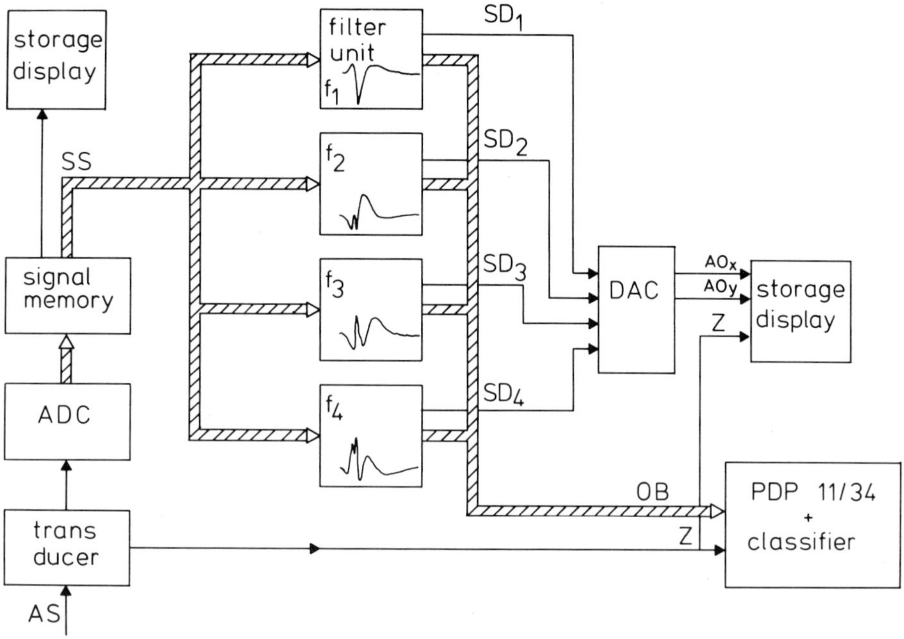

Fig. 6.7. Diagram of a hardware spike waveform analyzer. The basic features are (1) spikes, *Z*, are detected by level crossing; (2) the orthogonal templates, $f_1, \ldots f_4$, are realized as digital, finite impulse response filters; (3) the output of the templates, $SD_1, \ldots SD_4$, is sampled at the moment of level crossing (taking into account any filter delay); (4) the output values are displayed as in Fig. 6.6 for all template combinations; (5) moments of level crossing together with the output values of the matched filters, *OB*, are stored for off-line classification and spike waveform reconstruction. (Eggermont et al. 1983a)

When the separability is insufficient, the electrode will be moved in small (1 μm) steps or the preparation will be left to stabilize for some minutes. After the recordings have been done, a suitable combination of the α's for separation is selected, in principle separation can be done in a sequential way by using different (α_i, α_j) projections for different clusters. In this case the pair (α_2, α_3) provides the best separation for the three units that are present. This is shown in part b of Fig. 6.8 in the upper right corner – the classification part. Ellipses are drawn by indicating the long axis and one point on the short axis and units whose α_2, α_3 are within the ellipse are classified. This recording totalled about 4000 spikes, from which about 5% was left unclassified. These spikes probably represent superimpositions of the three spike waveforms. Parts c and d of Fig. 6.8 are shown in the lower half and comprise the single unit dot displays and the reconstructed spike waveforms on the basis of:

$$|\tilde{x}_i\rangle = \alpha_{ij}|f_1\rangle + \alpha_{i2}|f_2\rangle + \alpha_{i3}|f_3\rangle + \alpha_{i4}|f_4\rangle \;, \tag{6.16}$$

where α_{ij} is the jth component for the ith unit.

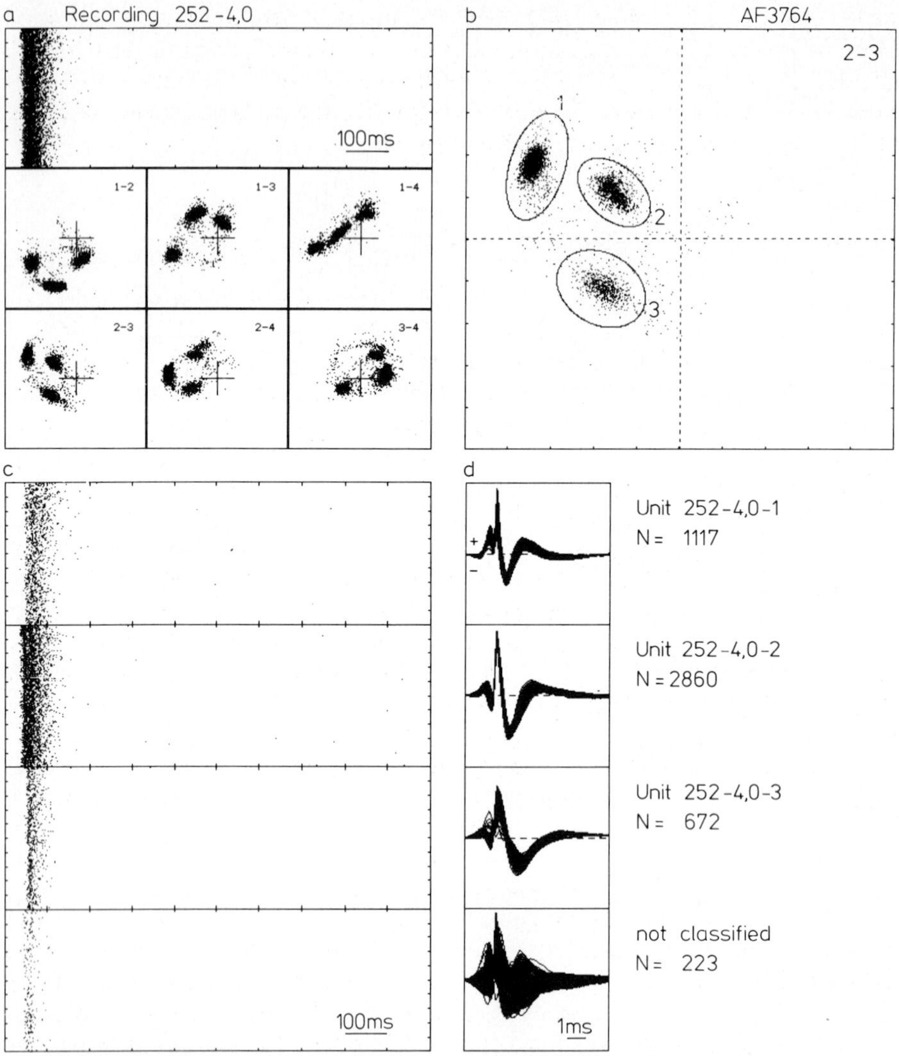

Fig. 6.8 a – d. Separation and reconstruction of a triple-unit recording. In **a** the multiunit dot display and the projections of the spike waveforms onto the templates are shown. **b** illustrates the separation: a particular clear projection is selected and ellipses are drawn around a cluster of dots. **c** illustrates the dot displays for each of the individual units as well as the nonclassified spikes. **d** shows reconstructed wave forms for each of the unit superimposed for quality estimation of the separation procedure. (Epping and Eggermont 1987)

This reconstution allows a visual control of the quality of the classification on basis of all four principal components, while in fact the separation is performed on the basis of only two.

Thus basically the spike separation procedure described above consists of three parts: spike detection and displaying, spike classification, and single-unit spike reconstruction. Most procedures do not allow off-line reconstruction of spike waveforms since the (α_i) are not stored but classification is done rather on-line and the results of the classification are stored (e.g., Abeles and Goldstein 1977), and therefore have to rely on other methods to be sure that the separation procedure was of sufficient quality.

6.2.6 The Performance of Spike Separation Procedures

What is the quality of a multiunit sorting device? This question was addressed by Heetderks (1978) and Wheeler and Heetderks (1982). They compared various methods when applied to the same dataset on basis of two probabilities. First of all the *sensitivity* of the test, i.e., the probability of correctly diagnosing spikes of waveform $|x_i\rangle$ to neuron i. When one has various units in the recording, the probability of a false alarm, i.e., diagnosing waveform $|x_j\rangle$ as belonging to unit $i \neq j$, has also to be taken into account. This is usually indicated by the *specificity* which equals 1 − the false alarm probability. A more or less reliable spike separation procedure requires that both the sensitivity and specificity of the test are at least 95%. This means that the distance – in terms of the classifying feature(s) – between two neurons should at least be larger than five standard deviations of the superimposed noise. It is hereby assumed that the size of the clusters depends only on the noise, and that waveform changes are negligible. The distance is a function of the number of features used and increases in a nonlinear, saturating, way with that number.

In the paper by Heetderks (1978), the performance of four multiple unit isolation techniques is compared as a function of the number of recording electrodes and features selected. A spike train, recorded with hook electrodes, from the cat medial articular nerve containing activity from eight units was used. The following features were used in the test: amplitude, filtered amplitude (projection on the first principal component as given in Abeles and Goldstein (1977) for cortical neurons), conduction latency between two electrodes, and a combination of conduction latency and filtered signal amplitude. On the required 95% confidence level none of the eight units could be reliably isolated using either amplitude or filtered amplitude as the single discriminating statistic. The use of conduction latency resulted in the isolation of six out of eight units at the indicated quality level. Finally, the combination of conduction delay and filtered signal amplitude correctly identified all units – with the proviso that two units were considered inseparable by all techniques considered. It must be remarked that the latter two were distinctly different on the oscilloscope trace and that very probably an amplitude-time window discriminator would have separated those accurately.

In a sequel to this study Wheeler and Heetderks (1982) compared for a 5-unit recording from the nerve cord between ganglia A1 and A2 from the cockroach by using two hook electrodes various single feature classifiers (e.g., amplitude, conduction delay), optimal linear filtering and template matching, and various principal component techniques. It appeared that single feature classifiers are useful only on waveforms relatively free of noise. For noisier recordings the method of choice is one of the principal component methods. These outperform all other techniques when the electrode signal contains more than 40% noise (RMS signal-to-noise ratio 0.4). Template matching theoretically is clearly the best technique; however, it does not outperform other techniques in a practical setting. A modification using different weighting factors as in Gerstein and Clark (1964) might show some improvement because in this way errors at the margins of the waveform can be avoided. The various principal component methods, based either on the correlation matrix or on the covariance matrix and used with or without signal normalization [cf. Eq. (H.5)] of the spike waveforms, did not show different results at the level of five classifying features. When the principal components calculated from the average covariance matrix were used with only one classifier, a much larger separation was obtained than for any of the other single feature classifiers: 21 standard deviations vs. about 6 for the amplitude or conduction delay, respectively 15 for the principal components calculated from the average correlation matrix. This advantage, however, became minimal when using two or more features.

Box I. Classification Theory

Classification theory (e.g., Duda and Hart 1973) is concerned with grouping objects into distinct classes on the basis of values taken by numerical variables defined as quantitative measures on the objects. Let x_{ik} ($i = 1, \ldots, n$; $k = 1, \ldots, p$) denote the value taken by the kth numeric variable on the ith object. Let d_{ij} denote the dissimilarity between the ith and j^{th} object. If d_{ij} is small one is inclined to assign the objects to the same class, if d_{ij} is large the objects will likely belong to different classes.

Commonly used measures of d_{ij} are the Euclidian distance:

$$d_{ij}^{(e)} = \left\{ \sum_{k=1}^{P} W_k (x_{ik} - x_{jk})^2 \right\}^{1/2} , \qquad (I.1)$$

where W_k is a set of weights applied to the numerical variables, the "City block" measure:

$$d_{ij}^{(c)} = \sum_{k=1}^{P} W_k \, |x_{ik} - x_{jk}| , \qquad (I.2)$$

and the Mahalanobis distance:

$$d_{ij}^{(M)} = (\vec{x}_i - \vec{x}_j)^T \, \Sigma^{-1} \, (\vec{x}_i - \vec{x}_j) , \qquad (I.3)$$

where Σ is the $p \times p$ covariance matrix of the numerical variables.

Box I (continued)

In case we have probability distributions with equal variance σ^2, the dissimilarity d′ is commonly expressed in units of the standard deviation σ. In Fig. I.1 an idealized distribution resulting from, e.g., a separation procedure of a double-unit spike train is given.

The projections upon two basis vectors are indicated by the two Gaussians. The height of the functions is related to the number of waveforms having this particular projection. There is some overlap between the distributions and any separation boundary will induce errors in the subsequent classification. This can be further elucidated by a cross-section through the distribution functions.

If we denote by α_i the probability of misclassifying a spike of type i, and if we denote by β_{ij} the probability of classifying a spike of type j as belonging to class i then *sensitivity* is defined as $(1-\alpha_i)$ and *specificity*

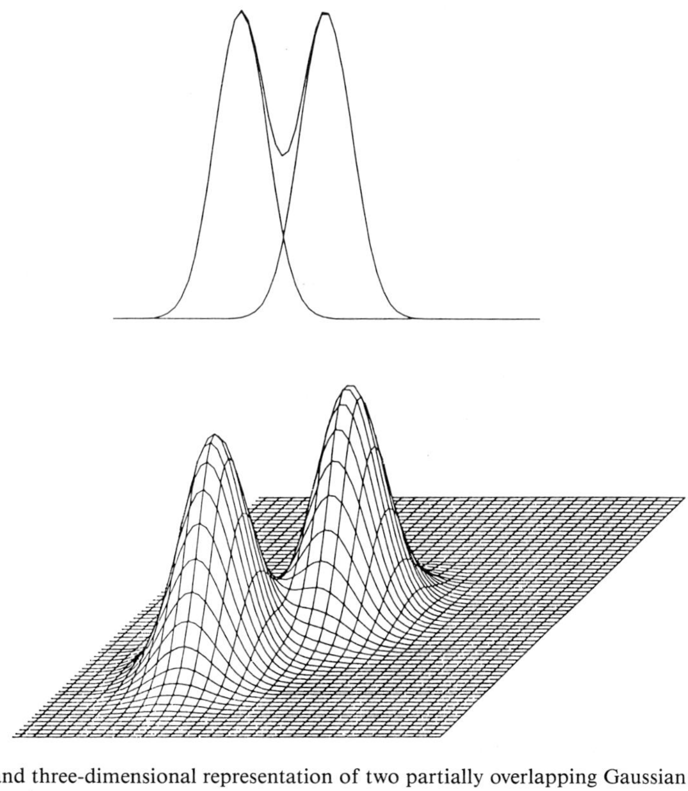

Fig. I.1. Two- and three-dimensional representation of two partially overlapping Gaussian distribution functions

Box I (continued)

as $(1-\beta_{ij})$. The dissimilarity is related to α_i and β_{ij}. Consider classification ellipses (in this particular case circles) drawn concentric around the center of gravity of the distribution then 5% false negative classification ($\alpha_i = 0.05$) requires a circle around the mean of the correct distribution with radius 2.5 σ. Assuming also a 5% false positive classification ($\beta_{ij} = 0.05$) again results in a classification boundary circle with radius 2.5 σ now centered around the mean of the other distribution. In this case the distance between the means d′ equals 5 σ, provided that the classification boundaries just touch each other. The situation indicated in Fig. I.1 has a $d' \cong 3.3$.

When the assumption of equal variance of the distributions is violated, then the noise causing the spike waveform to vary is not independent of the spike waveform. This situation is quite common but not easy to deal with. In this case classification boundaries are set heuristically.

6.2.7 Difficulties Encountered with Spike Separation Procedures

The two most common difficulties encountered are (1) spike waveforms that are completely or partially superimposed, and (2) spike waveforms from different neural units that are nearly or totally indistinguishable from each other. The first situation occurs frequently when transient stimuli such as a light flash or a tonepip are applied. In this case many neural units tend to fire nearly synchronously. This is especially a problem in areas where there is a strong retinotopic or tonotopic organization. One way out of this problem is to desynchronize the population by using continuous stochastic stimuli such as Gaussian modulated light or Gaussian white noise or the use of less "attractive" stimuli (e.g., by decreasing the stimulus amplitude). This, however, will not always result in sufficient numbers of spikes, especially in more central parts of the nervous system where such stimuli might be rather ineffective. When such superimposition occurs in a nerve fiber tract, recording with a number of electrodes along the tract might result in a separation of superimposed spikes (Roberts and Hartline 1975). We must take into account, however, that in nerves with very homogenous diameter of the fibers – such as in the auditory nerve – hardly any dispersion occurs, and that in such cases also this technique fails.

The second situation – nearly or totally identical waveforms from different units – occurs frequently in regions of the brain which are densely packed with cells or fibers from a rather homogenous population as, e.g., in hippocampus. A technique advocated to overcome this problem is to use closely spaced electrodes and hope that nearby units project differently on each recording electrode (e.g., McNaughton et al. 1983). A prerequisite, however, is that problem number 1, that of superimposed spikes, does not occur.

If one is not sure whether a single neuron is recorded from one, two or more units having the same waveform, an autocoincidence function or interval histogram usually reveals the presence or absence of a dead time (refractory period). If there is no dead time a single-unit recording can be ruled out.

7 Representation and Detection of Correlated Neural Activity

Three parameters are essential for representing multiple-neuron activity. These are the location in the brain ($\vec{r}$), the time of occurrence (t), and the value of the activity ($\vec{e}$). This last parameter can be an electric, magnetic, or optic quantity, representative of dipole moment, dipole strength, neural firing rate, type of neural unit, size of evoked potential, magnetic field, or optical activity. In principle therefore a representation requires a map of $\vec{e}$ onto a four-dimensional space-time continuum. This is not easy to represent in the sense that one cannot visualize such a representation in its totality. The solution then is generally to reduce the dimensionality of the individual representations and to show a sequence of lower dimensional projections or cross-sections.

The type of representation differs for recordings or continuous changes in the brain like those of graded (evoked) potentials recorded in the form of electric, magnetic, or optic signals, and for recordings of discrete changes such as sequences of action potentials in groups of neurons recorded by microelectrode arrays. First of all I will discuss briefly the representation of evoked potentials recorded from multiple sites on the scalp or from gross electrodes inserted into the brain, and then in much greater detail elaborate on the representation of multiple single-unit recordings. This will be followed by an extensive treatment of methods to detect correlated neural activity; we will see that an adequate representation greatly facilitates this detection.

7.1 Representation of Multiple-Site Recorded Continuous Neural Activity

The mapping of changing neural activity in a useful form first of all requires a reduction in the number of dimensions. In the case of stationary signals one could, for instance, calculate the spectrum over the time period under consideration and display these spectra as a function of recording site. This technique is used in the analysis of the EEG or ongoing cortical slow wave activity. In recording evoked potentials with an electrode in the cortex, preferably with multiple recording sites along the electrode, one can obtain a depth profile in the radial direction. Using a two-dimensional array of such electrodes then permits really probing the brain in all three spatial dimensions. Applying a current-source-density analysis (to be discussed later in this chapter) then permits identification of sources and sinks or at least concentrations thereof. In applications to humans one has to limit oneself to an array of electrodes attached to the scalp and one effectively covers only the tangential plane. The standard recording system uses 20 electrodes of which one electrode usually functions as a reference, sometimes one uses a reference based on the average activity of all electrodes (for details consult Nunez 1981). Thus one cannot adequately measure processes occurring in the radial direction from such data, unless one resorts to so-called inverse procedures which under a given set of assumptions allow one to calculate the location and moment of the equivalent dipole sources (e.g., Scherg and von Cramon 1985a, b). The problem of representation is thus generally reduced to the one of mapping $\vec{e}$ on a three-dimensional (x, y, t) space. This can in principle be done using color, but a perceptual meaningful assignment of the various parameters is not always possible (for a situation where a bivariate function, a time dependent complex spectrum, has been mapped on a two-dimensional (frequency-time) plane with some success see Johannesma et al. (1981). Reduced representations are therefore in use such as the chronotopogram (Lesevre 1982) and the brain map (e.g., Duffy 1982).

The chronotopogram reduces the spatial dimension by displaying only the activity of a linear array of electrodes selected from, e.g., a cross-montage. The electrical activity at each electrode with respect to a given reference is plotted as a function of time using spatial interpolation to estimate the activity values between electrodes. Thus a spatiotemporal map of electrical activity is obtained featuring isopotential lines. In Fig. 7.1 an example is shown for a cross-montage resulting in two chronotopograms. An averaged waveform for the central electrode is also shown to indicate the correspondence with the spatiotemporal map features.

The brain map or *topographic map* reduces the time dimension to a single point and electrical activity is plotted on a map of the scalp. Again, an interpolation technique is used to obtain values at points between electrode locations. A pseudocolor scale has become in vogue to use with these maps. Figure 7.2 gives the averaged evoked potentials recorded at each electrode site, the position of which is also indicated. To construct the brain map at the time given by the line transecting the waveforms the voltage at each electrode is calculated (lower left), interpolated using a grid and the map is plotted using a grey scale (lower right).

The actual selection between the spatiotemporal and the topographic map depends on the problem under investigation. An issue pertaining to both maps is *what* is represented? When using a set of electrodes spaced as in the standard 10–20 EEG system several difficulties arise. First of all there is a problem with spatial sampling. Due to volume conduction effects the point-spread function from the surface of the cortex to the scalp for evoked potentials is typically 2–5 cm in diameter (Nunez 1981). This means that a point source of activity in the cortex typically produces activation on the scalp in a circular area with that diameter. The electrodes are typically at distances of 4–5 cm, which means that the cortical activity present at the scalp is spatially undersampled, which introduces aliasing problems in the spatial interpolation between electrode positions. In laboratory studies the use of more electrodes is therefore advocated (e.g., Gevins 1984). In addition, electrical activity picked up by an electrode may be either due to local weak generators, distant strong generators

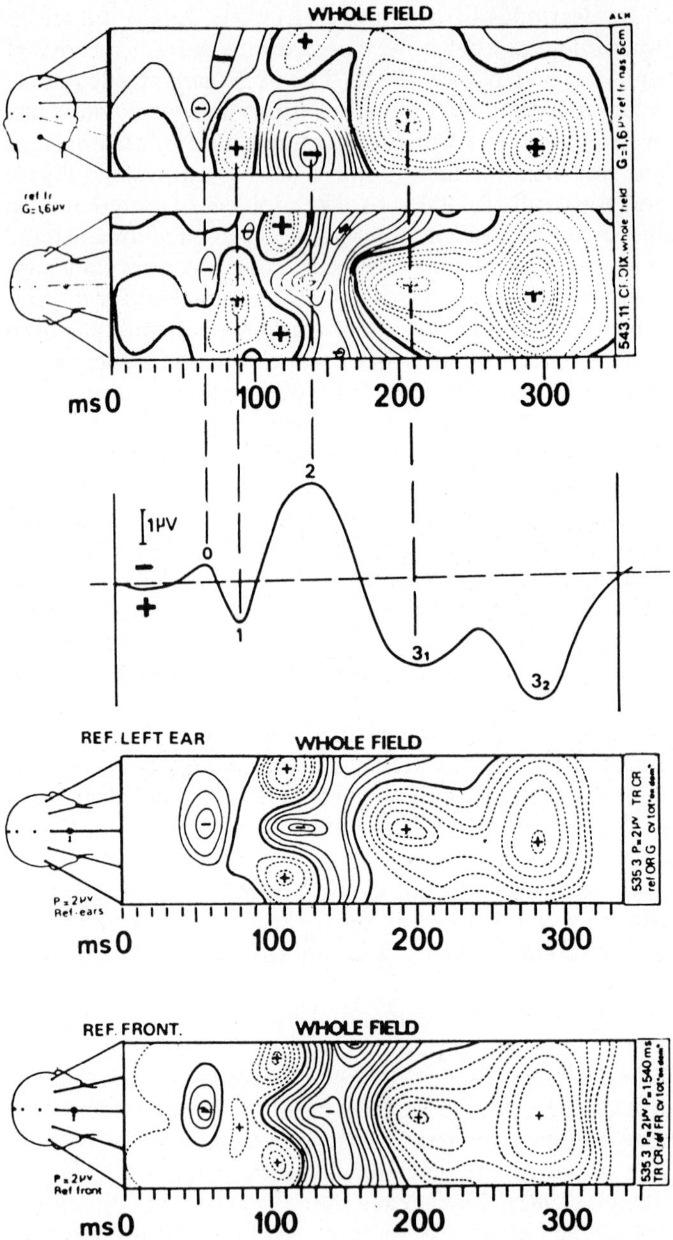

Fig. 7.1. Chronotopographical representation of the voltage recorded with coronal and sagittal electrode montages on the scalp. The voltage profile is indicated with *altitude lines*. The correspondence with the average evoked potential as well as the wealth of additional information can easily be seen. (Lesevre 1982)

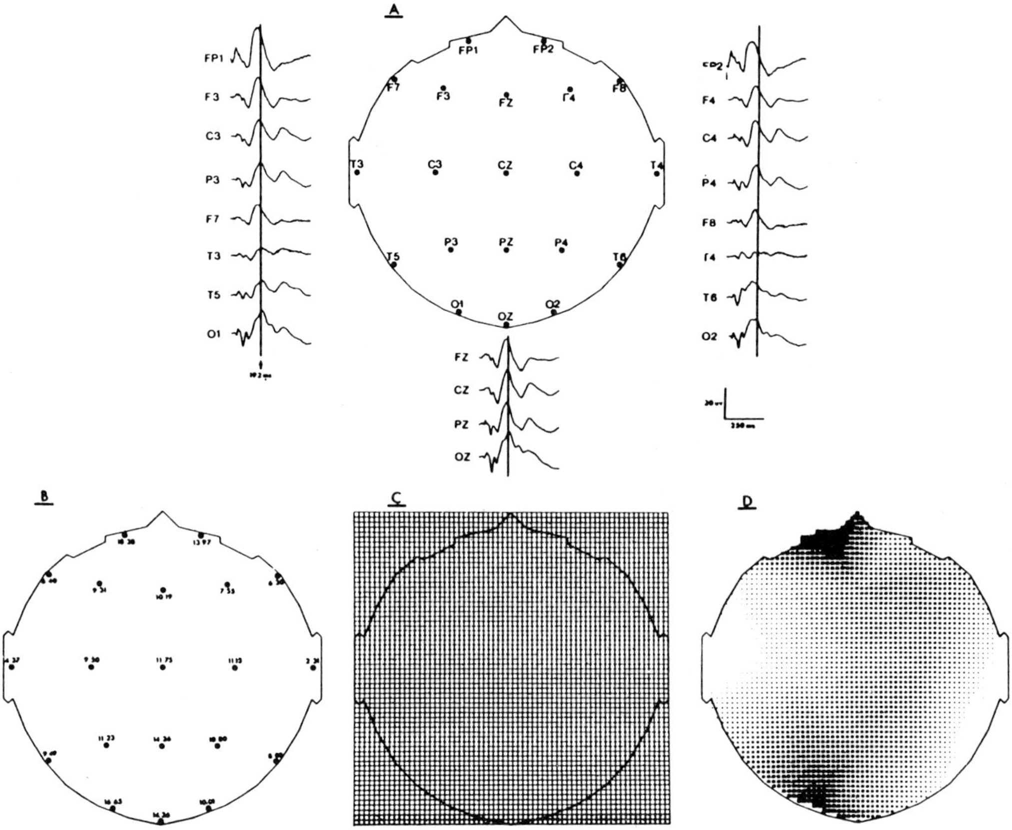

Fig. 7.2 A–D. Brain map. When the electrical activity of a complete (10–20 system) set of electrodes at one particular instant of time is represented on a two dimensional representation of the scalp a "brain map" results (Duffy 1982). **A** shows the electrode montage and the recordings at each electrode. **B** displays the numerical values at the selected time (peak of the EP); **C** shows a linear grid used for the interpolation, and **D** shows a map of the voltage distribution across the scalp

or weak distant generation observed through low resistance paths, e.g., skull openings, to the electrode. Thus the activity of neighboring electrodes may have considerable common input. It therefore makes quite a difference whether one represents in a brain map, e.g., RMS voltage, peak voltage, voltage gradients estimated by voltage differences between electrodes, or an estimate of a current-source density. The latter is constructed by applying the Laplacian operator to the potential recorded at a given point. In theory:

$$\Delta \Phi = \frac{\partial^2 \Phi}{\partial x^2} + \frac{\partial^2 \Phi}{\partial y^2}, \tag{7.1}$$

which can be approximated (Katznelson 1981) by using the potential at the point required, Φ_0, in combination with those at the four most nearby points:

$$\Delta \Phi \cong \sum_{i=1}^{4} \Phi_i - 4\Phi_0 \; . \tag{7.2}$$

If d is the interelectrode distance (assumed uniform), h is the scalp thickness and σ is the electrical conductivity of the scalp, then the vertical current density at point 0, J_0 can be estimated as:

$$J_0 \cong \frac{\sigma h}{d^2}\left(4\Phi_0 - \sum_{i=1}^{4} \Phi_i\right) . \tag{7.3}$$

Plots of this current source density transformation appear to be much more space-specific than are gradient or voltage plots. This transformation works best when recording from many electrodes. For 19 channels one has 9 current-source density channels whereas for 63 channels on has 43 (Gevins 1984).

The spatial smearing effect that is still present can be overcome by inverse filtering with the point-spread function (e.g., Freeman 1980). Using the optical

Box J. Stationarity of Random Processes

The signals recorded from within the brain or from the scalp are realizations of stochastic (point) processes. Stochastic processes are difficult to treat unless they are stationary. Several definitions of stationarity can be given. A strictly stationary process is a stochastic or random process whose probabilistic laws remain unchanged through a shift in time. In other words:

The stochastic process $\{X(t), t \in T\}$ is *strictly* (or completely) *stationary* if for *any* positive integer k and *any* points $t_1, \ldots, t_k$ in T the joint probabilities of $\{X(t)\}$ are unchanged by an arbitrary time shift τ. The process is stationary of order m if the above holds not for all k but only for $k \leqq m$. Clearly stationarity of order m implies stationarity of lower order $m' < m$, since the m-th order density determines all densities of lower order. The stochastic process is *weakly* (or covariance-, or wide sense-*stationary* if its expected value is constant and it has a finite second moment $E\{[X(t_1)-\mu][X(t_2)-\mu]\}$ that depends only on the time difference (t_2-t_1). Clearly, if X is stationary of order 2, then it is also wide-sense stationary; the converse, however, is not true: wide-sense stationarity involves only 1st and 2nd moments, not the complete density.

A Gaussian process is completely specified by μ and its covariance, and if it has weakly stationary properties it is automatically strictly stationary. Finally, different processes $\{X(t)\}$ and $\{Y(t)\}$ are called *jointly stationary* (in the strict sense) if the joint statistics of $\{X(t), Y(t)\}$ are the same as those of $\{X(t+\tau), Y(t+\tau)\}$ for any τ. Note that the processes X and Y might be individually, but not jointly, stationary. Two processes are jointly stationary in the wide sense if each is wide-sense stationary and their cross-covariance depends only on the time difference (t_2-t_1).

recording technique described in the previous chapter, the only smearing effect occurs due to some light scattering from nearby recording sites and by the spatial resolving power of the microscope. However, deconvolution techniques might sharpen up the picture also in these cases.

When estimating evoked potentials one generally uses ensemble averaging, and the brain maps or chronotopograms are constructed from such averaged responses. The assumption for this averaging procedure is that both the signal (single trial evoked potential) and the superimposed nose (the EEG, muscle potentials) are stationary and that the noise is additive and uncorrelated with the evoked potential. It is known that the normal background EEG is stationary for only a few seconds at a time and that stationarity is doubtful for EP's even over periods as short as 100 ms. Thus generally the assumptions underlying averaging are violated and more effort has to be put in single trial EP evaluation. The problem of stationarity also is vital in many of the approaches used in detecting correlated neural activity. In most cases there is at least a transient nonstationary period at the onset of stimulation. This may last from a few milliseconds to tens of seconds, depending on the part of the brain from which one is recording. Usually correlation analyses are carried out over the stationary part; however, one should always consider that in actual life stationary states hardly ever occur. Thus one might well assess correlations that are not very interesting from the biological point of view.

7.2 Representation of Multiunit Spike Trains

The most straightforward representation of multiunit spike trains is a registration of the electrode signal (e.g., Fig. 7.3). This representation is redundant and in most cases difficult to interpret. Under favorable conditions (e.g., second trace in Fig. 7.3) one may detect combinations of spike waveforms from different neural units that occur in a rather fixed pattern. One of the first reports about correlated neural activity (von Baumgarten and Schaeffer 1957) used this representation and indicated its value in cases with strong coupling between the neurons or in cases of weak coupling with rather common double occurrences.

Considering that spike waveforms for a given unit are fairly constant, one may form an abstraction of the multiunit electrode signal in the multiunit dot display. In a single-unit dot-display a dot is displayed every time a spike occurs. In the case of multiunit recordings one has to code the individual units as well. Care must be taken that each unit is identifiable and that the interrelation between the firing of units is preserved. One solution to this problem is (Eggermont et al. 1983b) to use different symbols such as circles, squares, triangles, etc. for the various units as in Fig. 7.4. One immediately spots the double occurrences, but the picture is too crowded about 50 ms after stimulus onset due to overlap of symbols. Another solution is to use colored dots and to represent each unit by its own assigned color, which permits a larger density of events to be represented (Epping et al. 1984). This color-coded dot display is introduc-

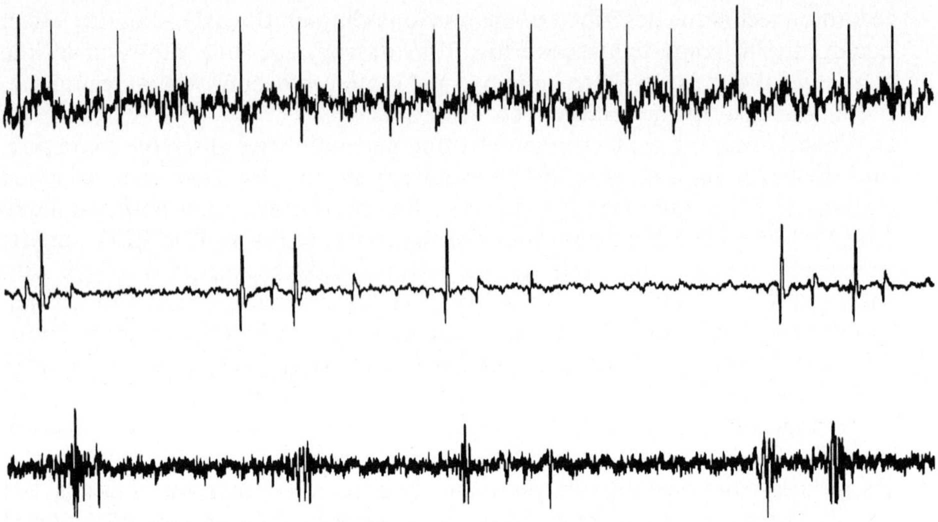

Fig. 7.3. Electrode signals for a single-unit, double-unit, and multiple-unit recording in the auditory mid brain of the frog, using tungsten micro electrodes with exposed tip of about 10 μm

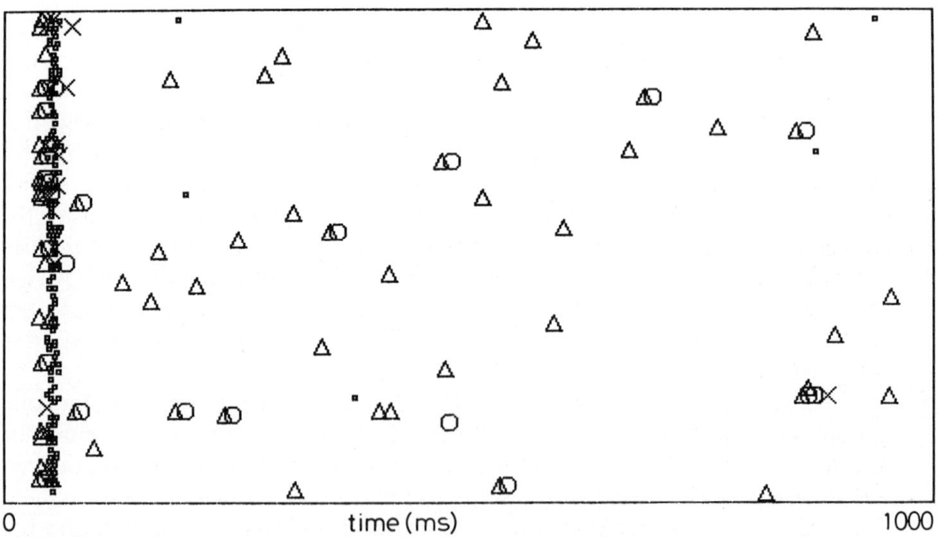

Fig. 7.4. Dot display representation for a four-unit recording. The stimuli were tonepips presented with frequency in random fashion at a rate of one per second. The time base is 1 s. One observes that all units, indicated with different symbols, have a latency of around 50 ms in their response to the tonepips. There is spontaneous activity as well; from this one observes some characteristic multiunit patterns: triangle-circle pairs are very frequent. (Eggermont et al. 1983c)

ed under the name of "neurochrome". Plate I (see pp. 280, 281) shows an example from an experiment involving multiple auditory neurons; rearrangement of the dots according to the frequency of the stimuli is shown as well: in such a representation the frequency sensitivity of the different units is immediately apparent.

7.3 Cross-Correlation

A redundant representation of the joint activity of two neurons is formed by the time-dependent cross-coincidence diagram (Fig. 7.5 a). On the vertical axis of the representation the times of occurrence of the spikes of unit A are indicated, and horizontally the lead or lag times τ for the firings of unit B with

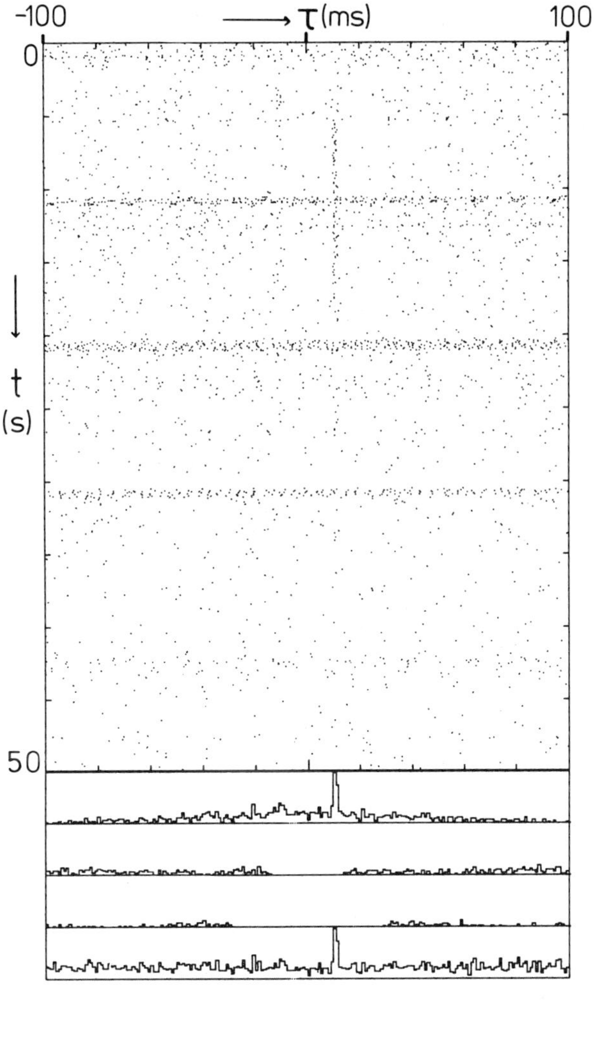

Fig. 7.5. Time-dependent cross-correlation. The *upper part* shows all coincidences between both point processes both as a function of the delay time τ as well as ongoing experiment time t. The *lower part* shows first-, second-, and third-order recurrence densities as well as the cross-coincidence histogram, which is the sum of all order recurrence time densities. It is clear that the entire correlation is due only to the initial one-third of the recording. (van Stokkum et al. 1986)

respect to unit A are given. Thus for each A-spike there is a sequence of intervals for spikes from train B that either precede or follow that spike. If there are instationarities in the firing of the A-train they will show up as horizontal high density regions of dots. If there are preferred intervals between the firings of the two units this will become immediately clear from the presence of vertical high density regions. Typically, the vertical axis comprises one stimulus sequence duration and is usually a factor 10–100 larger than the time span along the horizontal axis. In case the conditions for stationarity are met, one may average across all A-spikes to obtain a cross-coincidence histogram shown in the lower part of Fig. 7.5b.

If we indicate by $R_{AB}(t_{Aj,}\tau)$ the time-dependent cross-coincidence histogram, in which t_{Aj} indicates the instances where the spikes of train A occur, and τ indicates the intervals between the B-spikes and the A-spikes, then the average cross-coincidence histogram is formed in two steps. The first step consists of averaging:

$$R_{AB}(\tau) = \sum_{j=0}^{N} R_{AB}(t_{AJ}, \tau) \ . \tag{7.4}$$

The second step consists in forming bins of a certain width Δ and counting the number of coincidences in each bin.

An alternative to binning is to replace each dot in $R_{AB}(t_{Aj}, \tau)$ by a kernel function such as a gaussian or a half-wave cosine and to obtain a smooth estimate of the time-dependent cross-correlation function (e.g., Schneider et al. 1983).

When only the first-order intervals between the A- and B-spikes are taken into account one obtains a so-called cross-interval histogram (cf. upper part of Fig. 7.5b), which is related to the distribution of the forward- and backward-recurrence times (e.g., Perkel et al. 1967; Johnson and Kiang 1976) in case the two spike trains are independent. Consider a spike train A, then from a random instant of time the probability density $G_A(\tau)$ of encountering a spike at some later time is constant and given by:

$$\begin{aligned} G_A(\tau) &= \lim_{\Delta t \to 0} \frac{1}{\Delta t} \text{Prob}\{\text{event in A in } (t_0+\tau, t_0+\tau+\Delta t) \mid \text{random} \\ &\qquad\qquad\qquad \text{instant at } t_0\} \\ &= \mu_A^{-1} , \end{aligned} \tag{7.5}$$

where μ_A^{-1} is the mean firing rate of spike train A. If on the other hand we selected a particular event of the A-train as our starting point, then the probability density for an event at some later time is, of course, given by the autocorrelation function of A. In the particular application to two spike trains, events of spike train B are considered as random events with respect to train A. Because one is a priori not sure that the event is truly random, the cross-interval densities of order i indicated by $\eta_i(\tau)$ have to be compared with those for the recurrence times. In practize one either compares the distribution of the first-order cross-interval with that of the first-order recurrence time given by:

$$\int g_i(\tau)\,d\tau = \frac{1 - F_A(\tau)}{\mu_A}\,, \tag{7.6}$$

where $F_A(\tau)$ is the cumulative distribution function of the intervals of spike train A, or one compares the cross-correlation function of spike trains A and B

$$R_{AB}(\tau) = \sum_{\substack{i=-\infty \\ i \neq 0}}^{i=+\infty} \eta_i(\tau)\,, \tag{7.7}$$

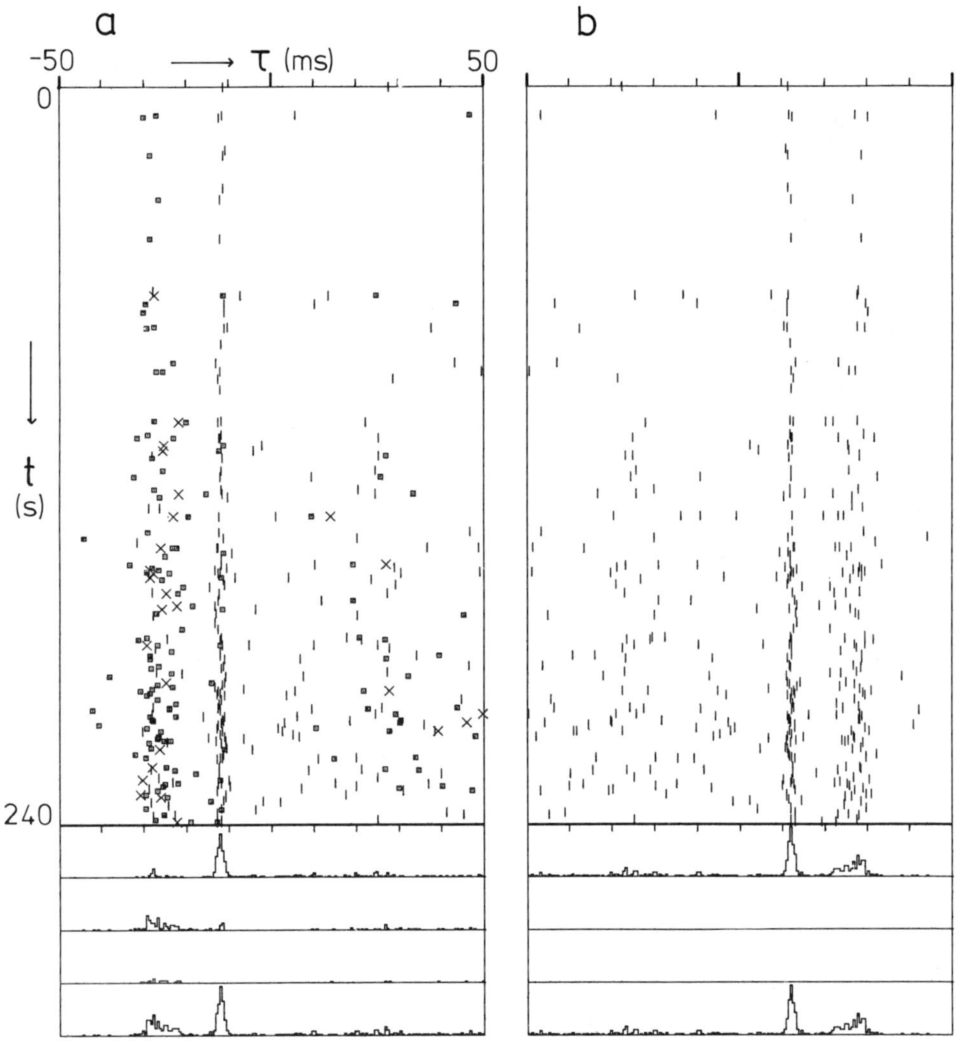

Fig. 7.6 a, b. Time-dependent cross-correlation between spikes and clicks with indication of recurrence times, both for spikes as trigger (**a**) and for clicks as trigger (**b**). The recurrence time histograms as well as the cross-coincidence histogram are shown. (van Stokkum et al. 1986)

with $G_A(\tau)$ as given in Eq. (7.5). In the case when the cross-interval histogram and the cross-correlation histogram of spike trains A and B are not significantly different from the predictions given in Eqs. (7.6) and (7.5) one may consider spike trains A and B independent.

Recently (van Stokkum et al. 1986), it has been shown that considering the time-dependent distribution of first to third order (a practical limitation) forward and backward waiting times (or cross-intervals) is of considerable advantage. First of all, it can be examined if the process is stationary and estimating of densities by obtaining cross-interval histograms is allowed. Secondly, the actual distribution of the cross-intervals of order one to three is more readily related to the underlying connectivity between the units, than is the cross-correlation function. An example is shown in Fig. 7.6. This example is taken from a multiunit recording from the auditory midbrain of the frog. A Poisson distributed click stimulus ensemble was presented with an average rate of $16\,s^{-1}$ and a minimum interval of 1 ms. In part a of Fig. 7.6 a cross-correlation is shown between spikes of a neural unit (trigger) and the clicks. In part b the clicks are used as a trigger and the spikes are plotted. One observes from part b that a click is followed after about 12 ms with a spike and that there is increased activity around 25 ms as well. Note that only first order waiting times are present. Turning again to part a, one observes that in most cases two clicks (one at 25 ms, the other 12 ms before the spike) precede the spike, suggesting a neural unit that is sensitive to this particular click combination. This method can of course also be used for two spike trains.

7.4 Joint Peri-Stimulus-Time Scattergrams

For use with a periodic stimulus a slightly different version of the joint-occurrence density $R_{AB}(t_{Aj}, \tau)$ was introduced by Gerstein and Perkel (1969, 1972) and denoted the joint PST scattergram. It represents in a two-dimensional plot the occurrence of three sets of events: a train of periodic stimuli and the occurrences of spikes for neural units A and B. The time base along the horizontal as well as vertical axis is equal to a stimulus period. Along the horizontal direction the B-spikes are plotted, in the vertical direction the A-spikes in such a way that for each B spike in a particular sweep all the A-spikes in that same sweep are indicated. Joint occurrences from different sweeps are superimposed in the scattergram. The principle is illustrated in Fig. 7.7a.

The most simple situation is obtained when either the A- or the B-train is spontaneously active but both respond to the stimulus with latencies t_A and t_B respectively. In the scattergram one obtains a dark region around the point (t_a, t_B) in the case of only one spike following the stimulus for each unit. In the case of both units responding tonically to the stimulus one will obtain a column at t_B and a horizontal bar at t_A with a length equal to the response duration. For two spontaneously active units without any stimulus dependence the scattergram will be randomly filled with dots in the case the neurons firing independently (Fig. 7.7c). If one (Fig. 7.7d) or both (Fig. 7.7e) neurons, in ad-

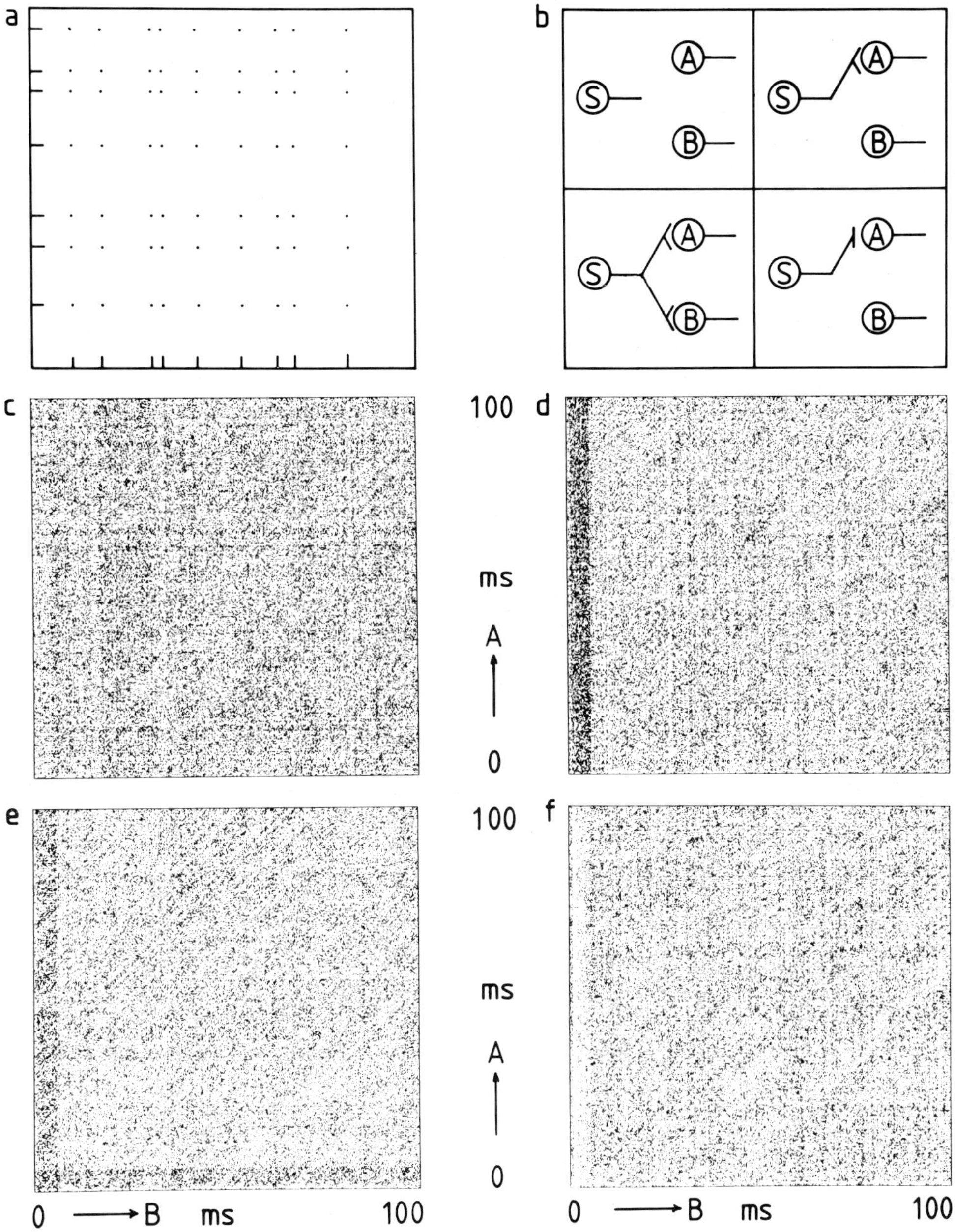

Fig. 7.7 a–f. Joint peri-stimulus-time scattergrams of simulated neural activity. **a** shows the principle of construction: on the horizontal axis one plots the spikes in train B that follow within say 100 ms of stimulus onset, along the vertical axis one plots all spikes of train A over (in this case) the same period of time (this need not be so in general), and indicate all joint occurrences by *dots*. **b** shows the wiring diagrams for the four model situations shown in parts (**c–f**), *S* indicates stimulus. **c** shows the scattergram in the case of independent, spontaneously active, spiketrains without stimulus effect. **d** shows independent spiketrains with the stimulus activating neuron A. **e** shows independent neurons with a common excitatory stimulus effect on both neuron A and B. **f** shows independent neurons with the stimulus-inhibiting neuron A

dition, are influenced by the stimulus, a combination of previously observed patterns occurs. If the stimulus inhibits spontaneous activity for one of the neurons a light region is observed (Fig. 7.7 f). The scattergrams become more interesting when spike trains A and B are not independent. If the A and B spikes tend to occur nearly simultaneously in a spontaneously active situation, then a dark band parallel with the main diagonal will become visible (Fig. 7.8 a). When one neuron is inhibiting the other, one observes for the spontaneous case a light region parallel to the main diagonal (Fig. 7.8 c). A systematic study of various scattergrams that occur for simulated two- or three-neuron networks is given by Gerstein and Perkel (1972). We will discuss a few examples. The first one illustrates the effect of common excitatory input upon two spontaneously active neurons. Let us assume that the common driver unit is also spontaneously active. This situation is often indistinguishable from the situation where one neuron excites the other: one observes a dark band coinciding with or parallel to the main diagonal (cf. Fig. 7.8 a). The case of two neurons with common and stimulus input is shown in Fig. 7.8 e and a case where the stimulus inhibits neurons A and B is shown in Fig. 7.8 d. Finally in a case where the stimulus excites neuron C, which in turn excites neuron A but inhibits neuron B, the joint PST of neuron A and B (Fig. 7.8 f) shows the dark and light regions parallel to the axis indicating the stimulus effect, and the diagonal showing the reciprocal shared input.

Correlations between spike trains can occur through a locking of the firings of both units to the stimulus. For instance, one neuron might fire with a latency of approximately 10 ms and the other with around 15 ms. If both neurons do not fire in exclusion, then the joint PST scattergram will show a dark band paralleling the diagonal at a "distance" of 5 ms. Such a case is shown in Fig. 7.8 e, where besides some complicated stimulus effects such as repetitive firings (or chopper-like activity) this nearly diagonal band is seen. By either randomly shuffling the stimulus intervals for one spike train or by shifting one spike train over one or more stimulus periods with respect to the other one, a new scattergram can be calculated that shows only the stimulus-locked activity. This is known as the "shift predictor" (e.g., Gerstein and Perkel 1972), and will contain all complicated stimulus effects but not the diagonal. When a histogram is made of the joint PST histogram as well as for the shift predictor, then the latter can be subtracted from the first to give a stimulus-corrected histogram (Gerstein 1988).

The density representation of the joint PST offers also an interpretation in terms of cross-correlation functions or renewal densities. When the ordinary PST density, estimated by the PST histogram is defined as:

$$P_A(t) = \lim_{\Delta t \to 0} \frac{1}{\Delta t} \text{Prob}\{\text{A}-\text{spike in } (t, t+\Delta t) \,|\, \text{stimulus onset at zero}\} \tag{7.8}$$

the joint PST-density as:

$$P_{AB}(t,u) = \lim_{\substack{\Delta t \to 0 \\ \Delta u \to 0}} \frac{1}{\Delta t \Delta u} \text{Prob}\{\text{A-spike in } (t, t+\Delta t) \text{ and B-spike in } (u, u+\Delta u) \,|\, \text{stimulus onset at zero}\} \tag{7.9}$$

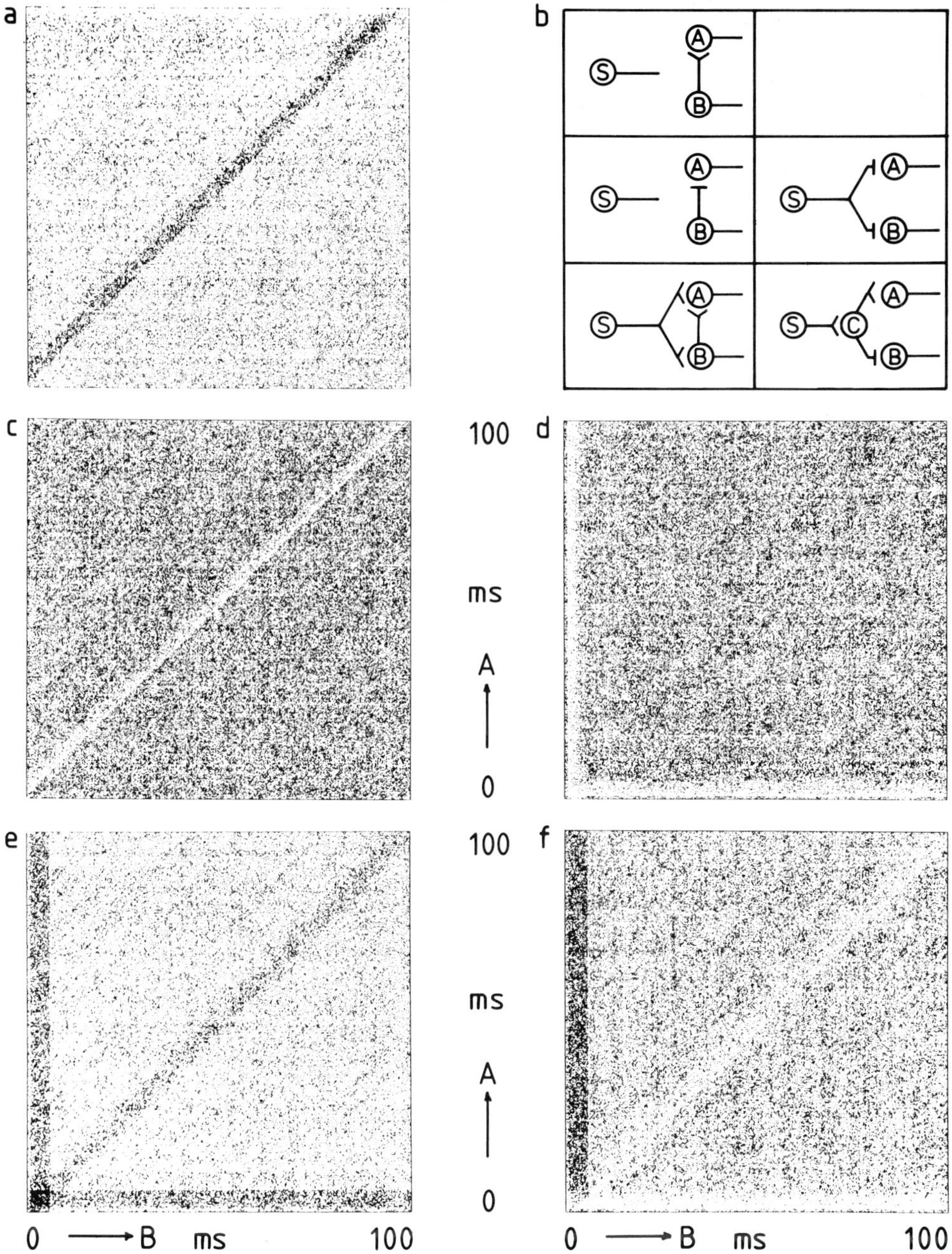

Fig. 7.8 a – f. More joint peri-stimulus-time histograms of simulated neural activity. **b** shows again the various model conditions. **a** shows the case for neuron B exciting neuron A and without a stimulus effect. **c** shows the same for inhibition. **d** shows common inhibition of both neurons by the stimulus. **e** shows a case where there is shared stimulus input with nearly identical latencies for both units, this results in a pseudo neural correlation effect along the diagonal. **f** shows a case where the stimulus activates a neuron C which reciprocally inputs on neurons A and B

and the conditional PST-density as:

$$P_{A|B}(t,u) = \lim_{\Delta t \to 0} \frac{1}{\Delta t} \text{Prob}\{\text{A-spike in } (t, t+\Delta t) \,|\, \text{B-spike at } u \text{ and stimulus onset at zero}\} \tag{7.10}$$

then

$$P_{AB}(t,u) = P_{A|B}(t,u)\,P_B(u) = P_{B|A}(u,t)\,P_A(t) \ . \tag{7.11}$$

If the A and B spikes are generated independently, then

$$P_{A|B}(t,u) = P_A(t) \quad \text{and thus}$$

$$P_{AB}(t,u) = P_A(t) \cdot P_B(u) \ . \tag{7.12}$$

That is, the density in the joint PST histogram equals the product of the marginal densities (i.e., the densities along the horizontal and vertical axis).

On this basis a control for independence could be constructed. This joint PST density under independence is, however, difficult to compare with a joint PST scattergram. In practice, a control situation is therefore constructed on the basis of either shuffling the intervals within a stimulus period or by shifting one spike train over a few stimulus periods and then constructing a new scatter diagram. If long-term trends are absent, both the shuffling and the shift control will yield the same results.

When the joint PST scattergram is projected onto a plane perpendicular to the diagonal, taking proper account of the boundary effects, the cross-coincidence histogram $R_{AB}(\tau)$ is obtained. A slightly different version is formed when on the vertical axis of the joint PST scattergram the time difference of the A-spikes with respect to the B-spikes is plotted. In this case we obtain a version of $R_{AB}(t_{Aj}, \tau)$ still being dependent of the time since the onset of the stimulus, that is in fact representing a region parallel to the diagonal in the joint PST scattergram.

The joint PST scatter diagram can be extended to three spiketrains and then results in the "PST cube" (Gerstein 1970) or in the "snowflake" (Perkel et al. 1975). The joint PST cube has three mutually orthogonal time axes, all one stimulus period long, representing the occurrences of spikes of units A, B, and C with respect to stimulus onset. An example is shown in Fig. 7.9a. The "snowflake" can be understood as resulting from the projection of the points in the PST cube onto a plane perpendicular to the main diagonal (Fig. 7.9b). Three time axes intersecting at 120° and representing $t_B - t_A$, $t_C - t_B$, $t_A - t_C$, i.e., the time differences in firing between unit B and unit A etc., form a hexagonal diagram (Fig. 7.9c, d), hence its name. In the snowflake a point is plot-

Fig. 7.9a–c. Joint peri-stimulus time scattergrams for three neurons. A trivial extension of the two neuron scattergram is the PST-cube (**a** from Gerstein and Perkel 1972). Another representation is to project all the points onto the plane perpendicular to the main diagonal, resulting in the "snowflake", **b** shows its construction and interpretation (Perkel et al. 1975). **c** shows a density representation (histogram) of the snowflake for three neurons from the auditory cortex. (Abeles et al. 1983). For further details see text

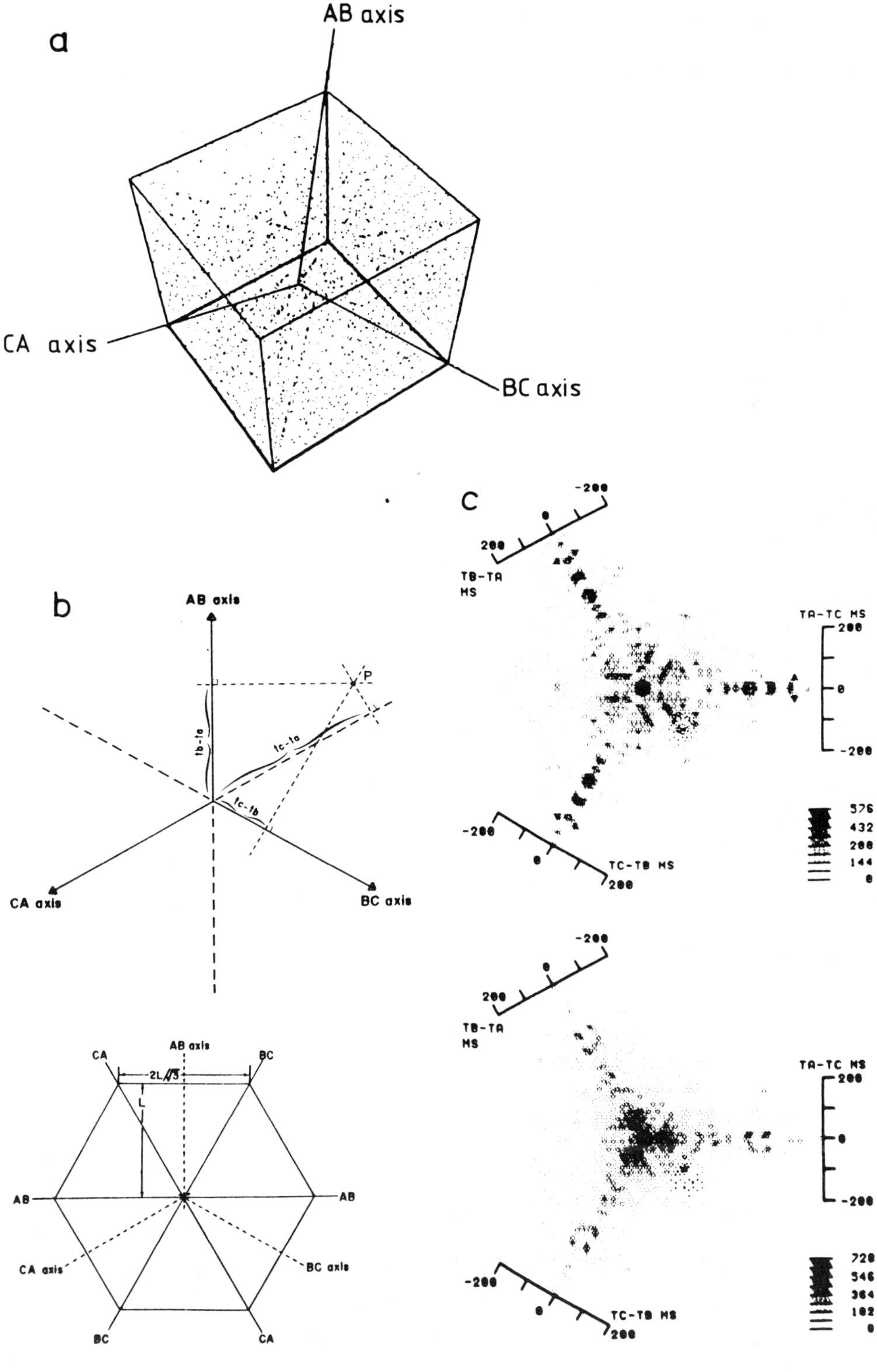
a
AB axis
CA axis
BC axis
b
AB axis
P
tb-ta
tc-ta
tc-tb
CA axis
BC axis
CA
AB axis
BC
2L/√3
L
AB
AB
CA axis
BC axis
BC
CA
c
-200
0
200
TB-TA MS
TA-TC MS
200
0
-200
-200
0
TC-TB MS
200
576
432
288
144
0
-200
0
200
TB-TA MS
TA-TC MS
200
0
-200
-200
0
TC-TB MS
200
720
546
364
182
0

ted for each triplet of spikes. For the formidable variety of intrinsic circuits among three neurons – Perkel et al. (1975) list 728 possibilities – this representation can be of considerable help. Abeles (1983) and Abeles et al. (1983) describe a slightly modified version in which the snowflake is partitioned into triangular bins. The number of dots in the bins is indicated by a gray code. The triple joint-PST density $P_{ABC}(t,u,v)$ can be expressed in two types of conditional densities:

$$P_{ABC}(t,u,v) = P_{AB|C}(t,u,v)\,P_C(v) \tag{7.13}$$

and

$$P_{ABC}(t,u,v) = P_{A|BC}(t,u,v)\,P_{BC}(u,v)\ , \tag{7.14}$$

where $P_{AB|C}(t,u,v)$ gives the rate at which joint firing of units A and B is related to events of unit C, and $P_{A|BC}(t,u,v)$ gives the rate of firing of unit A on the condition of the firing times of units B and C. Abeles (1983) argues that especially the simple renewal density $P_{A|BC}$ is very helpful for the interpretation of neural interaction, whereas P_{ABC} or $P_{AB|C}$ are often to complicated to interpret.

7.5 A Hierarchy of Multiunit Representations

Having discussed a large variety of ways to represent few- unit recordings and possible neural interactions, I will now briefly describe a hierarchical scheme of these representations (Fig. 7.10). This scheme consists of three levels; level I incorporates the running time of the experiment and is needed to judge stationarity of the recording, level II results after averaging stimulus periods, and level III focuses on time differences between the firings of the various neurons.

Basically it all starts with the multiunit electrode signal. After applying a multispike train separation procedure the spike train can also be represented as a symbol – or color-coded dot display which can be both a first- or second-level representation. On the first level we have also the double- and triple-unit joint occurrence diagrams, and the time-dependent cross-correlation for double and triple units in which the last one takes the form of a conditional diagram. On level II we obtain – through averaging over stimulus presentations – the Joint PST scattergram and the Joint PST cube. We may arrive at level III either directly from level I or from level II by averaging over time and so obtain the cross-coincidence histogram and the cross-coincidence plane or the snowflake.

The relations between the various representations can also be summarized in a more mathematical way:

The joint PST scattergram is an estimator of:

$$P_{AB}(t,u) = \lim_{\substack{\Delta t\to 0\\ \Delta u\to 0}} \frac{1}{\Delta_t \Delta_u} \text{Prob}\,\{\text{spike of neuron A in } (t, t+\Delta_t \text{ and spike of neuron B in } (u, u+\Delta_u)\,|\,\text{stimulus at zero}\}\ . \tag{7.15}$$

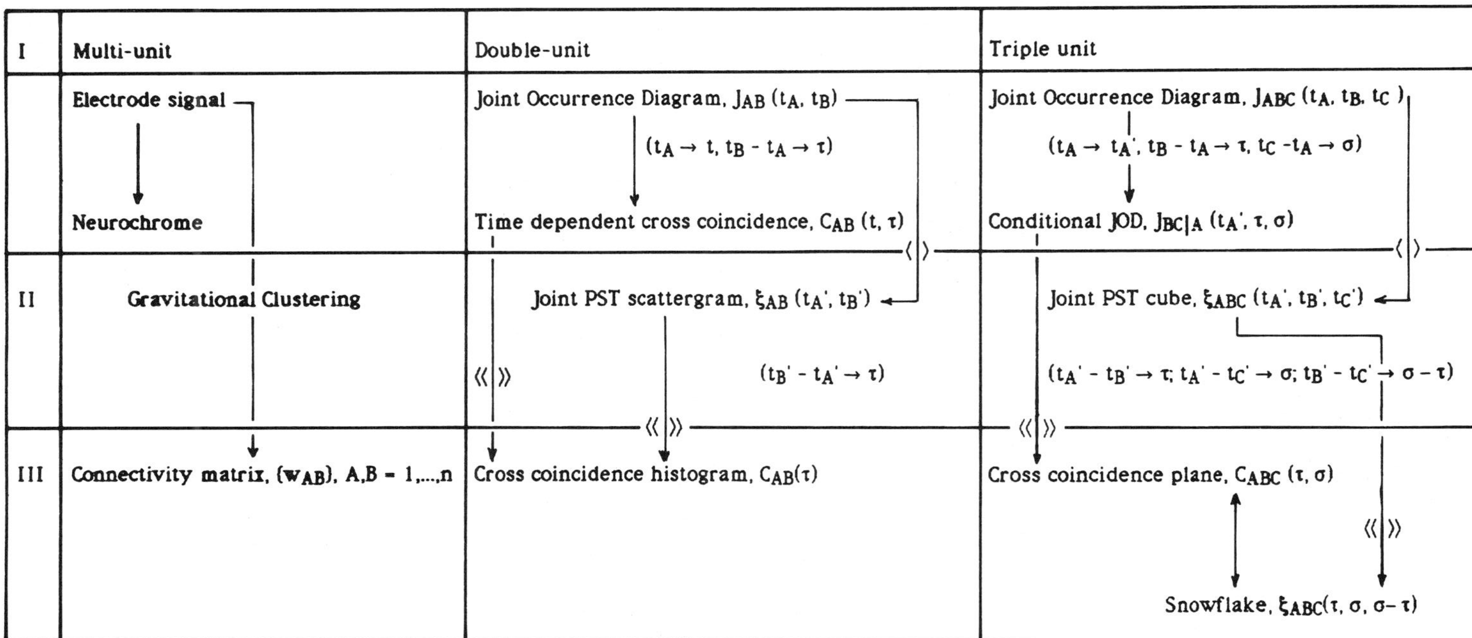

⟨ ⟩ denotes average over stimulus presentations
⟨⟨ ⟩⟩ denotes average over experiment time

Fig. 7.10. Hierarchy of multi-unit representations. *Arrows* indicate the way in which representations can be derived from each other; *in brackets* the required transformations are shown

One may derive conditional densities as:

$$P_{AB}(t,u) = P_{A|B}(t,u) P_B(u) = P_{B|A}(u,t)\, P_A(t) \ , \tag{7.16}$$

where $P_B(u)$ and $P_A(t)$ are the marginal densities of units B and A (estimated by their PST histograms). The correlation function is defined as:

$$R_{AB}(\tau) = \lim_{\Delta t \to 0} \frac{1}{\Delta t} \text{ Prob \{spike of neuron B in } (t_0+\tau,\ t_0+\tau+\Delta t) \,|\, \text{spike of neuron A at } t_0\} \ , \tag{7.17}$$

and therefore $R_{AB}(\tau)$ is related to $P_{B|A}(u,t)$ with $\tau = t-u$ and is obtained from $P_{B|A}$ through averaging over the complete stimulus presentation (if allowed). The time-dependent cross-covariance $C_{AB}(t,\tau)$ is related to $P_{AB}(t,u)$:

$$P_{AB}(t,u) = P_A(t)\cdot P_B(u) + C_{AB}(t,u) \ . \tag{7.18}$$

When the neurons are independent, then:

$$P_{AB}(t,u) = P_A(t)\cdot P_B(u)$$

and $C_{AB}(t,u) = 0$

Equation (7.18) provides an analogy with the relation between cross-covariance $C_{xy}(\tau)$ and cross-correlation $R_{xy}(\tau)$.

$$C_{xy}(\tau) = R_{xy}(\tau) - \mu_x \mu_y \tag{7.19}$$

when written as:

$$C_{AB}(t,u) = P_{AB}(t,u) - P_A(t)\cdot P_B(u) \tag{7.20}$$

and we can consider $P_A(t)$ and $P_B(u)$ as stimulus-modulated intensities of the two processes.

7.6 Correlation Analysis of Larger Numbers of Neural Units

The global analysis of neuronal ensembles obviously has to precede a detailed pair or triple correlation analysis as described in the previous pages. With the present possibilities of recording simultaneously and separably from 20–50 neurons (e.g., Krüger and Bach 1981; Gerstein et al. 1983), it becomes nearly impossible to analyze the underlying connectivity using standard correlation analysis. A 40-neuron recording for one stimulus condition and including the shift predictor analysis requires 1560 pair correlations to be computed and, worse, analyzed. Any new stimulus condition requires an additional 1560 correlation functions, and so on. The first attempt to analyze such large neuron clusters was (again) made by Gerstein et al. (1978) by simply comparing the number of coincident firings observed against that predicted in case of independence using a χ^2-statistic. The procedure searches for the highest χ^2 and then, for this neuron pair, searches for the neuron that combined with it gives the highest triple-coincidence score etc. In this way a list of ordered strings is obtained representing functionally associated neurons. To analyze these associations in more detail the standard correlation procedures can be applied to the reduced number of relevant neurons in the ensemble. The method is limited to excitatory couplings.

A more recent and very powerful method to analyze an ensemble of neurons is based on a pattern recognition paradigm called gravitational clustering (Gerstein et al. 1985b). The basic idea is to transform the N-simultaneously recorded point processes in time into an N-body problem in N-space by associating each of the N observed neurons with a particle in this space. Each particle is assigned a charge which increases with a fixed amount at the moment of each spike in the corresponding spike train. The charge decays with time in an exponential way. Forces between the charged particles will cause them to move:

$$\dot{\vec{r}}_i = \vec{F}_i / U$$
$$\vec{F}_i = q_i \sum_{j \neq i}^{N} q_j \, \vec{r}_{ij} \ , \tag{7.21}$$

where $\vec{r}_{ij}$ is a unit vector pointing from particle i to particle j. The force $\vec{F}_i$ thus is chosen independent of particle distance. A high viscosity in the N-space is assumed which prevents acceleration of the particles and makes their velocity proportional to distance. The result of this set of rules is that particles, corresponding to neurons which are firing in a statistically coherent manner, will aggregate into a cluster, different clusters signifying different coherence groups (assemblies?). It is reported that with typical neural interaction strength and range (related amplitude and time duration of the charge), useful clustering of the particles in the N-space occur after some 50–100 spikes of each neuron. This makes this technique an order of magnitude more sensitive than techniques discussed earlier. If a metric could be imposed on this process, the "final" distance between the particles is a measure of their association strength and a suitable two-dimensional representation thereof would form a map of the *connectivity or association matrix* (Aertsen et al. 1986). In order to detect inhibition, the sign of the force has to be reversed. However, the sensitivity of the procedure for inhibition is less than for excitation, a finding similar to that observed for the cross-correlation method (Aertsen and Gerstein 1985). An example from a model simulation of an inhibitory neuron pair (1, 2) among a group of ten independent neurons shows the power of the method. The trajectory indicating the distance between particles 1 and 2 decreases steadily with time while this is much less clearly so for the other independent neurons (Fig. 7.11). Conventional cross-coincidence histograms calculated over selected intervals are shown below and clearly show the inhibition of neuron 1 upon neuron 2.

A modification proposed by Gerstein and Aertsen (1986) takes care of the problem of how to identify the driver neurons and the follower neurons among the associated group. So far the corresponding particles are moving towards

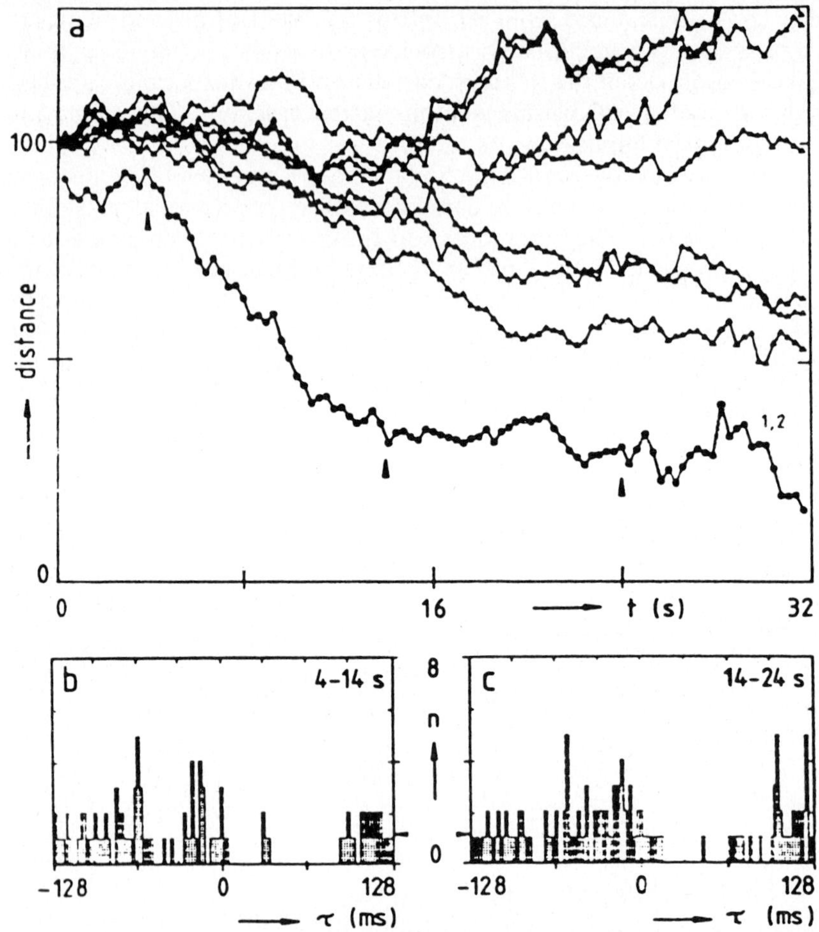

Fig. 7.11a–c. Gravitational clustering. **a** shows the pair-wise distances (inversely proportional to correlation strength) of the neurons in a simulated neural network as a function of time. One observes the gradual decrease in distance for the pair labeled *1, 2*; this indicates a functional correlation. **b** and **c** give the cross coincidence histograms for two 10-s periods or recording, indicated by the *arrowheads:* it is clear that one neuron inhibits the other. All the other pairs are independent in this simulation. (Gerstein and Aertsen 1986)

each other in a symmetrical way. The modification distinguishes an acceptor charge and an effector charge, both of them assigned to each particle. Acceptor and effector charge as a function of time can now be each other's mirror images, or, more generally, such that their cross-correlation becomes an asymmetric function of time. This breaks the symmetry in the force interaction and thus, although the number of calculations is doubled and the interpretation has become more difficult, this modification allows inference on the directionality on the connections.

The random motion of the trajectories of the independent neuron pair distances creating a signal-to-noise problem is, at present, one of the remaining problems of this technique (Aertsen et al. 1988). The form of the trajectory ensemble is suggestive of a diffusional type of drift. Anyhow, the technique holds great promise in arriving at a quick overview of the order in a neural assembly which can then be analyzed by more detailed techniques.

7.7 Are Higher-Order Correlation Analyses Necessary?

Gravitational clustering is capable of showing multiple interactions between neurons in an ensemble. An important issue in the quantitative analysis of such multiple-unit interaction is when and how the multiple correlation functions can be expressed in terms of pair correlation functions. The relationship between the shape of the pair correlation function and the underlying neural wiring diagrams has been studied for simulated networks by Moore et al. (1970) and offers some help in the interpretation of neural correlations in experimental situations. In Fig. 7.12 we show some of the basic neural interactions and their pair correlation functions: shared excitatory synaptic input resulting in a peak around $\tau = 0$ (Fig. 12a) shared inhibitory synaptic input resulting in a rather small peak around the origin (Fig. 12c); shared reciprocal synaptic input showing a central trough (Fig. 12d). It is noted that in practice most of the correlograms cannot readily be identified with one of these simulations or a combination of them: the inverse problem always is an order of magnitude more difficult than the forward one, moreover it does not necessarily have a unique solution. The problem that we face is whether pair correlations are sufficient to describe neural interaction in an assembly or whether we need more sophisticated measures. As an analogy, we consider the physical model system of an ideal gas where all the molecules or particles are independent. The equation of state is the well known ideal gas law pV = kT. For a nonideal, i.e., a real gas with interaction between the particles the ideal gas law no longer holds but is replaced by the "virial expansion" (Huang 1963):

$$\frac{pV}{kT} = \sum_{r=1}^{\infty} a_r(T) \frac{(\lambda^3)^{r-1}}{V}, \tag{7.22}$$

where $\lambda = (2\pi)^{1/2}/mkT$ is the thermal De Broglie wavelength and v is the volume occupied by one particle (v = V/N). The virial coefficients $a_r(T)$ are polynomials of so-called cluster integrals. The second cluster integral describes the interaction between isolated pairs of particles, the third cluster integral any interaction between triplets, etc.

Obviously in the description of the ideal gas we only need the first cluster integral: $a_1(T) = 1$, all the higher order $a_r(T)$ are zero and the ideal gas law follows. In a real but diluted gas the second cluster integral is needed, it appears that in this case pair interactions are sufficient to describe the equation of state of the gas (leading to the Van der Waals equation). When the gas

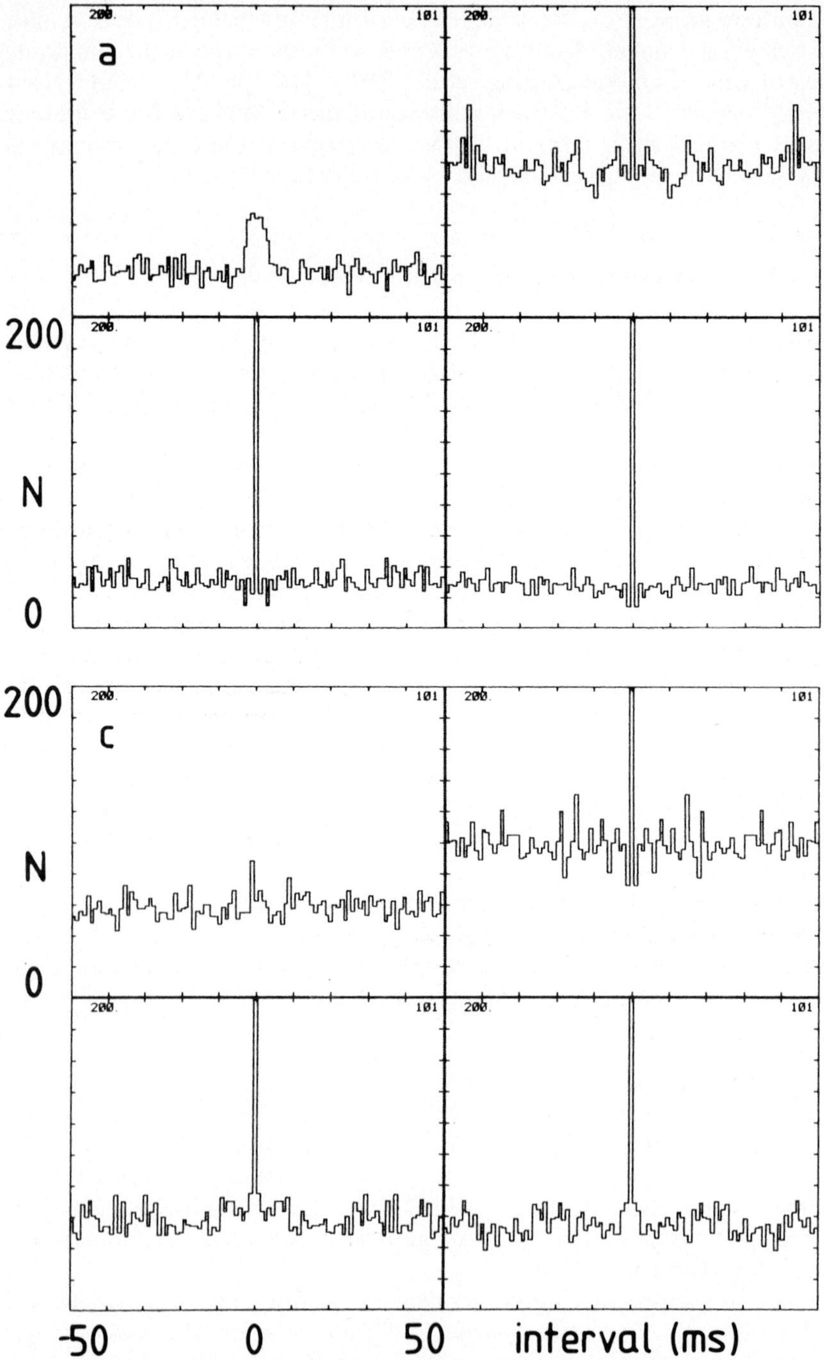

Fig. 7.12 a–d. Some forms of shared input. **b** is the key to the design of the other parts: for the three unit interactions the cross-correlation between the output neurons (*A* and *B*) as well as the auto-correlations for each of the three units are shown for delay times from −50 to +50 ms. Shared

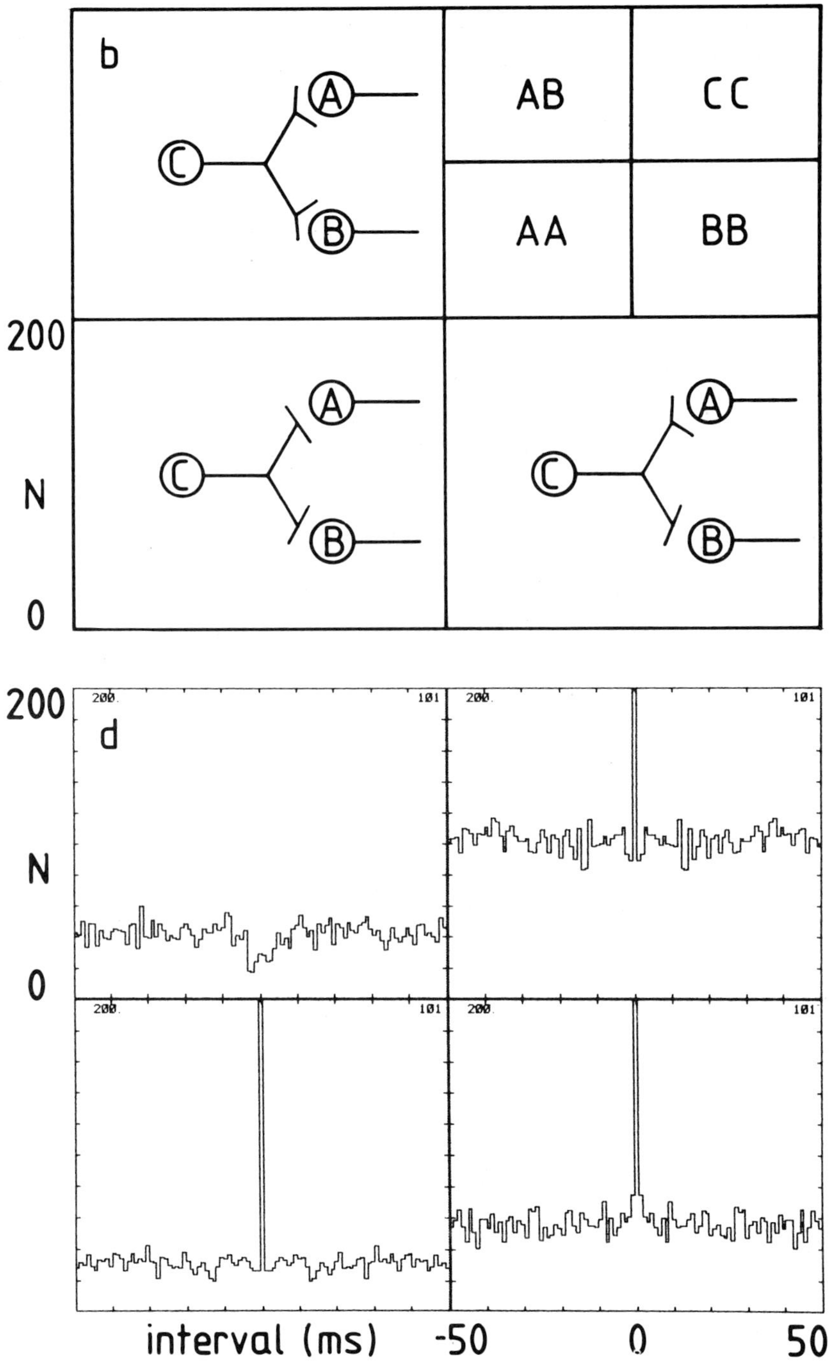

excitatory input results in a peak in the cross-correlation function around zero **a**. Shared inhibitory input of the same strength also results in a peak around zero but not as pronounced **c**. Shared reciprocal input produces a valley around zero **d**. All spike trains had Poisson characteristics becomes more and more dense, higher-order interaction terms are needed for its equation of state. It is my impression that an analog situation occcurs in the central nervous system, where obviously it is an interesting question which neuronal network property plays the role that particle density plays in the physics model.

Let us adopt a general formalism on the basis of the decomposition of higher-order joint densities into correlation functions (e.g., van Kampen 1981). The nth order joint density describing the firings of n-neurons is indicated by $P_{1,\ldots,n}(t_1,\ldots,t_n)$, where t_1 stands for the firings of neuron 1, etc. If the neurons fire independently we have simply the product of the marginal densities

$$P_{1,\ldots,n}(t_1,\ldots,t_n) = J_1(t_1)\cdot J_2(t_2)\ldots J_n(t_n)\ . \tag{7.23}$$

Define a hierarchy of correlation functions $g_1(t_1)\ g_{12}(t_1,t_2),\ldots,g_{1,\ldots,n}(t_1,\ldots,t_n)$ such that

$$P_1(t_1) = g_1(t_1)$$

$$P_{1,2}(t_1,t_2) = g_1(t_1)\,g_2(t_2)+g_{1,2}(t_1,t_2)$$

$$\begin{aligned}P_{1,2,3}(t_1,t_2,t_3) = {} & g_1(t_1)\,g_2(t_2)\,g_3(t_3)+g_1(t_1)\,g_{2,3}(t_2,t_3)+g_2(t_2)\,g_{1,3}(t_1,t_3)\\ & +g_3(t_3)\,g_{1,2}(t_1,t_2)+g_{1,2,3}(t_1,t_2,t_3)\ .\end{aligned} \tag{7.24}$$

This can be used as a test to decide whether higher-order correlations are necessary. In our notation $g_{1,2}(t_1,t_2)$ is the time-dependent correlation function estimated from $C_{1,2}(t_{1,j},\tau)$ where $\tau = t_2 - t_1$. In the stationary case then, once we know the marginal densities related to $g_1(t_1)$ and the cross-coincidence function $C_{ij}(\tau)$ which relates to $g_{ij}(t_i,t_j)$, we can predict whether we need to incorporate higher-order correlation terms. This procedure runs as follows: the triple joint occurrence density $J_{1,2,3}(t_1,t_2,t_3)$ is calculated from the three point processes, and on basis of the $g_1(t_1)$ and $g_{ij}(t_i,t_j)$ a predicted $\tilde{J}_{1,2,3}(t_1,t_1,t_3)$ is calculated. When $J_{1,2,3}$ equals $\tilde{J}_{1,2,3}$ then obviously $g_{1,2,3}(t_1,t_2,t_3)$ is zero within statistical limits and pair correlations are sufficient to describe the behavior of the triple unit. This, however, does not imply that for $n>3$ the same holds!

7.8 Correcting Cross-Correlations for Effects of Stimulus Coupling

A peak in the cross-correlation function, or a diagonal band in the joint PST scattergram can be the result of a neural interaction between units but also the result of a coupling of the firings of both neurons to the stimulus. As we will see in the next chapter, the interpretation of an (averaged) cross-coincidence function will require knowledge of the autocoincidence functions of both spike trains (cf. Fig. 7.13) and an estimate of the amount of correlation due to this stimulus coupling. In our present discussion, I will follow largely Eggermont et al. (1983a):

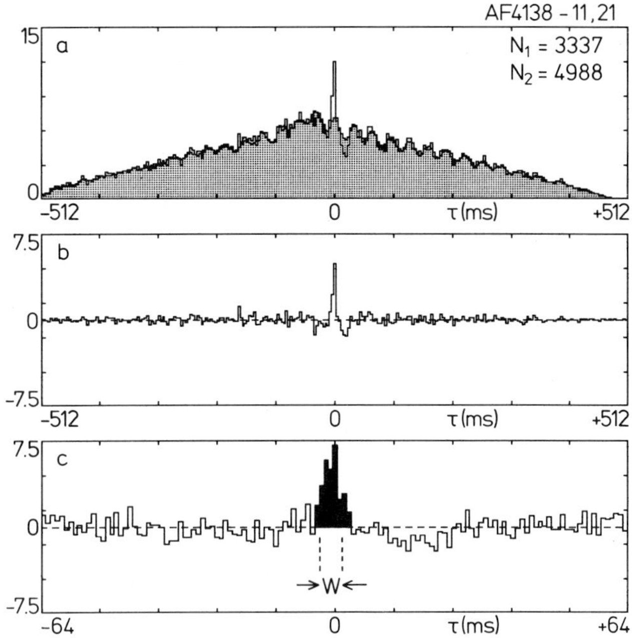

Fig. 7.13 a – c. Stimulus- and neural- cross-correlation in a neuron pair from the auditory midbrain of the grass frog. **a** shows the sum of stimulus and neural correlation, **b** the result after subtraction of the stimulus coupling part of the correlation; **c** shows the same on an enlarged time scale: this is probably a case of shared input. (Epping and Eggermont 1987)

Consider two spike trains $z_1(t)$ and $z_2(t)$ of duration T represented as sequences of δ-functions:

$$z_k(t) = \sum_{i=1}^{N_k} \delta(t-t_{ki}) \; ; \quad k = 1,2 \tag{7.25}$$

with cross-correlation function:

$$R_{z_1 z_1}(\tau) = \frac{1}{T} \sum_{i=1}^{N_1} \sum_{j=1}^{N_2} \delta[\tau-(t_{1j}-t_{2j})] \; . \tag{7.26}$$

For a binwidth Δ a cross-coincidence function:

$$C_{z_1 z_2}(\tau, \Delta) = \frac{1}{\Delta} \int_{\tau-\Delta/2}^{\tau+\Delta/2} R_{z_1 z_2}(\sigma)\, d\sigma \tag{7.27}$$

can be defined, which in experimental situations takes the form:

$$C_n(\Delta)\, C_{z_1 z_2}(n\Delta, \Delta) \quad n = 0, \; \pm 1, \; \pm 2, \ldots$$

(see also Aertsen et al. 1979).

One may present the stimulus ensemble to the animal twice; in case of a double-unit recording resulting in spike trains $z_1(t)$ and $z_2(t)$ respectively $z_1'(t)$ and

Table 7.1. Overview of possible coincidence histograms for a double-unit recording with two presentations of the stimulus ensemble

	$z_1(t)$	$z_2(t)$	$z'_1(t)$	$z'_2(t)$
$z_1(t)$	C_{11}	C_{12}	$C_{11'}$	$C_{12'}$
$z_2(t)$		C_{22}	$C_{21'}$	$C_{22'}$
$z'_1(t)$			$C_{1'1'}$	$C_{1'2'}$
$z'_2(t)$				$C_{2'2'}$

$z'_2(t)$ for the second one. For those four spike trains one may compute ten cross-coincidence histograms as shown in Table 7.1. We will use the notation C_{ij} for $C_{z_i z_j}$.

On the main diagonal two estimates for each of the units autocoincidence functions are found. Under long-term stationary firing conditions $C_{ii} \cong C''_{ii}$, $i = 1, 2$ and one may use $C_i = 1/2\ (C_{ii} + C'_{ii})$ as an estimator of the autocoincidence function of spike train i. For the off-diagonal elements it is noted that $C_{k1}(\tau) = C_{1k}(-\tau)$ therefore Table 1 is symmetric, provided that the sign of τ is adjusted. We define C_{12} and $C_{1'2'}$ as the simultaneous cross-coincidence histograms of spike trains z_1 and z_2 and again one may use $C_{ij} = 1/2\ (C_{ij} + C_{i'j'})$ as an estimator for the cross-coincidence function.

The *nonsimultaneous* cross-coincidence histograms $C_{12'}$, and $C_{1'2}$ are thought to represent only the correlation between spike trains z_1 and $z_{2'}$ or between $z_{1'}$ and z_2 which is due to stimulus coupling. Note that this method of estimating the correlation induced by the stimulus is a generalization of the shift predictor for a strictly periodic stimulus train as used by, e.g., Perkel et al. (1967) and discussed in previous parts of this chapter.

Another cross-coincidence histogram is obtained by taking both spike trains for the same unit to obtain $C_{11'}$, and $C_{22'}$. These histograms represent the amount of stimulus coupling of the units individually and have been called "existence function" (Aertsen et al. 1979) because any significant peak around $\tau = 0$ indicates a stimulus coupling effect and hence the existence of a stimulus-response relation.

There is no model-free approach for the separation of correlation produced by direct neural interaction and that caused by stimulus coupling. The model that is generally supported considers the effect of applying a stimulus as *additive*, i.e., the presentation of the stimulus only adds spikes but has no effect upon the strength and type of neural interaction (Perkel et al. 1967). Assuming that this model holds, $z_1(t)$ is composed out of *spontaneous* and neurally induced activity $z_{1n}(t)$ and direct *stimulus-induced* activity $2_{1s}(t)$ in an additive way:

$$z_1(t) = z_{1s}(t) + z_{1n}(t) \tag{7.28}$$

$$z_2(t) = z_{2s}(t) + z_{2n}(t)\ . \tag{7.29}$$

The estimators of the simultaneous and nonsimultaneous cross-correlation functions between the two spike trains are:

$$C_{12}(\tau) = E[z_1(t)\, z_2(t+\tau)]$$
$$= E[z_{1s}(t)\, z_{2s}(t+\tau)] + E[z_{1n}(t)\, z_{2n}(t+\tau)] \qquad (7.30)$$

and

$$C_{12'}(\tau) = E[z_1(t)\, z_2'(t+\tau)] = E[z_{1s}(t)\, z_{2's}(t+\tau)] \;. \qquad (7.31)$$

Since obviously spontaneous activity from both neurons is not correlated when the firings are a complete stimulus sequence apart (which will be in the order of many seconds). Under stationary conditions

$$E[z_{1s}(t)\, z_{2s}(t+\tau)] = E[z_{1s}(t)\, z_{2's}(t+\tau)] \quad \text{and thus:}$$

$$C_{12}(\tau) - C_{12'}(\tau) = E[z_{1n}(t)\, z_{2n}(t+\tau)] \qquad (7.32)$$

can be used as an estimator for correlation due to neural interaction. An example of such a procedure for two neurons from the frog's auditory midbrain is shown in Fig. 7.13, which in the upper part shows a superimposition of $C_{12}(\tau)$ and $C_{12'}(\tau)$, the latter one shaded. Under the assumption of additivity $C_{12}(\tau) - C_{12'}(\tau)$ is shown in the middle part, and on an expanded time scale in the lower part of the figure. The symmetric peak surrounding the origin suggest a shared excitatory input situation.

In general, the right hand side of Eq. (7.32) may still be dependent on the type of stimulation used, for instance, through spikes caused by the stimulus due to unobserved neurons that contribute as well to $z_1(t)$ as to $z_2(t)$. Thus in an additive model one can still expect a stimulus dependent $C_{12}(\tau) - C_{12'}(\tau)$ that is not induced by using the wrong model.

In practice, several situations will generally occur: the simultaneous and nonsimultaneous histograms are similar giving rise to a nonsignificant difference, the simultaneous histogram differs from the nonsimultaneous one but both showing a peak or through, and the simultaneous histogram shows a clear peak or through but the nonsimultaneous histogram is flat. In the first case neural interaction cannot be shown to exist, there is no neural correlation. This, however, does not warrant the conclusion that the neurons are independent. It may simply be so that the stimulus lock of the neuronal firings is so strong that this overwhelms any neural interaction additive to the stimulus-coupled correlation. In such cases a weaker or another stimulus usually shows a difference for $C_{12}(\tau)$ and $C_{12'}(\tau)$. Clearly the additive model does not hold in such cases. The situation in which the nonsimultaneous histogram is flat is indicative for spontaneously active neurons that are not influenced by the stimulus. An example of such situations is shown in Fig. 7.14.

A more general model incorporating also a multiplicative stimulus effect may be formulated as:

$$\begin{aligned} z_1(t) &= \{\alpha_{1s}(t)+1\}\, z_{1n}(t) + z_{1s}(t) \\ z_2(t) &= \{\alpha_{2s}(t)+1\}\, z_{2n}(t) + z_{2s}(t) \;. \end{aligned} \qquad (7.33)$$

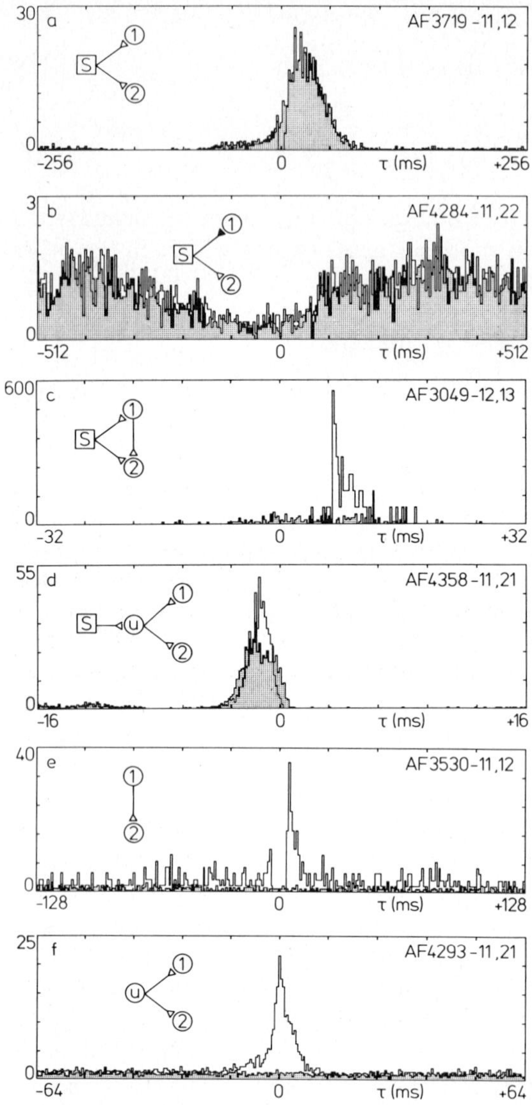

Fig. 7.14a–f. Manifestations of cross-correlation. Simultaneous and non-simultaneous (*shaded*) cross-correlation histograms are superimposed, the inserts show possible functional wiring diagrams. **a** is an example of a fully stimulus-related cross-correlation with probably shared input. **b** is a stimulus-related cross-correlation with reciprocal input. **c** represents unidirectional excitatory input. **d** correlation due to neural shared input as well as stimulus related shared input. **e** unidirectional shared input under spontaneous conditions. **f** neural shared input only under spontaneous conditions. (Epping and Eggermont 1987)

If $\alpha_{is}(t)$ is only slowly varying with respect to the firing rate of $z_1(t)$ and only a function of the type and level of the stimulus, the evaluation of $C_{12}(\tau)$ and $C_{12'}(\tau)$ results in:

$$C_{12}(\tau) - C_{12'}(\tau) = \{1 + \alpha_{1s} + \alpha_{2s} + \alpha_{1s}\alpha_{2s}\} \, E\,[z_{1n}(t)\, z_{2n}(t+\tau)] \ , \tag{7.34}$$

Thus the estimate of the contribution of neural interaction to the cross-correlation will depend in magnitude upon the type and level of stimulation. In case very different types of stimulation are used, e.g., tonepips presented once per second and continuous wide band noise, one will, in addition, expect different time courses for the modulator action and we will need incorporation of terms such as $E[\alpha_{1s}(t)\,\alpha_{2s}(t+\tau)]$. This will result in a change in the time course of the peak in the cross-coincidence histogram as well.

It must be remarked that this treatment of stimulus corrections is far from complete; it indicates only that the additive model that is tacitly used in nearly all reports probably is not universally correct. Estimating the model for the stimulus effect depends on the outcome of the neural correlation procedure which estimates the type of neural interaction. This may be a circular problem for which at present is not solution.

Recently, an unbiased, model-free method for obtaining the significance of a cross-correlation between two spike trains has been published (Gerstein 1988). This method is based on the joint PST histogram, formed by binning the joint PST scattergram. The bin width is chosen such that in each stimulus trial at most one spike per bin is obtained. When there is no interaction between the neurons, the estimator of the expected JPST histogram, and related to the shift predictor for the pair correlation, is the cross-product matrix of the individual PST histograms (Palm et al. 1988). This predictor can be subtracted from the JPST histogram, and the result divided by the standard deviation in each of the bins. This will result in a normalized JPST histogram. In order to arrive at a significance test we have to test the normalized JPST histogram against a theoretical predictor that is adapted for each experimental condition, using the predictor results in the "surprise" (Palm 1981; Palm et al. 1988), the negative logarithm of the probability of finding a particular value. This calculation is performed at every bin of the JPST histogram under the assumption that events in the bins are occurring independently. Since the summation in the direction of the main diagonal over the JPST histogram results in the cross correlogram, this method automatically results in a significance test as well as a model-free stimulus correction procedure for the cross-correlogram (Gerstein 1988).

In conclusion: When stimulus-dependent neural correlations are obtained there are a number of possibilities: the interneuronal coupling itself changes due to application of a stimulus (facilitation, short-term learning), the finding is artifactual due to the selection of the wrong model, or the finding can be explained on the basis of a pool of neurons affecting the firing behavior of the pair or triplet under study in a stimulus-dependent way. First of all, however, we have to investigate whether in physiological data stimulus-dependent neural correlations exist at all. For that purpose the next chapter will give a survey of experimental findings in the vertebrate central nervous system.

8 Correlated Neuronal Activity Observed in the Vertebrate Brain

Sensory nervous systems generally consist of a number of nuclei connected by nerve tracts and projecting to the sensory cortex. The internuclear connections provide divergent and convergent excitatory connections to the cells of the next nucleus or to those of the primary cortex. Because of the divergent connections many cells within a sensory nucleus will experience a shared or common input. Most sensory nuclei are composed of principal – or relay – neurons and interneurons (e.g., Shepherd 1979). Interneurons in most cases impose an inhibitory action upon the principal cell activity. This inhibition can be of the feedforward type (principal cell and interneuron share input from sensory afferents) or of the recurrent or feedback type (the interneuron receives its input directly or indirectly from the principal cell). Both types of inhibition have been demonstrated in the thalamus (e.g., Dubin and Cleland 1977).

A priori then one expects that internuclear correlation studies will show the excitatory character of the neural connections and will provide information about the degrees of divergence and convergence. Intranuclear correlation studies should reveal dominantly shared input and inhibitory connections. From the thalamic level upward one would expect an increasing influence of the state of the animal on the strength and time-course of neuronal correlation. Especially for the cortex it has been shown that neurons tend to fire in a synchronized way when the subject is in deep sleep or under anesthesia. When the neurons are disturbed by sensory input they become desynchronized (Noda and Adey 1970, 1973). As we have seen, the state of the animal might depend on the stimulus that is used; in this way the observed neuronal correlation might become stimulus-dependent. We will explore this systematically for both the visual system, the auditory system and the somatosensory system. After having obtained a general idea about what can be deduced from correlation studies (an elaboration thereupon will follow in Chap. 9) we will survey studies in the motor system and nonsensory (at least not primary sensory) parts of the central nervous system.

8.1 The Visual System

The visual system offers a rare opportunity to study internuclear connections with success. Because a neuron's receptive field actually is a small area on the retina, an electrode can be inserted into this area and a retinal ganglion cell in it can be impaled. The probability of finding a functional connection between that retinal ganglion cell and another cell, e.g., in the LGN is then quite high. Besides this method we will also encounter double-unit recordings on one or two electrodes inserted in a nerve fiber tract or nucleus.

8.1.1 The Retinal Ganglion Cells

A favorite retina for investigation has always been that of the goldfish, and the first attempt to detect correlated neuronal activity between ganglion cells was carried out in this structure. Schellart and Spekreijse (1972), searching for the origin of the stochastic nature of ganglion cell activity determined cross-correlation histograms for five pairs of ganglion cells recorded in the optic nerve using a single electrode, but could not demonstrate any correlation. Arnett (1978), however, in a correlation analysis of 37 unit pairs found evidence for a common input to pairs of cells. Twenty-four pairs showed a sufficient level of spontaneous activity to permit a correlation analysis; for these cells also the receptive field size and type were determined. It appeared that units with nonoverlapping RF's gave flat cross-correlations (6 out of the 24 cases). In four cases, units with RF overlap did not show a cross-correlation either. Nine pairs had "like-type" receptive fields (e.g., both on-center or both off-center) and in the cross-correlation histogram showed a broad peak around the origin. Five pairs showed a pronounced dip around the origin and were subsequently shown to have complementary receptive fields. These results were interpreted as evidence for a noise source common to the ganglion cells. This was confirmed and elaborated upon by Johnson and Levine (1983), who studied 26 unit pairs on single electrodes. By comparing the delay in the cross-correlation histogram with the difference in response latencies of the two ganglion cells, they were able to conclude that the common noise causing the correlation should be located at a late stage in the retinal network. This was further investigated by Ginsburg et al. (1984) in 31 pairs recorded on single electrodes in 28 retinas. Pairs of like-response type showed a central peak in the cross-correlation histogram, pairs of unlike responses showed a dip. In 16 pairs out of these 31, the cross-correlation histogram under stimulus conditions appeared to differ from that under spontaneous discharge. This was explained by assuming that stimulation changes the gain of the retinal pathways. It must be remarked that the stimulation used was highly nonstationary: on-periods of 9–30 s separated by at least 20 s of darkness. As we have shown in the previous chapter, an average cross-correlation histogram over a stimulation period where profound changes in adaptation take place must be interpreted very carefully.

In the rabbit retina Arnett and Spraker (1981) studied 35 pairs of retinal ganglion cells using single or dual electrode recordings. Thirteen out of 14 like-type pairs had a peak near the origin in the cross-correlation histogram, and all 21 unlike-type pairs showed a dip around the origin. The authors suggested that a common noise source in the bipolar cells could explain these results.

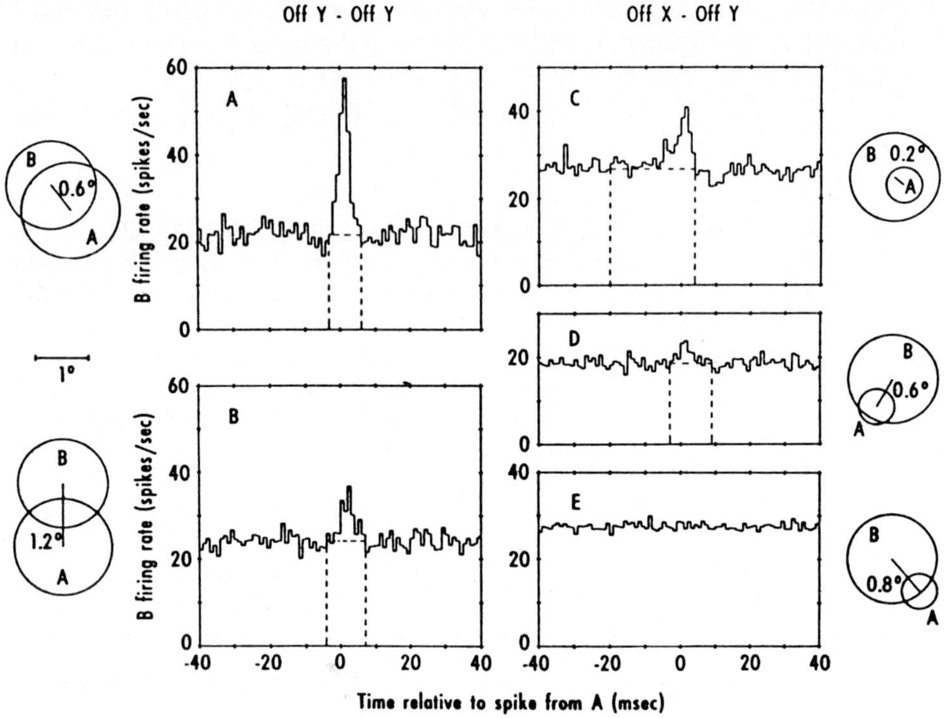

Fig. 8.1 A–E. Correlation strength dependence on the amount of overlap of receptive fields. The receptive field sizes and the amount of overlap are indicated beside the correlograms. It seems that the actual size of the overlap area is the determining factor for the strength of the correlation. (Mastronarde 1983 a)

Recently, Mastronarde (1983 a, b, c), investigating neural correlations for cat ganglion cells, confirmed these general results, and in addition showed that the strength of the correlation depends on the type of ganglion cell, the type of receptive field and the amount of overlap of the receptive fields. It was found that X- and Y-type ganglion cells can share a common input. Because of differing RF-sizes for X and Y cells the correlation between two nearby Y cells is stronger than between an X and Y-cell. By having one electrode in the optic nerve and another one in the retina often a whole series of cells with RF overlap to the optic nerve fiber could be recorded from. Examples of the effect that the amount of overlap has on the correlation for like-type cells are shown in Fig. 8.1, and on correlations for unlike RF-type X- and Y-cells in Fig. 8.2. It is observed that the correlation peaks and dips are sharper for situations where both cells are Y-type ganglion cells. Under mesopic stimulus conditions (constant illumination) the correlation between nearby cells was attributed to shared input from spiking amacrine cells with a low rate of spontaneous activity.

Under scotopic conditions the ganglion cells showed correlations that seemed to originate in individual quantal events in the rods. These are 40–50-ms-

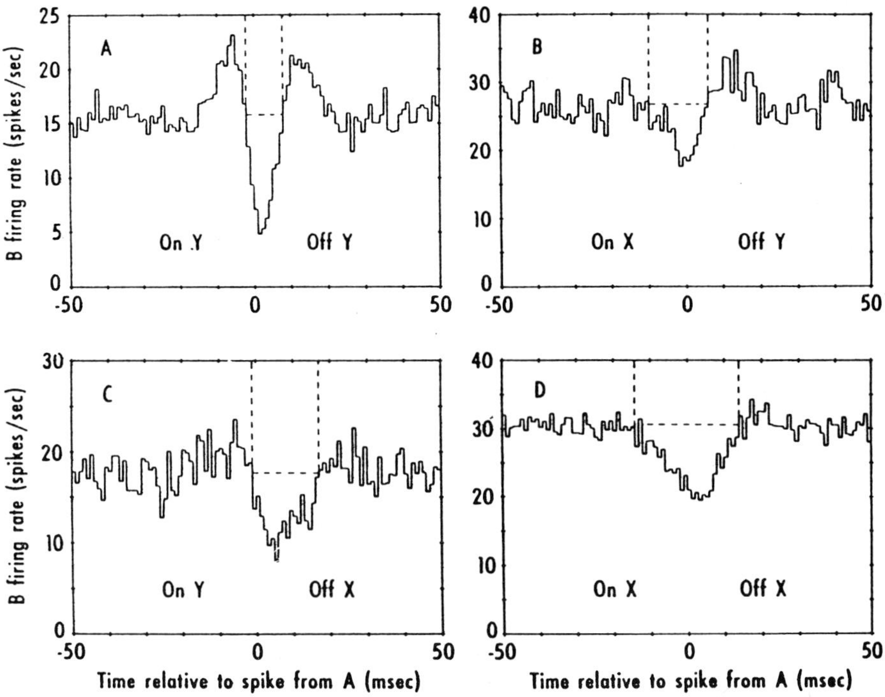

Fig. 8.2 A–D. Correlations between various on-off pair combinations for X and Y cells. The reciprocal input gives rise to decremental cross-correlograms. (Mastronarde 1983a)

lasting effects that may explain the very broad peaks and dips observed in the cross-correlation histograms. A possible pathway could be through rods – rod bipolars and AII amacrine cells (Sterling 1983). Because under scotopic conditions the inhibitory surround mostly disappears, cells from a larger area could contribute inputs which also may give rise to broadened cross-correlation peaks. Besides these shared input correlations between ganglion cells manifesting themselves through rather broad peaks and valleys, often very sharp, superimposed peaks could be seen. Mastronarde interpreted these as a direct interaction between ganglion cells. Y-cells of the like-type showed this interaction most clearly. It appeared that interactions between neighboring X and Y cells or between two X cells were weaker than between Y cells. Gap junctions between neighboring cells were seen as a likely mechanism for this interaction. Summarizing, the main effects observed by simultaneous recording from retinal ganglion cells can be interpreted as evidence for common input to cells with overlapping receptive field (regardless whether they are X or Y cells), and an ephaptic interaction between neighboring like-type cells. Note that, because of the shared input, X and Y ganglion cells do not represent completely independent parallel systems.

8.1.2 From Ganglion Cell to Lateral Geniculate Nucleus

The connectivity between optic nerve fibers and cells in the lateral geniculate nucleus has been studied restrictively in the cat. As expected, it was found that on-center retinal ganglion cells always project upon on-center LGN cells and that off-center retinal ganglion cells exclusively project upon off-center LGN cells. It was demonstrated (Cleland et al. 1971) that 5 out of 62 LGN cells had only one retinal input cell, the other 57 LGN cells all had multiple inputs (generally two to three). In an extended study (Levick et al. 1972) it was found in 8 out of 105 LGN cells that more than 99% of the spikes were related to a single retinal ganglion cell, i.e., a one-to-one correspondence. Eleven cells had two retinal inputs, four cells had three retinal inputs and only one cell with six retinal inputs was found. Thus a functionally effective excitatory input to each LGN neuron is provided by only a small number of retinal afferents: from one to six. Levick et al. note that retinal inputs with inhibitory actions were sought but never found. Therefore they explained the suppressive surround in LGN cells (cf. Chap. 3) by recurrent inhibition through interneurons. Eckhorn and Pöpel (1972) recorded simultaneously from cells in the optic tract and the LGN, either with a dual electrode (58 pairs) or a single electrode (44 pairs). In 73 pairs the units were both optic tract fibers. Of the remainder, ten pairs showed evidence for a strong monosynaptic coupling between an optic tract fiber and a principal (P) cell, in two cases there was a weak monosynaptic coupling between a fiber and a principal cell. In two cases an excitatory coupling between two principal cells was concluded, furthermore one common inhibition and one additional inhibitory connection between P-cells was reported.

Further evidence for inhibitory interaction within the LGN came from Dubin and Cleland (1977), who used electrical stimulation of the retinal inputs to the LGN and recorded simultaneously from retinal and LGN cells. It appeared that so-called intrageniculate interneurons are organized similarly to principal cells: they receive at most two to three inputs from retinal cells through monosynaptic connections. Evidence was found for common input to principal cells and interneurons, which is suggestive for feedforward inhibition. These intrageniculate interneurons have the same type RF as the principal cells but could be distinguished on the basis of differing spike waveforms. This then indicates that the interneurons in the LGN actually produce spikes; we will encounter different views in the next paragraph (see also LeVay 1971).

In conclusion, LGN principal cells and intrageniculate interneurons are monosynaptically excited by only a few retinal ganglion cells, probably two or three and not more than six. There is no sign of monosynaptic inhibitory connections between retina and LGN cells.

8.1.3 Inside the Lateral Geniculate Nucleus

Projections from both eyes arrive in the LGN, however, they end in different laminae so that one does not expect a functional connection between them.

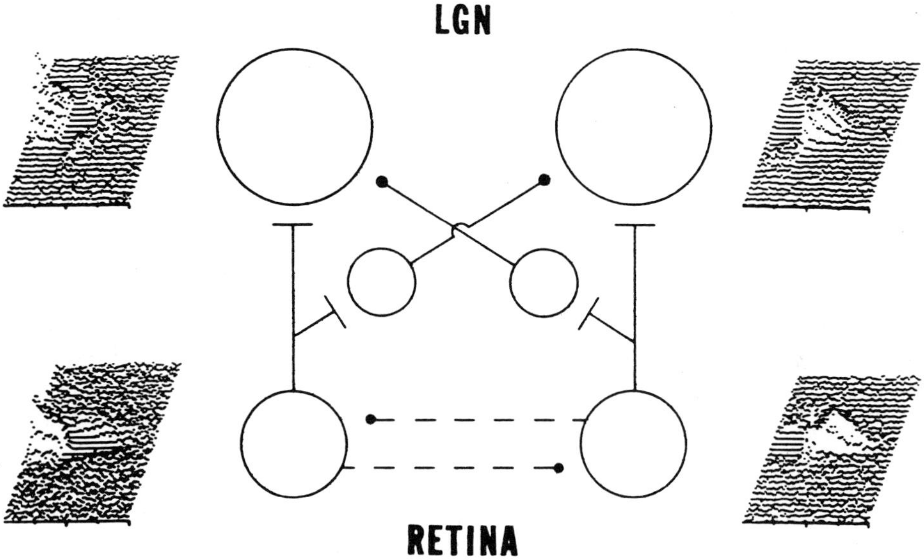

Fig. 8.3. Proposed LGN wiring diagram, on the basis of the fact that the spatiotemporal organization in the LGN response planes (*upper left and right*) is generated at the retinal level (*lower left and right*). (Stevens and Gerstein 1976). See also Fig. 4.7

Arnett (1975) observed 13 neurons pairs in the LGN where one of the cells received input from the ipsilateral eye and the other from the contralateral eye and did not find any correlation. Thirteen additional pairs from one eye with nonoverlapping RF's did not show correlations either. On the other hand, 13 pairs with ipsilateral and overlapping RF's showed a correlation in 10 cases, and 17 pairs with contralateral and overlapping RF's showed correlation in 11 cases. Like-pairs had cross-correlation histograms showing a peak near the origin, unlike-pairs a dip. It was concluded that most of the observed functional interactions observed took place already in the retina. No signs of monosynaptic inhibition were found.

Using single electrodes, Stevens and Gerstein (1976) recorded from 31 pairs. In 11 pairs neural correlation was found, two of these pairs showed overlapping, like-type receptive fields and had a peak in the cross-correlation histogram (CCH). In seven cases a dip in the CCH was found for overlapping, unlike RF-types. In the retina dip-like CCH's could be explained by a common input to ganglion cells showing an otherwise complementary RF center. This explanation could still hold for LGN cells as in the interpretation of Arnett (1975). Stevens and Gerstein, however, elaborate on two other possibilities based on the action of interneurons in the LGN. If the functional interaction took place in the retina one would at least expect that the RF's for LGN neurons are similar to those for retinal ganglion cells. This obviously is not the case: there is a much more potent and more extensive inhibitory surround in LGN cells. Stevens and Gerstein (1976) considered a mutual feedforward inhibition and a mutual recurrent inhibition (cf. Dubin and Cleland 1977; Lindstrom 1982). Based on their observations of stimulus dependence of the CCH as well as an excitatory overlap of the RF's of neurons showing decremental CCH's, these authors concluded that the most likely model to explain their dip-like CCH's was the feedforward inhibition model (Fig. 8.3).

8.1.4 The Striate Cortex

Simultaneous recordings of retinal ganglion cells and cells in the striate cortex (Lee et al. 1977) showed that the input to most simple cells consists of either on-center or off-center fibers, but not both. In 29 pairs with at least partially overlapping receptive fields, six cases showed an excitatory drive with a peak in the CCH around 6.5 – 7.5 ms. Considering that in retinal ganglion cell-LGN pairs the peak observed in the CCH is at 3 – 4 ms, the authors concluded that there is a direct monosynaptic activation from optic radiation afferents.

The recording from more than one unit simultaneously from the striate cortex has a relative long history – taking the first single-unit recording as the onset of history – since in 1963 Griffith and Horn recorded from 11 pairs in a conscious cat. Nine pairs showed a significant correlation that pointed to a common input or a reciprocal input. The next significant fact in this type of recording occurred in 1976 when Kimura et al. introduced glutamate iontophoresis to increase the spontaneous activity of simultaneously recorded neurons. The full reports were published by Toyama et al. (1981 a, b) and comprise 208 pairs of striate neurons. Common input was found in 105 pairs; in most cases (101/105) the peak in the CCH was situated within ± 0.3 ms from zero time, and its width was usually less than 0.9 ms. It was found that cell pairs featuring common excitation were mainly located in layers III – V of the striate cortex. It is known that the major LGN projection to the striate cortex is to layer IV, where the X-cells project to the lower part and the Y-cells to the upper part. There is also an X- and Y-cell input to the upper part of layer VI and a projection from the C-laminae of the LGN surrounding layer IV. Thus the region of functional LGN input overlaps with that estimated on basis of anatomical connections (see Gilbert 1983 for a review). An example from a recording is shown in Fig. 8.4. The recording method employs two independently movable electrodes, one is inserted in an oblique way and impales the cell indicated by the two concentric circles in the hatched area, the other is lowered perpendicular to the surface in the same column as the impaled cell. For all cells encountered while moving the perpendicularly inserted electrode the CCH is obtained and the type of cell is indicated (S – simple, C – complex, H – hypercomplex, E – e-on or e-off type). The range of common excitatory input is shown by the open rectangle, in general the strength of the correlation decreases with distance between the cells in the pairs, as exemplified by the peaks in the CCH's. Common excitation occurred in all cell types, except the hypercomplex cells. Evidence for monosynaptic excitation (delay in the CCH 0.6 – 0.9 ms) was found in 22 pairs and a correlation with a delay of 2 ms in

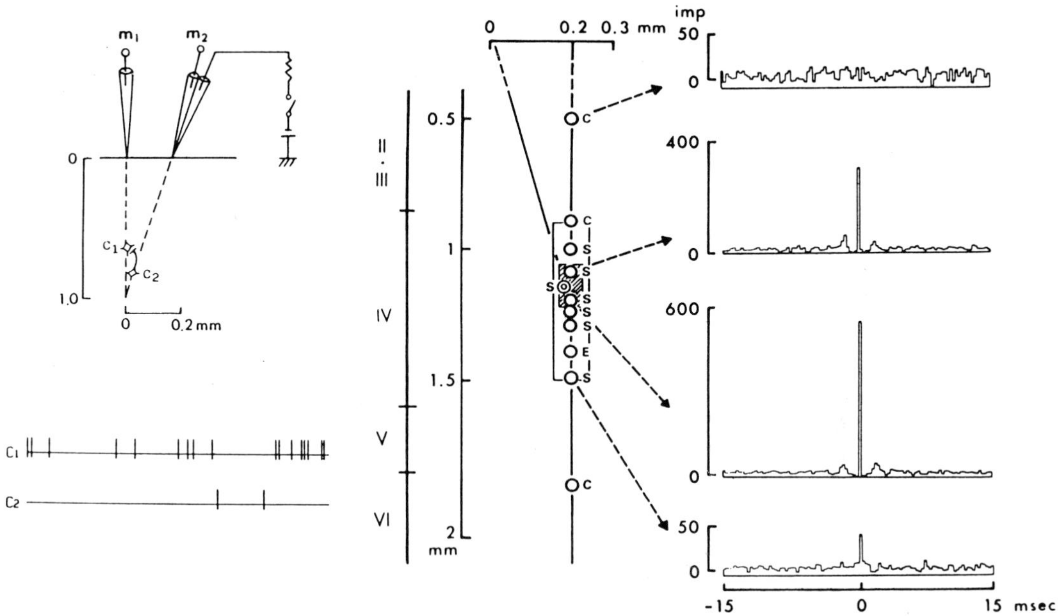

Fig. 8.4. Correlation between two simple (*s*) cells in the visual cortex usually results from shared excitatory input. The extent of interaction is about 300 μm both ways in the column, the strength decreases with distance. (Toyama et al. 1981 b)

two pairs. The CCH peaks lasted from 5 – 7 ms up to 15 ms. This delayed excitation was mostly found in the upper part of the supragranular layer and occurred in 14 out of 45 complex-complex cell pairs and in 5 out of 13 complex-hypercomplex cell pairs. No evidence was found for an excitatory interaction between simple cells and complex cells. Another example of a recording is shown in Fig. 8.5 and indicates the wide extent over which common input acts (open rectangle) and the narrow range (shaded) for excitation. It is noted that, in the latter case, the CCH peaks are less well resolved than for common input.

Delayed inhibition suggestive of monosynaptic nature (delay in the CCH 0.6 – 0.9 ms) was found in 23 neuronal pairs, the trough in the CCH could last as long as 80 ms. This intracortical inhibition was only observed in combinations of an e-on or e-off cell to a simple cell (8/16) or to a complex cell (3/17), or from a simple cell to a complex cell (12/36). This would corroborate the general assumption that the simple cells (including Toyama et al.'s e-on and e-off cells) correspond to the stellate cells. Of these stellate cells the smooth type is generally accredited with inhibitory action, while the spiny stellate cells as well as the pyramidal cells receive two types of synaptic endings, one presumably excitatory onto their spines, and the other inhibitory onto their somata, dendritic shafts, and axon initial segments (Gilbert 1983). Figure 8.6 shows an example of the ranges of inhibition (dark) and common input (open)

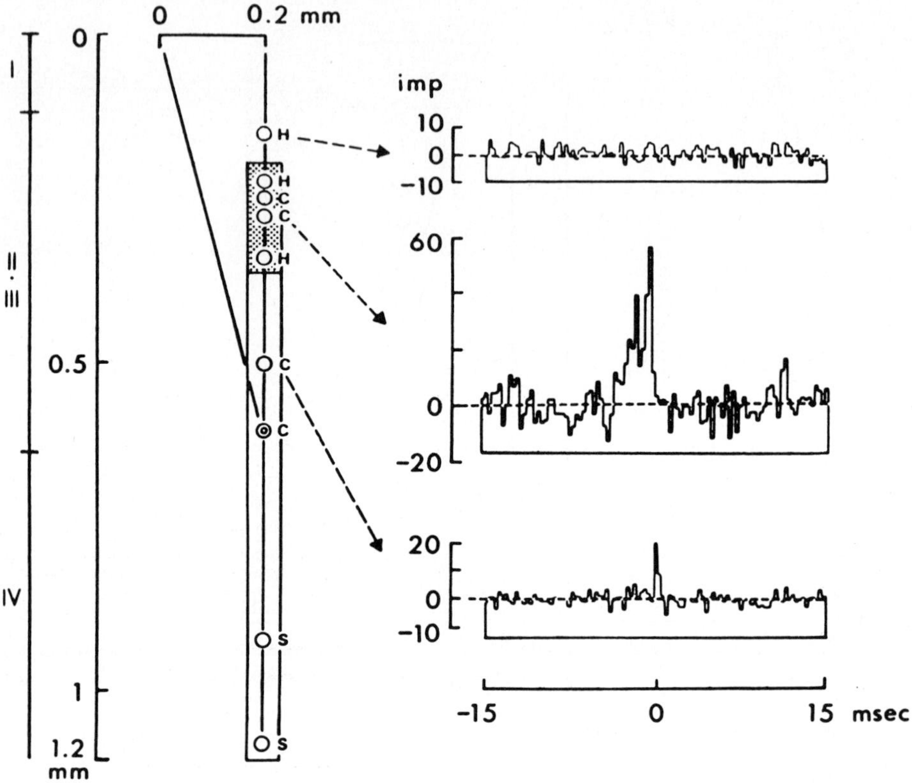

Fig. 8.5. Correlation between a complex (*C*) cell and other complex or hypercomplex (*H*) cells in the visual cortex results from common input (larger cell distance) as well as monosynaptic excitation (short distance). (Toyama et al. 1981 b)

that were encountered. The findings presented by Toyama et al. (1981 a, b) and reviewed in Toyama and Tanaka (1984) depend to a large extent on the topical application of glutamate. For instance, the sharp peaks in the cases with common input were barely detectable when only spontaneous activity was used but became prominent by increased activity of the cells either by glutamate or by photic stimulation.

The fact that in this analysis no evidence was found for an excitatory interaction between simple cells and complex cells and that simple as well as complex cells receive common input from the same LGN cells, therefore contradicts a simple hierarchical model as well as a simple parallel model in which simple and complex cells receive respectively X- and Y-cell input through separate channels (cf. Gilbert 1983 for a review).

Van Kan et al. (1985), using two independent electrodes, recorded from 30 neuron pairs in anesthetized paralyzed cats of which 13 pairs, showed an overlap in receptive field. Only in five cases was a significant correlation found, in all these cases the interelectrode distances were less than 150 μm and

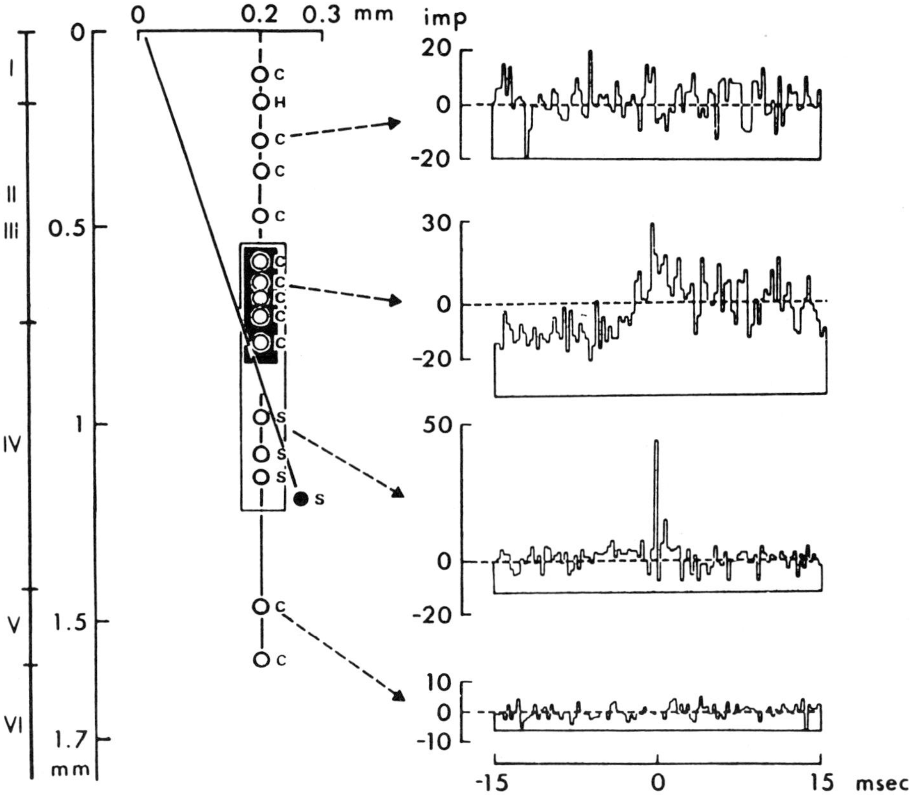

Fig. 8.6. Correlation between a simple (*s*) cell and complex (*c*) cells usually results from intracortical inhibition. (Toyama et al. 1981 b)

orientation and direction preference were the same as was the dominant input from the same eye.

Michalski et al. (1983), using multielectrode bundles, recorded from 286 pairs in the cat striate cortex. Only 13 pairs failed to show any interaction. In 58 pairs there was evidence for a direct excitatory synaptic connection on the basis of the very narrow peak in the CCH near zero. In one case direct inhibitory interaction and in 14 cases a mixed inhibitory, excitatory interaction was observed. Shared input was concluded for 131 pairs of which 43 had highly asymmetric peaks. In only six cases was the shift predictor (estimating the amount of stimulus lock) equal to the CCH. In 56 cases the correlation appeared to be stimulus-dependent. In 49 cases the recording was from pairs in separate cortical columns. Neural correlation was observed much less frequently in the intercolumnar pairs than within columns. Direct synaptic interactions were observed only twice in the intercolumnar pairs, in both cases the preferred orientation of the neurons was very similar.

From this overview of intracortical interaction it is again surprising that intracortical inhibition is so difficult to detect. Even in Toyama et al.'s data the results are not overwhelmingly clear in this respect. One possibility for the lack of signs of inhibition in the CCH – dips or troughs – is that "excitatory" peaks can occur in cases of inhibition, e.g., as a result of a rebound in firing that follows the hyperpolarization of the postsynaptic cell. These excitatory peaks will be much more easy to detect than the inhibitory troughs. Thus broad peaks may be indicative of inhibitory connections (Michalski et al. 1983). In a simulation study Aertsen and Gerstein (1985) clearly demonstrated that it requires a much stronger inhibitory synaptic connectivity to produce a dip in the CCH than is required for an excitatory connection to produce a comparable peak. This is especially obvious in cases with low spontaneous activity. Basically, when the background activity is raised – as in the case of glutamate application or sensory stimulation – the probability of detecting a significant dip becomes higher. However, we should still keep in mind an alternate possibility: spikeless neural interaction. This is expected in neurons without or with a small axon, such as many interneurons. Such inhibitory interactions obviously cannot be detected using correlation between spike trains.

In addition note that both Krueger (pers. commun.) and Eckhorn (pers. commun.) report very broad ($\cong 100$ ms) shared input for recording sites separated laterally by more than 160 μm. This is unlike Toyama et al. (1981 a, b), who basically recorded from within one column.

8.1.5 Geniculo-Striate Connections

Tanaka (1983, 1985) recorded from 243 geniculate-striate cell pairs consisting of 231 LGN cells and 100 striate cells. It was found that 82 pairs with overlapping RF showed incremental correlations with positive delays. Evidence was found for 37/42 pairs containing X-geniculate cells and for 28/36 pairs containing Y geniculate cells to have monosynaptic excitatory connections. In general the convergence number to an e-on/off or simple cell was larger than 10. Complex cells usually had a convergence number of more than 30. Convergence from 2–5 LGN cells could actually be demonstrated in 17 out of 32 striate cells, each of which were tested in pair with 3–14 LGN cells. Monosynaptic connections were found between LGN cells and *all* types of striate cells: X-geniculate-simple cells (26/71), X geniculate-standard complex (17/45), Y geniculate-simple (5/20), Y geniculate-standard complex (20/33), and Y geniculate-special complex (11/29). None of the 16 X geniculate-hypercomplex cell pairs showed significant correlations (Tanaka 1985). Excitatory connections to e-on/off or simple cells thus came mostly from X-cells, and those to hypercomplex cells came mostly from Y-cells. But for standard complex cells inputs from X-cells occurred as frequently as those from Y-cells.

The descending system from the visual cortex to the LGN, being mainly from layer VI, has been studied by Tsumoto et al. (1978). From the 211 pairs investigated, 18 pairs gave evidence for an excitatory cortico-geniculate projection (the RF centers were less than 1.7° apart). In 15 out of these 18 pairs the peak in the CCH was at 1–5 ms, in three cases in the range of 13–18 ms. In eight pairs an inhibitory connection, mostly with peaks around 2–7 ms, was found.

Such evidence for inhibition within the LGN has also been found by Dubin and Cleland (1977), who stimulated the cortical inputs to the LGN electrically. This produced excitation to so-called perigeniculate interneurons which also provide recurrent inhibition. These interneurons appear to be binocular, have nonconcentric RF's and are of the on-off type.

The fact that Tsumoto et al. could establish excitatory interactions on the basis of spike train correlations is evidence for the spike-producing character of these perigeniculate interneurons.

8.2 The Auditory System

In the auditory system there are more stages peripheral to the thalamus that are accessible to recording and in addition have spike producing neurons in more nuclei than in the visual system. Still the number of correlation studies in the auditory system lags behind that for the visual system.

8.2.1 The Auditory Nerve

The auditory nerve contains a large proportion ($\cong$95%) of cells that exclusively innervate the inner haircells, generally around 20 nerve fibers synapse on a single haircell. Both large, medium, and small-sized synaptic terminals have been found on the same haircell and this correlates with fibers having a high, medium, or low spontaneous rate (cf. Chap. 2). Provided that transmitter release at the various synaptic terminals is independent, one does not expect any correlation between the firings of the nerve fibers under conditions of spontaneous activity. Because phase-lock is one of the properties of auditory nerve fibers for frequencies up to 4–5 kHz, a stimulus correction procedure has to be carried out when testing for independence under stimulus conditions. Johnson and Kiang (1976) tested for this presumed independence in the cat by comparing the forward and backward recurrence times for one spike train with reference to the events in another simultaneously recorded spike train. They used two independently movable electrodes and did not try to position them close together. In 21 fiber pairs with spontaneous activity ranging from 7.5–131.5 spikes per second, the forward and backward recurrence times were the same within statistical limits. Under stimulation with high-frequency continuous tones – which do not produce phase-lock – 11 pairs with sometimes nearly identical characteristic frequencies also fired independently. For 32 pairs stimulated with low-frequency continuous tones the forward and backward recurrence times were also indistinguishable when a bin size of one stimulus period was used. A better method to correct for the phase-lock would have been to calculate the recurrence time distributions after a shift of the firings of one unit over several periods of the stimulating tone (compare the shift-prediction procedure in the cross-correlation analysis as discussed in Chapter 7). Nevertheless it seems fair to conclude that auditory nerve fibers are acting independently.

8.2.2 The Dorsal Cochlear Nucleus

The dorsal cochlear nucleus consists of principal neurons – often called fusiform cells or giant cells – and various types of interneurons, known as small cells or granule cells. The principal cells receive direct input from auditory nerve fibers as well as from interneurons, as was established in morphological studies (Mugnaini et al. 1980). Electrophysiologically one distinguishes so-called type II/III cells with a V-shaped tuning curve, of which the excitatory area is flanked by inhibitory (or suppressive) sidebands, and type IV cells with broad tuning curves which are mainly inhibitory. Voigt and Young (1980) and Young and Voigt (1981) studied 41 units pairs in the dorsal cochlear nucleus of decerebrated, unanesthetized cats, mainly using a single electrode. In the single electrode recording situation the best frequencies of the neurons were the same; when two electrodes were used (with separation of about 400 μm) the best frequencies of the neurons in the pair were less than half an octave apart. In 25 out of 41 pairs a clear correlation was observed, very likely this was a neural correlation since either the correlation was for spontaneous firing or when stimulating with high-frequency tones, which rules out phaselock. In one case reported, however, the characteristic frequencies were in the 2-kHz range with potential phase-locking to the stimulus. From both papers one obtains the impression that no shift predicter was used to correct for correlations due to stimulus coupling through phase-lock. Seventeen pairs consisted of a type II/III unit and a type IV unit, 20 pairs were composed of two type IV units, and in four pairs both units were type II/III. Most of the type IV units were sufficiently active under no-sound conditions to be able to detect inhibition. In 12/17 pairs combining a II/III type unit with a type IV unit an inhibitory through was observed in the CCH, suggesting a monosynaptic inhibition of a type II/III unit onto a type IV unit. In the remaining five pairs no signs of correlation was observed. Fifty percent of the type IV-only pairs had a flat CCH, the other half showed broad peaks surrounding zero-time delay and therefore being indicative for common input.

Thus because of the high spontaneous activity of the type IV cells a monosynaptic inhibition could be demonstrated by cross-correlation methods (Fig. 8.7). Voigt and Young (1980), however, remark that when barbiturate anesthesia is administered to decerebrated cats, the inhibitory region of a type IV unit in the vicinity of the characteristic frequency changes into an excitatory one. They conclude that a type IV unit is subject to both an excitatory drive (from the auditory nerve) and an inhibitory drive (from the low-spontaneously active interneurons) which is sensitive to anesthesia.

The type II/III low-spontaneously active neurons were linked to the so-called small cells and granule cells, i.e., to interneurons which have an in-

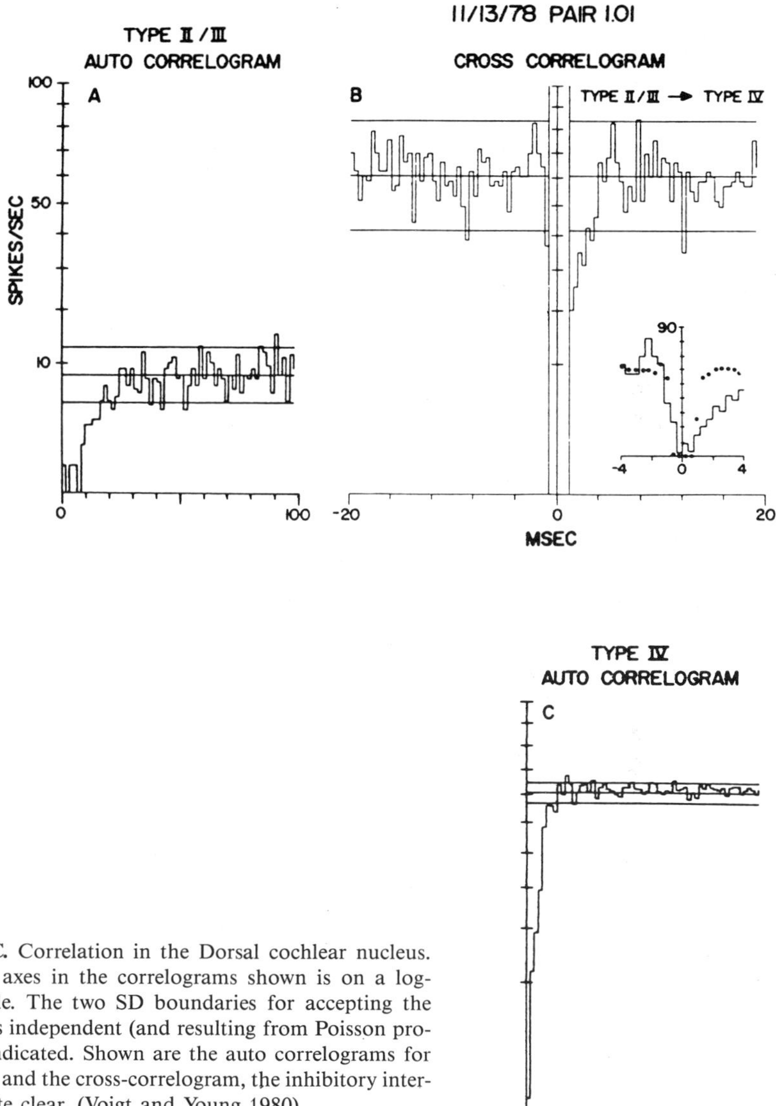

Fig. 8.7 A–C. Correlation in the Dorsal cochlear nucleus. The vertical axes in the correlograms shown is on a logarithmic scale. The two SD boundaries for accepting the spiketrains as independent (and resulting from Poisson processes) are indicated. Shown are the auto correlograms for the two cells, and the cross-correlogram, the inhibitory interaction is quite clear. (Voigt and Young 1980)

hibitory action on the type IV cells which most likely correspond to the principal cells. While the principal cells are known to receive a direct input from the auditory nerve, the low spontaneous activity of the interneurons argues against a direct input from the cochlea. This is corroborated by the absence of signs for common input. A possibility is that these low-spontaneously active interneurons receive input from higher-spontaneously active interneurons which do not show signs of inhibiting with type IV cells.

8.2.3 The Auditory Midbrain

The output from the principal cells of the dorsal cochlear nucleus passes through the dorsal acoustic stria directly to the inferior colliculus. Another, ventral, pathway takes its course from the ventral cochlear nucleus via the superior olivary complex and the lateral lemniscus to the inferior colliculus. To date there are no correlation studies reported from the mammalian inferior colliculus, but a comprehensive study has been carried out in its presumed homolog: the torus semicircularis in the grass frog (Eggermont et al. 1983a, b, c; Epping et al. 1984; Epping and Eggermont 1987). These authors used two recording electrodes, but generally unit pairs were encountered on each single electrode as well. A total of 264 pairs were studied, of which 140 were recorded on a single electrode and 124 on two electrodes. Since spontaneous activity in the frog is generally not higher than one spike s^{-1} most correlation studies were carried out under stimulus conditions. We have adopted the term "synchrony" to indicate a correlation in the presence of stimulation, regardless of whether it is of neural origin or stimulus-induced. The term correlation is reserved for of specific neural origin, i.e., synchrony which "survives" after an appropriate correction procedure for stimulus-induced covariation of firing. About 60% of the pairs exhibited synchronized behavior, with no differences between unit pairs recorded on one electrode or two electrodes. The majority of the synchronization correlograms were of the incremental type, only in 15 pairs was the decremental CCH found (six on one electrode, nine on two electrodes). Neural correlation was observed in 37 pairs (21 on one electrode and 16 on two electrodes) with no noticeable difference between the one- or two-electrode recordings, except that no neural correlation was observed for interelectrode distances of more than 300 μm. Most of the neural correlation could be ascribed to common input: in 15 cases for one-electrode and 16 cases for two-electrode recording. Unidirectional – and always excitatory – neural interaction was observed in only six unit pairs recorded on the same electrode. About half of the neural correlations were found to be stimulus-dependent.

An example of a double-unit recording showing stimulus-dependent neural correlation is shown in Fig. 8.8. The recording was from two electrodes, one of the units (Fig. 8.8a) was inhibited by tonepips with frequencies below 800 Hz over a time period of approximately 250 ms, the other unit (Fig. 8.8b) was excited by frequencies below 800 Hz and had a latency of approximately 50 ms. Both units were spontaneously active and the simultaneous and non-simultaneous (shaded) cross-coincidence histograms for that condition are

Fig. 8.8a–h. Unit pair from the auditory midbrain in the frog with stimulus-dependent neural correlation. **a** and **b** show dot displays of both single units to tonepip stimuli. Note the complementary response properties: neuron A is inhibited by frequencies below 800 Hz while unit B is activated in this frequency range. **c–h** show the simultaneous and nonsimultaneous (*shaded*) cross-correlation functions for various stimuli. The neural correlation appears to be dependent on the type of stimulus used. (Epping and Eggermont 1987)

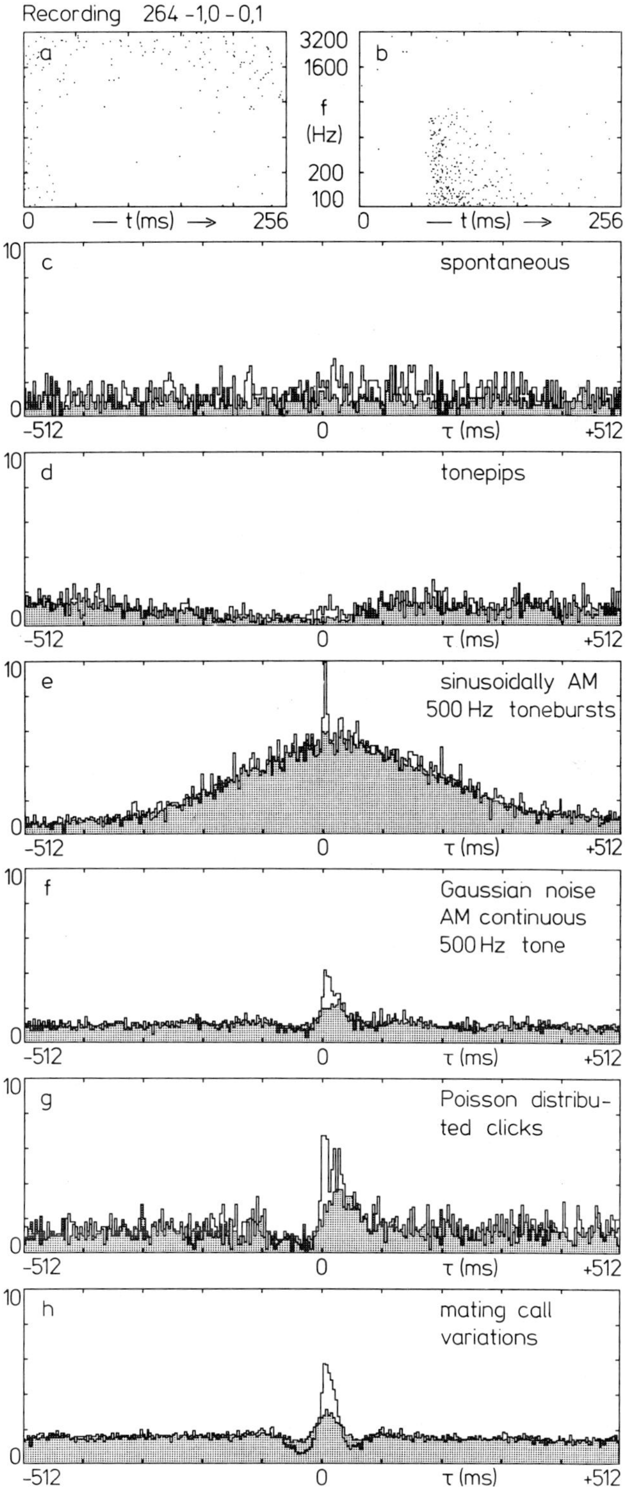
Recording 264 -1,0 - 0,1
a
b
3200
1600
f
(Hz)
200
100
0 — t (ms) → 256
0 — t (ms) → 256
c spontaneous
d tonepips
e sinusoidally AM 500 Hz tonebursts
f Gaussian noise AM continuous 500 Hz tone
g Poisson distributed clicks
h mating call variations
10
0
-512 0 τ (ms) +512

shown in Fig. 8.8c. There is some evidence that for $\tau > 0$ the simultaneous CCH is systematically higher than the nonsimultaneous CCH. Under stimulation of 48 ms duration, random frequency tonepips presented once per second (Fig. 8.8d), there is evidence for a stimulus-induced decremental CCH. Furthermore around $\tau = 0$ the simultaneous CCH is clearly above the nonsimultaneous CCH suggesting a common input. When 0.5-s sinusoidally modulated 500 Hz tonebursts are used as stimulus (Fig. 8.8e), both a clear and broad stimulus-induced synchrony and a sharp neural correlation become visible. The triangular-shaped nonsimultaneous CCH can be conceived as resulting from a correlation of two rectangles of neural activity each approximately 0.3 s long. The very sharp peak around $\tau = 0$ again suggests common input. A Gaussian noise-modulated continuous 500 Hz tone (Fig. 8.8f) again shows evidence for neural correlation, in this case however the peak is asymmetric and mainly present for $\tau > 0$. This is more indicative for a monosynaptic excitatory interaction. The same can be concluded from stimulating with a Poisson-distributed click train ensemble (Fig. 8.8g). In addition, the asymmetric stimulus coupled inhibitory effect in both Figure 8.8f and g remains clearly visible. Finally (Fig. 8.8h), using an ensemble of frog mating calls and synthetic variations thereof again we observe a peak to the right of $\tau = 0$ but now more symmetric and surrounded by two decremental regions (the simultaneous CCH is here below the nonsimultaneous CCH). The overall conclusion is that the width and position of the neural correlogram peak clearly depends on the type of stimulus used: the deterministic stimulations (Fig. 8.8d, e, h) produce rather symmetric peaks in the neural CCH suggestive of common input but with a difference in conduction time to both neurons. The more statistically structured stimuli (Fig. 8.8f, g) suggest more unilateral interactions.

While the investigation of neural correlations might shed some light on the neural circuitry in the areas under study, the synchronization of neural activity under stimulus conditions might be of more importance to the animal. Epping and Eggermont (1987) could show that two neurons recorded simultaneously and having no specific firing preference while presented with the stimulus ensemble consisting of natural and various synthetic mating calls, in principle could operate as a highly selective neuronal group (Fig. 8.9). When only the firings giving rise to a sharp peak in the neural CCH (Fig. 8.9a) – not unlike that in Fig. 8.8e – were taken into account, there were only responses to two very much alike vocalizations (Fig. 8.9d, e) instead of six or seven different ones (Fig. 8.9b, c). Any neuron in a higher part of the nervous system, performing such a "coincidence-detector" task would turn out to be a highly selective neuron and could possibly qualify as a feature detector.

8.2.4 The Medial Geniculate Body

The medial geniculate body (MGB), the auditory center in the thalamus, was extensively studied by Heierli et al. (1981, 1987), who recorded from 950 neuron pairs in the paralyzed and weakly anesthetized cat. About 40% of the

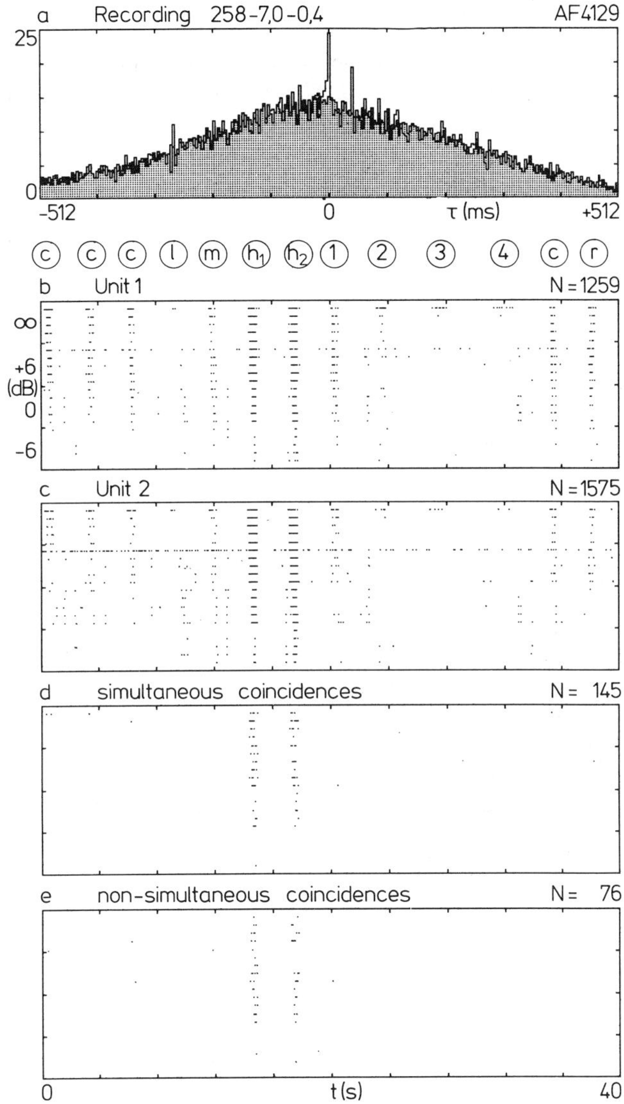

Fig. 8.9 a–e. Unit pair in the auditory midbrain of the frog showing enhancement of selectivity to mating call variations in double-unit coding compared with single unit coding. **a** shows the simultaneous and nonsimultaneous (*shaded*) cross-correlation histograms. **b** and **c** show the dot displays of both single units to a stimulus ensemble of mating calls and variations thereupon. The stimulus consists of three normal mating calls. **c** Synthesized versions with carrier 201 Hz (*l*), 557 Hz (*m*), 1067 Hz (h_1), and 1542 Hz (h_2), and temporal variations with original interval length (*1*), two (*2*), three (*3*) and four (*4*) times the interval length, followed by a normal call (*c*) and a time-reversed call (*r*). Moreover a noise background was added reducing the signal-to-noise ratio (SNR) in four steps of 6 dB. The total stimulus ensemble consisted of five identical sequences. The responses to identical calls, indicated above the dot displays were arrayed vertically. For convenience response to calls in equal amounts of noise were grouped together, no noise background (SNR = ∞) shown in the first five rows, maximal noise background (SNR = −6 dB) represented in the last five 5 rows. **d** shows the conditional dot display of unit 1 spikes in close correlation (−2 ms, 2 ms) with unit 2 spikes

neuron pairs fired independently, 20% showed a CCH indicative of common input, in 15% evidence was found for an excitatory synaptic interaction, and only in 3% of the cases was an inhibitory synaptic interaction found. The remaining neuron pairs were reported to show a weak or ambiguous interaction. All these results were obtained for spontaneous activity. When using a stimulus (clicks, noise, and tonebursts), there are indications that interaction effects could change: 93 pairs were evaluated with no effect in 31%, a qualitative change in 53% and a quantitative change in 16%. Not much difference in interaction type distribution was found for the various parts of the MGB.

8.2.5 The Geniculocortical Projection

The projection from the medial geniculate body to the primary auditory cortex was studied by Creutzfeldt et al. (1980) in the awake, restrained guinea pig using simultaneous recording from a thalamic and a cortical cell. Typically, first the thalamic cell was recorded, its characteristic frequency determined and then another electrode was inserted in the auditory cortex at the place where units with this characteristic frequency were likely to be found. From 69 pairs studied, nine combinations showed a positive correlation of their spontaneous activity. The peak was usually delayed by 3–5 ms and at most one cortical spike followed a thalamic spike. Those units in thalamus and cortex that formed a correlated pair had identical characteristic frequency and tuning curves of the same shape. When natural stimuli were used, cortical cells tended to be more phasic in their response and more selective than the thalamic cells that projected onto them. The relatively rare finding of interconnected pairs may be explained by the fact that anatomically the projection of the medial geniculate body to auditory cortex appears to be more diffuse than that from LGN to striate cortex.

8.2.6 The Primary Auditory Cortex

Dickson and Gerstein (1974) studied 61 neuron pairs in the auditory cortex of the paralyzed cat under spontaneous activity and 107 pairs under stimulus conditions. In 51 out of these 107, both neurons had a phasic response to tonebursts, in 32 cases only one neuron was affected by the stimulus and in 21 cases both neurons were unresponsive to the stimulus. Only 19 pairs showed a correlation completely due to stimulus coupling. From the total of 168 pairs studied, 54 showed a strong neural correlation (but only 5 thereof were other than common input), 31 showed weak neural correlation and the remaining 74 pairs did not show neural correlation. In 48 pairs which showed neural correlation and which were tested for more than one stimulus type, a dependence thereupon was found in 15 cases. Most of this dependence was attributed to a change in the firing rate of both neurons giving a better opportunity to detect any correlation. Only in five cases was a change in the CCH peak observed while the neurons' firing rates stayed the same. Thus there appears to be little direct stimulus dependence of neural correlation in the auditory cortex. The majority of the observed correlations was due to shared input. In fact, 93% of the pairs recorded on a single electrode showed this, and 35% of the dual electrode pairs. Dendritic trees in the auditory cortex have a radius of approximately 160 μm, thus a given axon bundle could be responsive for shared input for neurons within a range of 300 μm. Since a unilateral neuronal interaction was concluded only in 5% of all pairs, the source of the shared input in all probability is outside the cortex and presumably can be attributed to incoming axons from thalamic neurons. A corroboration for most of these findings comes from the investigations of Frostig et al. (1983), who studied 14 recordings each with 3 to 6 cells and yielding a total of 117 possible neuron pairs. Fourteen percent of the pairs gave a flat CCH, 12% showed evidence for a direct excitatory interaction, and 74% was indicative for common input (66%) or reciprocal input (8%). Again the fact that shared input within such a local group of neurons was a factor 6 more frequent than direct excitatory connections suggests that the source of shared input is outside the local group. The three cell correlation technique, using the snowflake representation, showed that within a local group the shared inputs of the various pairs were different and also independent. 88% of the pair correlations was reported to be stimulus-dependent; no evidence was given about changes in firing rate of the individual neurons under the various stimulus conditions except in one example, so that a firm establishment of this conclusion cannot be made.

Frostig et al. (1983) find their results at variance with the standard concepts of cortical organization, especially the fact that the percentage of direct connections found was small and their inability (like others) to show evidence for inhibitory connections. Since signs of common input in CCH's are always much more pronounced than those of unidirectional excitation, these can easily be missed: cells might share inputs in addition to having a direct interconnection. For the inability to detect inhibitory interactions the authors list as the major causes: low background firing rates, small neurons causing the inhibition being missed by the electrode, and inhibition being rare within a local group and mediated through long-range connections, e.g., intercolumnar. These findings, however, support the view of the cortex as a statistical, weakly coupled, machine (Braitenberg 1978a; Palm 1982a; Aertsen et al. 1988).

8.3 The Somatosensory System

The somatosensory cortex shows "dominance" regions just as the visual cortex, but in this case they represent interleaving regions to which the rapid (RA) and slowly (SA) adapting cutaneous receptors project. Metherate and Dykes (1985), recording with independent but closely positioned (150–500 μm) multibarrel electrodes from the middle layers of somatosensory cortex could show that neurons which receive input from the same modality (both RA or SA) show correlations when receptive fields overlapped. When the input to both neurons was RA and SA respectively no correlations were observed. They applied L-sodium glutamate both to isolate otherwise silent neurons and to enhance the background activity of the neurons to about 15 spikes s^{-1}. From the 53 pairs studied, 30 had overlapping receptive fields, but only 19 pairs had well-defined background activity so that they could be tested fully. Of these 14 showed a neural correlation; 5 out of these 14 were interpreted as receiving common input, the other were probably due to mono- or oligosynaptic connections. Although the authors mention decremental CCH's, the figures shown point only to incremental CCH's. Thus despite the application of glutamate there were probably no clear inhibitory interactions detected. The results suggest that up to this level the SA and RA regions do not have synaptic interactions. It appears that the cortical SA and RA bands also differ pharmacologically: GABA-containing neurons control the RF-size for neurons in RA but not in SA regions.

8.4 Simultaneous Versus Sequential Single-Unit Recording

While searching for single units, one either finds very few neurons (high impedance electrodes) or one finds a number of units that can be separated using the techniques described in Chapter 6. Thus the first advantage of simultaneous recording is a higher sampling gain of single units. Even with the coarser electrodes, however, one usually encounters much fewer neurons in a given track than one calculates from the density of cells and the recording volume seen by the electrode. There are some obvious reasons: interneurons often do not produce spikes, neurons are damaged by the electrode (maybe even in the absence of injury discharges), neurons are silent under most stimulus conditions, and in the cortex neurons may be inhibited most of the time. The "spike-view" of the brain in addition to electrode bias may certainly produce facts that are not readily integrated with, e.g., intracellular recordings.

What can be measured by simultaneous recordings supplemented by correlation studies? First of all the results are mostly applicable to local circuits, since most of the multielectrode studies using large (>300-μm) electrode distances do not show much correlated activity. For nearly all regions common input seems to be the dominant result; the peaks in the CCH are in addition better defined than for unidirectional neuronal interaction. Even reciprocal input usually presents itself very clearly, especially in cases with sufficient spontaneous activity. This basic background firing level allows detection of inhibition when it is large enough (dorsal cochlear nucleus) or artificially enlarged (visual cortex by glutamate or photic stimulation). The correlation studies carried out in the visual and auditory system so far have provided us with *constraints* for local and global circuit models (e.g., Mastronarde 1983 a, b; Toyama et al. 1981 a, b), with *evidence* for cooperative action of nerve cells in localization tasks or identification tasks (Gerstein et al. 1985 a), and with *indications* for coding strategies in nervous systems, e.g., in the formation of feature detectors (Epping and Eggermont 1987).

We will pursue these topics when we survey correlation studies in the motor system and shall describe the effects of the state of the subject on the outcome of correlation studies especially in the non-specific areas of the brain.

8.5 The Motor System

8.5.1 The Motor Cortex

Simultaneous recording from more than one single unit in the motor cortex started with the observations of von Baumgarten and Schaeffer (1957). They observed tendencies for synchronized activity between neighboring (i.e., recorded on the same electrode) cells. Allum et al. (1982) were able to make recordings from 14 neuron pairs from layer V of the awake, behaving, monkey. They studied the correlation in the firing activity when the monkey was exerting an isometric force. In four cases the CCH showed a trough distinct from the origin, indicative for inhibitory correlations between the neurons, in seven cases the CCH showed a peak surrounding the origin and most likely due to common input, in the remaining three cases a flat CCH was found. Because the event causing these correlations had a specific onset, the possibility that these correlations were due to this behavioral event rather than a neural connection remains possible, since the authors did not obtain an estimate for this type of coupling. Murphy et al. (1985 a, b), also recording in awake monkeys from this part of the cortex, provided such a shuffled correlation, thus their data allows a more extended classification and interpretation. They recorded from 237 pairs, 82 of which were on the same electrode. Clear signs of unidirectional neural correlation were found in 42 out of 82 pairs recorded on the same electrode and 40/155 pairs on dual electrodes. Exclusive synaptic interaction was found in 39/82 pairs, the other cases showed signs of inhibition or shared input as well. About 24% of the 237 pairs showed shared input: 27 pairs on the same electrode and 30 pairs on dual electrodes. The majority (76%) of shared input was of the incremental type, in 10% of the cases a valley was observed in the CCH, the remainder showing a mixed pattern.

Excitatory (25/42), inhibitory (9/42), and complex (8/42) interactions were observed between cells recorded from one electrode. An inhibitory interaction was recorded only once for the 40 dual electrode pairs showing synaptic interaction, the other 39 pairs showed excitatory interaction. This suggests, according to Murphy et al., that inhibitory connections are found mainly within a functional column, or at best between cells in adjacent columns. Shared input was found for electrode distances (horizontal) of up to 3 mm, evidencing the axon branching from the input source. This input source was considered to be of the cortico-cortical type as well as thalamocortical. In the latter case, both the visual trigger input for the task as well as the feedback from proprioceptors are possible candidates.

Event (movement) related correlations were found in 72/237 pairs and were in 27/72 cases due to event-coupling only.

A very clear boundary between shared input and synaptic interactions was found. Peak (valley) width for shared input had long durations, within 80–480 ms, while synaptic interactions showed peak (valley) widths of 10 ms at the most. This is a relation just opposite the findings of Toyama et al. (1981 a, b) in visual cortex: in that case common input peaks had a width of less than 0.9 ms, while unidirectional excitation had peak widths of 5–15 ms and unidirectional inhibitions even of 40–80 ms under photic stimulation.

8.5.2 Respiratory Areas in the Brainstem

Interactions between respiratory neurons in the brainstem are part of the production of the respiratory rhythm. Three types of correlation exist: short delay (peak at 2–4 ms), intermediate delay (peak 5–150 ms), and long delay (peak 0.5–6 s). Harper and Sieck (1980) recorded from 40 pairs of neurons in the nucleus parabrachialis medialis using a bundle of fine wire electrodes and found in 22 pairs short delay correlation, in 16 pairs intermediate delay correlation and in 23 pairs long delay correlation. The three types can occur in the same pair; the long delays were only found, however, in awake state or REM sleep state. The short time scale correlations were further investigated by Feldman et al. (1980), who recorded from three medullary regions the correlations between two expiratory neurons or two inspiratory neurons. Using both single-electrode pair recording and dual-electrode recording, differences between adjacent and distant neuronal correlations could be compared. Distant in all these cases refers to electrodes on opposite sites of the medulla. When single-electrode recording was used the spikes were separated using a window discriminator. Usually the population of larger spikes derived was from a single unit, but the population of smaller spikes usually comprised several units. The cross-correlations calculated in these cases therefore were single-unit-multiunit. Sixteen out of 26 inspiratory neuron pairs and 28 out of 45 pairs of expiratory neurons recorded on the same electrode showed correlation peaks in the 3-ms range. These findings did not occur with distant electrodes: only one inspiratory neuron pair (out of 43) showed a clearcut peak in the CCH and none of the 97 expiratory neuron pairs showed such a peak. It was concluded that local clusters of expiratory or inspiratory neurons had a high incidence of local interaction or shared input and that this interaction underlies the production of the respiratory rhythm. Inhibitory interactions were not found, which was attributed in part to the dead time (duration of the longer spike) in the correlation procedure. Since spike activity was quite high, it is probably also legitimate to conclude that it is due to the rarity of such interactions. Graham and Duffing (1981) did not observe any correlation in 22 pairs of expiratory neurons in recordings with two electrodes which were 250–350 μm apart. This suggests that the interconnected local circuits that were found for pairs on one electrode have a very restricted range of action, or that other, more experiment-related factors play a role.

8.5.3 Intercostal Motoneurons and Skeletal Motoneurons

Motoneurons are generally very large and have a large number of synaptic contacts. Sears and Stagg (1976), on the basis of a motoneuron surface of approximately $165 \times 10^3\ \mu m^2$ and a synaptic density of 20 per $100\ \mu m^2$, estimated about 33×10^3 synapses per motoneuron. These synapses are from a variety of afferents including muscle spindle afferents, and afferents from propriospinal and supraspinal sources (e.g., in case of intercostal motoneurons from respiratory centers in the brainstem). The larger motoneurons, giving larger spikes, are α-motoneurons and when recorded simultaneously show a short-term synchronization evident as a peak in the CCH surrounding the origin and extending to ± 3 ms at its base. This supports the conclusion that there is joint firing due to common input.

The muscle spindle afferents appear also to be synchronized to within a few ms, with the cardiac pulse being a strong contributing factor in this. It was found in addition (Kirkwood and Sears 1982a), by recording intracellularly from motoneurons and extracellularly from muscle spindle afferents, that one afferent connects about 42%–48% of the motoneurons. In a subsequent study (Kirkwood and Sears 1982b) it was shown that the CCH's of motoneuron discharges showed peaks with halfwidths of about 1 ms that were much shorter than the mean half widths of the recorded epsp's and only slightly longer than the rise time of the epsp's. They concluded that the shape of the CCH peak should be approximated by a linear combination of the epsp and its first time derivative (see Chap. 9). They also showed that the CCH's calculated for multiunit efferent discharges were consistent with those of the single efferent discharges.

This short-term synchronization of motoneurons was found for separations up to three or four segments (Kirkwood et al. 1982a), indicating that the majority of axons responsible branch to make strong synapses on motoneurons only over a restricted range.

Generally, there was a decrease in peak amplitude with increase in segment separation but not in peak duration, thought to result from a decrease in the number of common inputs. On the basis of an estimated conduction velocity – from shifts in peaks in the CCH – the axons were assumed to originate from bulbospinal respiratory neurons. The influence of the muscle spindle afferents was estimated to produce less than 10% of the observed correlation (Kirkwood et al. 1982b). This fairly simple shared input scheme for the α-motoneurons deserves some further elucidation. There is an important influence of the state of the animal on the firings. The narrow peak described so far is observed only under moderately deep anesthesia. For lightly anesthesized cats this narrow peak appears to be superimposed on a common input type broad peak (± 20 ms around $\tau = 0$), and sometimes high-frequency oscillatory CCH's appear (Fig. 8.10). The change in width is also observed in the epsp's. The broad peak synchronization was also promoted by spinal cord lesions and therefore it was attributed to the activity of interneurons. This activity would involve a presynaptic synchronization, i.e., the interneurons proba-

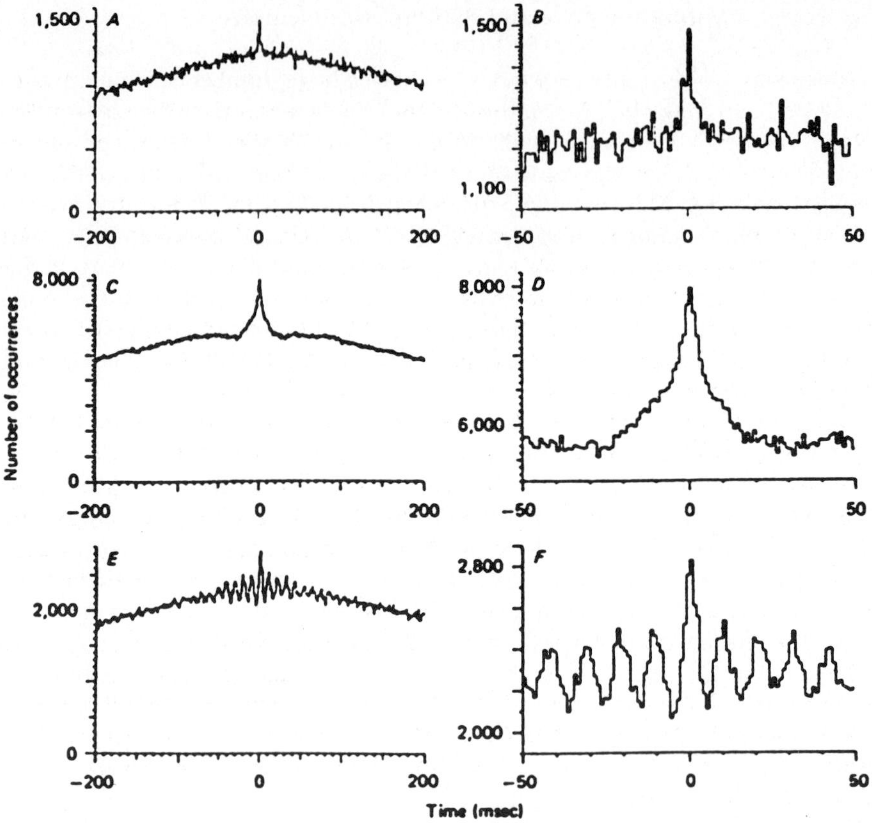

Fig. 8.10 A–F. Effect of anesthesia upon the shape of the cross-correlation between intercostal motoneurones. Under moderately deep anesthesia there is only a narrow shared input type correlogram (**A, B** enlarged time scale), for lighter anesthesia this peak becomes considerably broader and/or is accompanied by oscillations in the correlogram (**C–F**). (Kirkwood et al. 1982a)

bly share a common input that is passed in a slightly desynchronized way to the motoneurons.

Thus whatever the source, correlated activity in motoneurons is the result of shared input. Shared input usually gives the strongest peaks when neurons are anatomically or functionally close. In this case closeness relates, according to Kirkwood et al. (1982a), to the location of the region of the muscle that is innervated by the motoneurons rather than to the closeness of the motoneurons in the spinal cord.

Synchronization of the strong broad peak type may be undesirable for motoneurons: it may cause tremor (Shiavi and Negin 1975) in skeletal muscle which requires a smooth action. On the other hand, it might be contributing to a cooperative action in muscle carrying out ballistic movements (eye movements, coughing movements for intercostal motoneurons). Gamma-motoneurons in skeletal muscle in the cat gastrocnemius soleus show a high degree of synchrony reflected in a central peak in the CCH (Ellaway and Murthy 1985a, b). In fact, 170 out of 188 neuron pairs showed this synchrony in the decerebrated, spinal cat. However, only six out of 47 pairs showed this correlation in the decerebrated cat with intact spinal cord. The peak in the CCH was also absent in the intact anesthetized cat, but again present in the anesthetized, spinal cat. Thus, descending input from the brain is disrupting the – undesirable – synchrony in the firings of γ-motoneurons. The source of the synchrony was concluded to be afferents entering the spinal cord at the same segment that contains the gastrocnemius motoneuron pool.

In human tibialis anterior, muscle activity from single motoneurons was isolated and estimates of the CCH showed that the units tended to fire independently (Shiavi and Negin 1975). Furthermore, the joint interval histogram showed that there was no correlation between intervals of order 1 to 10. The pooled interval histogram of recordings with five to six motoneurons showed no tendency of these units to form a regular process; except for a preponderance of short (<10 ms) intervals the pooled interval distribution was exponential. It was concluded that this asynchrony is necessary and the independence sufficient for a normal smooth contraction.

8.6 Comparison Between the Sensory and Motor Systems

Going from sensory periphery to sensory cortex in the visual system the correlation between simultaneously recorded neuronal activity is dominated by common input in the case of overlapping receptive fields. Mostly the projections to higher centers show a very small amount of convergence up to the level of the thalamus, whereas primary targets in the cortex usually share input from 10–50 afferents. This shared input is very well synchronized; peaks in the CCH of visual cortex never are wider than 1 ms in this case. Inhibitory interactions have been demonstrated in the dorsal cochlear nucleus and the auditory and visual nuclei in the thalamus. In the primary visual auditory and somatosensory cortex common input dominates, followed by unidirectional excitation and inhibition. In the motor cortex common input and excitation are equally represented and inhibitory interactions have been demonstrated as well. A clear distinction with regard to the time course of the interactions is present, however. In the motor cortex it is the excitatory interactions that show a sharp peak (valley), albeit not as narrow as the common input in the visual system as found by Toyama et al. (1981a, b). The common input correlations in the motor cortex never showed peak widths less than 80 ms. Going to more peripheral levels of the motor system, short-term synchronizations show up under special conditions in skeletal muscle motoneurones and intercortal motoneurons: peak width is less than 10 ms under these conditions. This suggests that these common input correlations derive from only one neuron in the motor cortex or otherwise are synchronized by brainstem neurons. In intact animals the cortical efferents seem to be capable of effectively desynchronizing the motoneurones.

Table 8.1. Occurrence of correlation types in sensory and motor cortex

	Auditory	Visual	Motor
Common input	50% – 74%	46% – 50%	33%
Excitatory	10% – 12%	12% – 20%	30%
Inhibitory	–	0.5% – 12%	11%
Mixed exc./inh.	–	0% – 7%	10%
Total pairs	147	494	82
Total correlation	60% – 86%	59% – 89%	87%

Based on the data from Dickson and Gerstein (1974) and Frostig et al. (1983) for the auditory cortex, Toyama et al. (1981 a, b) and Michalski et al. (1983) for the visual cortex, and Murphy et al. (1985 a, b) for the motor cortex we can make a further comparison on the occurrence of the various correlations in these three cortical areas (see Table 8.1). The percentage of neural correlations found ranged from 60% – 86% for the auditory cortex, 59% – 86% in the visual cortex, and 87% in the motor cortex. Assuming that the way of recording, including sample bias, was most likely the same in all three cases, this indicates that the percentages of neural correlations are the same in these areas. Common input appeared to be more frequent in the auditory and visual cortex than in the motor cortex. Intracortical excitation on the other hand was occurring about twice as much in the motor cortex than in sensory cortex. When inhibition was detected using glutamate iontophoresis, the percentages were the same (12% in visual cortex, 11% in motor cortex). This holds also for the category mixed excitatory/inhibitory correlations. Without the application of glutamate, inhibitory interactions were not found (auditory cortex) or is only one case (visual cortex). The percentage of common input plus excitatory types of correlation again were very much the same for the three cortical regions: 60% – 86% for the auditory cortex, 50% – 70% for the visual cortex, and 66% for the motor cortex. All these correlations were for neurons recorded within the same cortical column or recorded on one electrode, and seem to indicate some organizational differences between sensory and motor cortex. Such differences could reside in the strength or extent of the neural input or to the strength or extent of the internal coupling, as well as in the relation between those two factors.

8.7 Correlated Neural Activity and the State of the Animal

From our survey of the peripheral motor system, the effect of anesthesia level and transection of the spinal cord upon the correlations observed was very clear. It indicates that although the electrodes may be probing local circuits, there is always a global effect that influences the action at the more local levels (e.g., Bach and Krüger 1986). In this paragraph the influence of sleep in its various stages, wakefulness, and anesthesia upon the presence and strength of neural correlations will be investigated. The regions where this effect has been studied comprise the reticular formation, the hippocampus and various parts of the cortex.

In the midbrain reticular formation synchronized firing for cells 150 μm apart had been noted already by von Baumgarten and Schaeffer (1957). Peripheral stimuli disturb the spontaneous activity patterns (Amassian et al. 1959) but do not seem to have an effect on the degree of synchronization (MacGregor et al. 1973). MacGregor et al. could show that in four out of nine unit pairs from the reticular formation in the unanesthetized rat there was a slow rhythmic interdependence, even for neurons separated about 1 mm from each other. This was considered consistent with the view that the entire system is being driven by some common external rhythmic source. In an extended study MacGregor et al. (1975) recorded 32 pairs in and around the reticular formation region in the midbrain of the unanesthetized and artificially inspired rat and found that from nine neuron pairs recorded on the same electrode six showed short latency correlated firing. From 23 unit pairs recorded with dual electrodes, five showed a short latency correlation in firing. The tendency for midbrain neurons to exhibit correlations in firing with other midbrain neurons was apparently increased when the cells were exhibiting rhythmic firings (the question though is what comes first). Slow rhythmic activity in the CCH was found to have a period that was an integer multiple of 0.64 s. Neuron pairs recorded on a single electrode showed these broad rhythmic correlations with a peak centered around zero, pairs on distant electrodes had a peak distant (up to 100 ms) from the origin. Siegel et al. (1981) also observed short-term correlations with a peak of 1 – 3 ms wide in about 50% of the 20 unit pairs studied with a single electrode. They took this as evidence for a common synaptic input to a large proportion of adjacent reticular cell pairs.

In somatosensory cortex, Li (1959) observed for electrodes less than 0.5 mm apart that during arousal the units seldom fired synchronously. The most apparent correlation in neural activity was found during sleep or anesthesia. This was found to corroborate the statement that "groups of nerve cells very often tend to act in unison when there is nothing to prevent them" (Adrian 1935 quoted by Li). Noda et al. (1969) recorded from 22 pairs of hippocampal pyramidal neurons in the CA_2 and CA_3 area in the cat. Using a single electrode they found that the discharges of two nearby units showed a correlation peak (less than 5 ms wide) during intermediate and deep sleep. During wakefulness a less prominent peak was occasionally seen. During arousal or REM sleep no correlation in firing was found. There was also an increased probability of joint appearance of long silent periods in neuronal pairs during intermediate and deep sleep. This pattern was interpreted as due to the inhibitory action of the basket cells onto the pyramidal cells. Basket cells receive inputs from pyramidal cell collaterals and are contacting in turn 200 – 500 pyramidal cells. This constitutes a global recurrent inhibition loop. Other cell types mainly excite pyramidal cells. It was assumed that the main change occurring during these states of wakefulness and sleep was in the proportion of excitation and inhibition received by the pyramidal cells.

Basically the same was found for the association cortex (Noda and Adey 1970, 1973). During slow-wave sleep or anesthesia, high correlations in the firing rate of simultaneously recorded neurons were obtained. Neurons discharged, however, quite independently during arousal, wakefulness or REM sleep, suggesting that desynchronization of neighboring cortical units occurs whenever the brain is processing information in a manner likely to result in a movement or to lead to memory storage. An isolated piece of suprasylvian-gyrus cortex – except from blood supply – in the unanesthetized, unrestrained cat "behaves" in a quite independent way. The spontaneous activity of such a piece of cortex is heavily correlated between neurons, the CCH showing a broad peak with a width of some 400 ms around zero (Burns and Webb 1979). No effect of sleep state on these correlations could be demonstrated. In intact cortex none of the 11 observed pairs showed such a peak, but light sleep produced near synchronous firing among neighboring units. Burns and Webb (1979) stated that "it is not unreasonable that when the brain is processing information, neighboring cortical neurons are very unlikely to be found doing the same thing at the same time." Holmes and Houchin (1966) observed in the cortex of the rat that spontaneously active units under deep urethane anesthesia show a high correlation in their firings with zero lag for neighboring units, with, in addition, a strong periodic component (period around 2 s) present. Under light urethane anesthesia the activity of cortical cells becomes desynchronized (see Fig. 8.10). This also happens during sensory input under deep urethane anesthesia. Light nembutal anesthesia does not show a correlation in activity either, and deep nembutal anesthesia disrupts all spontaneous activity.

It is thus apparent that an active brain or a brain exposed to sensory input is not a global correlating brain since it shows less correlation between neuronal activity patterns than a sleeping brain. Also during REM sleep neurons tend to fire independently. Intermediate and deep sleep stages seem to be more or less comparable to the stage that a piece of cortex, deliberately isolated from input, experiences. The fact that the neurons are highly correlated in their firings even when recorded with distant electrodes may suggest that they take part into one large assembly. Does sleep unite the cortex into one large assembly? During active data processing the large assembly is then broken up in a large number of smaller assemblies, which probably overlap and ignite each other in ever-changing sequences. One may also say that during activity a particular neuron may take part in a large number of assemblies, but during sleep only in one or very few. During active processing most of the neuron's firing is not associated with that of a nearby neuron since both may belong for the majority of their activity to different assemblies.

In this respect it is also conceivable that iontophoretic application of large amounts of, e.g., glutamate reduces the number of assemblies in which a particular neuron is taken part by joining assemblies together. This might increase the probability of encountering correlated neurons, but only shows up in the number of inhibitory interactions.

8.8 Correlated Neural Activity and Cognition

While we have seen an effect of such extensive state changes as sleep and wakefulness, it is highly interesting to find out whether within the waking state different correlation patterns are occurring during different cognitive tasks. Gevins et al. (1981) explored the cross-correlation between macropotentials – either evoked or spontaneous EEG – from the scalp in humans. Most of this activity reflects postsynaptic potential changes rather than spike activity, although spike activity may be related to it. Cross-correlations were computed for averaged evoked potentials from 44 pairs of electrodes formed out of the 15 electrodes pasted on the subject's scalp. All pairs showed significant correlation coefficients at zero lag ranging from 0.38 to 0.98 depending on the electrode sites. Correlations were higher after stimulus presentation than before. This was assumed consistent with the idea that stimulus-related neural processing should be accompanied by increased correlation between the activity at various sites. However, this correlation between averaged evoked potentials did not reveal any difference between two distinct visuomotor tasks imposed on the subject. In contrast single-trial cross-correlations to which nonlinear, distribution-independent, multivariate pattern recognition techniques were applied, revealed dynamic patterns of correlation clearly distinguishing between the two tasks. The complex correlation patterns obtained shifted in the anterior-posterior and left-right axis between successive 175-ms intervals. This indicates that many areas in both hemispheres are involved in these tasks. It also indicated that during purposive behavior many cortical and subcortical areas are related, and that increasing functional interrelation between neural areas may be reflected in increasing correlation of their low frequency-evoked potentials. Similar suggestions can be derived from the findings of Freeman and Viana di Prisco (1986), who interpreted changes in the spatial EEG pattern recorded from the olfactory bulb in rabbits during learning to respond to odors, as reflecting the process of synaptic modification and nerve cell assembly formation.

The report by Gevins et al. (1981) seems to contradict the observations that synchrony between individual neurons – neighboring or distant – tends to be very high during sleep or anesthesia and virtually absent during arousal or REM sleep. One may dismiss this by saying that these cats and rats were not involved in cognitive processing, but it is very likely that correlations between scalp recorded potentials would be found also in these animals. Furthermore it suggests that the procedure of averaging waveforms may not always be the most optimal approach to indicate cognitive processing, in fact it may even destroy the information about it.

We may well ask at this moment what information we want to extract from the presence or absence of synchronous firing in small neuronal groups. Furthermore, what is the important part of this synchrony: that during stimulation or the execution of a motor act or that part which remains after applying the shift predictor? The first situation gives us the correlations the animal might use in its identification and localization task or in the precise control of muscle movements. The second situation relates to functional wiring diagrams in local circuits or to the flow of neural activity in more global structures. Both correlations are needed to understand at least a little about the functioning of a brain (see also discussions in von Seelen et al. 1988). In the next chapter we will deal exclusively with the identification or wiring diagrams and connection strength between neurons as derived from the correlation of pre- and postsynaptic neural activity. In later chapters we will investigate how correlation, initially due to stimulation or carrying out motor acts, might lead to changes in connectivity between neurons. This will be discussed with respect to theories and experimental facts about plasticity, development, learning, and memory.

9 System Identification from Neural Correlation

The analysis of correlated neural activity can form the basis for constructing wiring diagrams of neural circuits (e.g., Stevens and Gerstein 1976; Toyama and Tanaka 1984). A wiring diagram is to be read as a scheme for the connectivity of the network, where additional graphical elements such as thickness or color of the wires may be used to indicate specific aspects of the connection, such as strength, sign, and latency (e.g., Gerstein et al. 1985a). In general the "wiring" diagram approach tends to emphasize the gross aspects of the connectivity, not so much the detailed aspects such as precise time course of the interaction or the shape of the postsynaptic potentials. A more general use of input and output signals is to identify the system that is in between: this in general leads to a (non)linear model that describes the transformation from the input signal to output signal. From such a model the connection between two (or more) neurons can be derived. A necessary requirement for applying these techniques is that the system is time-invariant and has finite memory. These conditions may create some problems in biological systems. Thus the results and methods discussed in this chapter pertain only to nonlearning systems with time-invariant synaptic strengths.

Consider a system of two neurons, A and B, that are monosynaptically connected via an axon carrying a spike train x(t) from A to B, the output of neuron B is spike train y(t) (cf. Fig. 9.1). We assume that the intracellular potentials of the cells can be measured: A(t) respectively B(t). Thus four signals are available: two continuous ones and two discrete ones. The aim is to find the properties of the synapse in between. This of couse is a simplification, if the axon of neuron A branches and several branches connect with neuron B, we assume their properties combined into this one, idealized synapse. In principle, four system identification problems can be encountered in such a case:

1. Identification on the basis of two continuous signals, A(t) and B(t). In this case the spike generating mechanism of neuron A is incorporated into the synaptic properties. When neuron A does not use action potentials to com-

Fig. 9.1 General model for system identification in the nervous system. The membrane potentials of the two neurons are indicated by $A(t)$ and $B(t)$, and the output spike trains by $x(t)$ and $y(t)$

municate with neuron B, but instead has dendrodendritic or dendrosomatic synapses (cf. Shepherd 1979) then this method is the most direct one to estimate the synaptic properties.

2. Identification on the basis of A(t) and y(t), i.e., continuous input and discrete output. The estimation of the synaptic properties includes the spike-generating mechanism of cell B, and in case A and B connect through spike trains also the spike-generating mechanism of neuron A.
3. Identification on the basis of x(t) and y(t), i.e., two discrete signals. The estimation of the synaptic properties includes the spike-generating mechanism of neuron B.
4. Identification on the basis of x(t) and B(t), i.e., discrete signal input and continuous signal output. This is the most direct way to estimate the connectivity between a spiking neuron A and neuron B.

In general, the input and output signals will be stochastic signals and the system in between will be nonlinear. In addition to the assumed finite memory and time invariance properties, it is furthermore necessary for a good characterization of the system that the bandwidth of the input signal is large compared to that of the system. For reasons of mathematical convenience, it is preferred to use Gaussian wide band noise to describe the continuous signal and Poisson processes to describe spike trains. For most neurons, however, neither the input signals at the soma nor the spike trains in the axon resemble the ideal input signals. Therefore in the practical estimation procedure of a synapse one could try to impose the ideal input signals by either injecting a Gaussian current into the soma of cell A or electrically exciting the axon of cell A by a Poisson train of electric pulses. This can be done in some invertebrate preparations (e.g., Brillinger et al. 1976) or in vertebrate motoneurons (e.g., Kirkwood and Sears 1978), but in the vertebrate CNS one generally is already quite lucky to find two spiking neurons that are monosynaptically connected and produce a sufficient number of spikes to allow for systems analysis. However, for the sake of simplicity, let us assume that the input signal can be controlled and let us discuss the four situations distinguished above.

9.1 Identification on the Basis of Continuous Input and Output Signals

If Gaussian white noise (GWN) could be injected in the soma of cell A, and one could register the intracellular potential of cell B, then one could estimate the so-called Wiener Kernels by cross-correlation (e.g., Marmarelis and Marmarelis 1978). For a system that is nonlinear up to third degree, e.g.,

$$B(t) = \sum_{n=0}^{3} K_n(A)(t) \tag{9.1}$$

with K_n-bounded homogeneous polynomial functionals of order n one finds the Wiener-kernels to be (apart from corrections for the diagonal terms):

$$
\begin{aligned}
w_0 &= \langle B \rangle \\
w_1(\tau) &= R_{AB}(\tau)/P \\
w_2(\tau_1, \tau_2) &= R_{AAB}(\tau_1 \tau_2)/2P^2 \\
w_3(\tau_1, \tau_2, \tau_3) &= R_{AAAB}(\tau_1, \tau_2, \tau_3)/6P^3 \ ,
\end{aligned}
\tag{9.2}
$$

where P is the power of the Gaussian input, and where $R_{AB}(\tau)$ is the first-order cross-correlation function, $R_{AAB}(\tau_1, \tau_2)$ is the second order, and $R_{AAAB}(\tau_1, \tau_2, \tau_3)$ is the third-order cross-correlation function:

$$
\begin{aligned}
R_{AB}(\tau) &= \lim_{T\to\infty} \frac{1}{2T} \int_{-T}^{T} B(t)\, A(t-\tau)\, dt \\
R_{AAB}(\tau_1, \tau_2) &= \lim_{T\to\infty} \frac{1}{2T} \int_{-T}^{T} B(t)\, A(t-\tau_1)\, A(t-\tau_2)\, dt \\
R_{AAAB}(\tau_1, \tau_2, \tau_3) &= \lim_{T\to\infty} \frac{1}{2T} \int_{-T}^{T} B(t)\, A(t-\tau_1)\, A(t-\tau_2)\, A(t-\tau_3)\, dt \ .
\end{aligned}
\tag{9.3}
$$

Note from Eq. (9.2) that the Wiener kernels, through P, are dependent on the properties of the input ensemble. This is because the Wiener kernels are defined such that the Wiener functionals generated by them (and used in the functional expansion of the nonlinear system) are orthogonal with respect to the Gaussian input (e.g., Marmarelis and Marmarelis 1978).

From the Wiener kernels, one may calculate the so-called Volterra kernels (e.g., Eggermont et al. 1983b). For our third-order system [Eq. (9.1)] this gives:

$$
\begin{aligned}
v_1(\tau) &= w_1(\tau) - 3P \int w_3(\tau, \sigma, \sigma)\, d\sigma \\
v_2(\tau_1, \tau_2) &= w_2(\tau_1, \tau_2) \\
v_3(\tau_1, \tau_2, \tau_3) &= w_3(\tau_1, \tau_2, \tau_3) \ .
\end{aligned}
\tag{9.4}
$$

For more general systems more complex relations can be given (e.g., Aertsen and Johannesma 1981a). These Volterra kernels which are independent of the input stimulus describe the contributions of respectively the linear part, the quadratic part, and the third-order nonlinearity to the systems output. In this particular case the second- and third-order Wiener kernels completely describe the second- and third-order nonlinearity, but the first-order Wiener kernel does not completely determine $v_1(\tau)$. The Volterra representation of our system as given in Eq. (9.1) will be three parallel systems: the linear part, quadratic part, and third-order part, the outputs of which are added. With these Volterra kernels it is possible to describe B(t) for any input A(t):

$$
\begin{aligned}
B(t) &= v_0 + \int v_1(\tau)\, A(t-\tau)\, d\tau + \iint v_2(\tau_1, \tau_2)\, A(t-\tau_1)\, A(t-\tau_2)\, d\tau_1\, d\tau_2 \\
&= \iiint v_3(\tau_1, \tau_2, \tau_3)\, A(t-\tau_1)\, A(t-\tau_2)\, A(t-\tau_3)\, d\tau_1\, d\tau_2\, d\tau_3 \ .
\end{aligned}
\tag{9.5}
$$

Therefore we say that by v_0, $v_1(\tau)$, $v_2(\tau_1, \tau_2)$ and $v_3(\tau_1, \tau_2, \tau_3)$ the system under consideration has been identified.

This method can be used in cases where the input is a sensory stimulus and the output is nonspiking activity from a single cell or groups of cells. Examples are a visual noise and recordings from horizontal cells or the EEG recorded from the scalp (see, e.g., Marmarelis and Marmarelis 1978) or pseudorandom auditory noise stimulation and EEG recordings (e.g., Dobie and Wilson 1984; Wilson and Dobie 1987). In cases where the EEG is used as the response signal the analysis is usually restricted to a first-order cross-correlation, this can be interpreted as a kind of evoked response and might resemble the evoked response to a light flash or a click stimulus.

In case the system is assumed to be linear and the mean output signal in the absence of input is zero, all system kernels collapse, except the first one which simply equals the impulse response. In that case one may estimate the systems impulse response $h(\tau)$ from the equation:

$$R_{AB}(\tau) = \int_{-\infty}^{\infty} R_{AA}(\tau)\, h(t-\tau)\, dt \ . \tag{9.6}$$

The solution is easily obtained in the frequency domain provided that $\hat{R}_{AA}(\omega)$, the spectrum of the input signal, is nonzero over the frequency range of interest:

$$\hat{h}(\omega) = \frac{\hat{R}_{AB}(\omega)}{\hat{R}_{AA}(\omega)} \ . \tag{9.7}$$

In the case where the input signal is GWN, one has $\hat{R}_{AA}(\omega) = 1$, and therefore

$$\hat{h}(\omega) = \hat{R}_{AB}(\omega), \quad h(t) = R_{AB}(t) \ . \tag{9.8}$$

9.2 Identification on the Basis of Continuous Input and Discrete Output

In this case the cross-correlation in Eq. (9.6) reduces to so-called triggered averaging (e.g., Kirkwood 1979), i.e., for every action potential $y(t)$ a portion of the continuous input signal that precedes it is sampled and added to the sum of the previous samples. For a system that is linear or for a system containing a static nonlinearity this procedure results in an estimate of the impulse response of the linear part of the system, provided GWN was used as the input signal.

One can also interpret this average as the optimum signal to elicit an action potential from the system. In case the probability of spike generation is proportional to the value of $B(t)$ it can be shown (Johannesma 1980) that:

$$R_{Ay}(\tau) = R_{AB}(\tau) \ . \tag{9.9}$$

For a neuron which fires whenever the potential $B(t)$ crosses the voltage level b (in either direction) one obtains:

$$R_{Ay}(\tau) = b\, R_{AB}(\tau) \ . \tag{9.10}$$

When the neuron fires only for positive going crossings of the trigger level one finds (de Boer and Kuyper 1968):

$$R_{Ay}(\tau) = b\, R_{AB}(\tau) - \alpha \left(\frac{2}{\pi}\right)^{1/2} \dot{R}_{AB}(\tau) \ , \qquad (9.11)$$

in which $\dot{R}_{AB}(\tau)$ is the time derivative of $R_{AB}(\tau)$. Thus the cross-correlation between $A(t)$ and $y(t) = \sum \delta(t - t_1)$ is a distorbed estimate of $R_{AB}(\tau)$ because, in addition, it contains a term proportional to the time derivative of $R_{AB}(\tau)$. For normalization purposes α is taken such that:

$$\alpha^2 = \frac{\int R_{AB}^2(\tau)\, d\tau}{\int \dot{R}_{AB}^2(\tau)\, d\tau} \qquad (9.12)$$

This method of spike-triggered averaging, also called reverse correlation method (de Boer 1967, 1968, 1969, 1973), has been applied to identify the filtering properties of the basilar membrane-haircell system in the vertebrate inner ear. The correlation procedure uses Gaussian wide band noise as the signal $A(t)$ and the spikes from single auditory nerve fibers constitute the output signal $y(t)$. The filter properties that are derived are a convolution of the mechanical filtering by the basilar membrane and the low pass filtering action of the haircell-auditory nerve fiber synapse. A more exhaustive discussion on the use of reverse correlation methods in auditory research can be found in Eggermont et al. (1983c).

Brillinger and Segundo (1979) studied the spike-generating mechanism in neurons of the sea slug *Aplysia* by injecting current into the cell body and monitoring the spike train generated in the axon of that neuron (see also Bryant and Segundo 1976). Assume $A(t)$ is Gaussian current and $a(t)$ is the time function that converts synaptic current into a potential at the trigger zone (the system kernel). Under the conditions of low firing rates, a fixed threshold b, and an exponential decay of the membrane potential to its resting value, it can be shown that:

$$R_{Ay}(\tau) - \phi(b, P) \int_0^\infty a(\sigma)\, R_{AA}(\tau - \sigma)\, d\sigma \ , \qquad (9.13)$$

where $\phi(b, P)$ depends on the threshold b as well as on the noise power. When $A(t)$ is GWN this reduces to:

$$R_{Ay}(\tau) = \phi(b, P)\, a(\tau) \ . \qquad (9.14)$$

Brillinger and Segundo (1979) found that the estimated system kernel, $a(\tau)$, was usually narrower than the postsynaptic potentials observed in the same neurons under natural stimulation conditions (usually by a factor of 2), and that $a(\tau)$ could be biphasic. Assuming that Eq. (9.11) also holds under the conditions in this experiment, this suggests a rather strong contribution of the derivative of the epsp's to $a(\tau)$.

Kirkwood and Sears (1978) used this method when recording the intracellular potential of a motoneuron that shared an afferent with another motoneuron from which spikes were recorded. By averaging the intracellular potential around the time of spike occurrence in the other neuron they obtained an "average common excitation potential". This was considered a measure of the strenght of the shared input to these cells, it could be demonstrated that the shape of this potential was (in an intricate way) related to the shape of the epsp of that motoneuron that was impaled by the intracellular electrode.

This method can be extended to nonlinear systems (which most synapses are) in a way analogous to the continuous input-continuous output case. The calculation procedure, however, is generally more simple. Applications, to date are described for stimulus-response relations in the visual (Marmarelis and Marmarelis 1978), auditory (Eggermont et al. 1983b) and vestibular (O'Leary and Honrubia 1975) systems.

9.3 Identification Based upon Discrete Input and Discrete Output

9.3.1 Linear System Analysis

This is the feature of interest when recording multiple spike trains in the CNS. The problem focuses around the question: is the estimate of $R_{xy}(\tau)$ related to $R_{AB}(\tau)$? In order words, can one estimate the epsp size and shape from cross-correlation between spike trains? This problem has been attacked directly by Kirkwood and Sears (1978, 1982a, b) and Kirkwood (1979) for motoneurons, where they were able to obtain estimates of both $R_{Ay}(\tau)$ and $R_{xy}(\tau)$. By comparison they found that:

$$R_{xy}(\tau) = \varepsilon R_{Ay}(\tau) + \delta \dot{R}_{Ay}(\tau) \; . \tag{9.15}$$

A relation like (9.15) would arise in a situation where x(t) is a spike train with Poisson distribution and B(t) would result from that by replacing each spike by a γ-function-shaped epsp, and an output spike would be generated upon upward going crossings of a fixed threshold level. This situation is obviously comparable to the one discussed in Section 9.2 [see Eq. (9.11)]. Since $R_{Ay}(\tau)$ estimates the epsp, it follows that the cross-correlation peak between pre- and postsynaptic spike trains will be larger the steeper the rise of the epsp. From their experimental data, Kirkwood and Sears estimate that $\varepsilon \cong 2\delta$, thus the contribution of the derivative can, indeed, be quite large.

Several other opinions regarding the relation between $R_{xy}(\tau)$ and epsp size and shape have been expressed: Moore et al. (1970) assume that $R_{xy}(\tau) \cong \varepsilon R_{AY}(\tau)$, and Knox (1974) inclines to $R_{xy}(\tau) \cong \delta \dot{R}_{Ay}(\tau)$. The formulation of Kirkwood and Sears therefore takes an intermediate position. From a direct comparison of epsp's and the time course of the CCH for cat lumbar motoneurons, Fetz and Gustafsson (1983) concluded that in most cases simple epsp's produced CCH peaks that resembled their time derivative (cf. Fig. 9.2) but with an onset of the CCH peak delayed by about 0.5 ms. Complex, compound epsp's and ipsp's produced CCH's that deviated significantly from a prediction based upon the psp-derivative. Surmeier and Weinberg

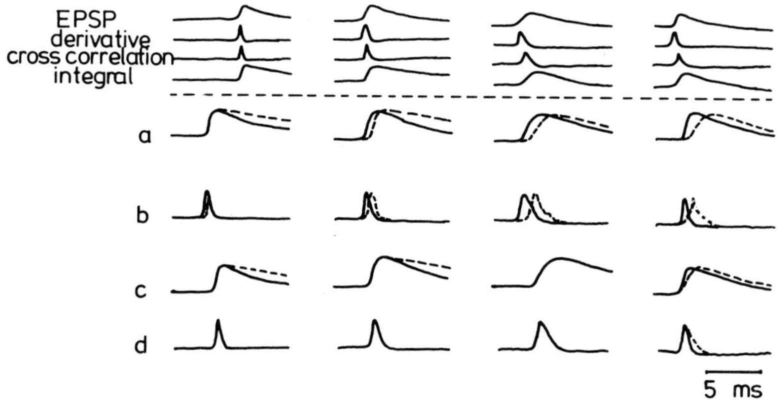

Fig. 9.2 a – d. Comparison of epsp shape and cross-correlation histogram (CCH) in cat lumbar motoneurons. There is a close resemblance between the time derivative of the epsp and the CCH (**b, d**), or alternatively between the epsp and the integral over the CCH (**a, c**). Epsp measures *full lines*; CCH measures *dashed lines*. (After Fetz and Gustafsson 1983)

(1985) concluded from a model study of cuneate nucleus neurons that the CCH peak had a strong resemblance to the underlying epsp and little to the time derivative of the epsp.

All that can be said is that model studies and experimental results seem to depend in their outcome on the particular part of the nervous system that is studied.

For more general nonlinear models it is insufficient to obtain only $h(\tau)$, and an expansion based upon kernels, analogous to the Volterra-Wiener expansion discussed for the continuous input cases, is needed. Brillinger et al. (1976) give a heuristic discussion to this problem. The zero-order kernel, output without input, gives the mean spontaneous rate. The first-order kernel relates to the effect of a single presynaptic spike; it is the best "linear" predictor of the average change of the instantaneous rate at time t in spike train y(t) when a single spike occurs at time zero in train x(t). The first-order kernel can be found from:

$$R_{xy}(t) - m_y = h(t) + \int h(t-\tau)\{R_{xx}(\tau) - m_x\} dt \ , \tag{9.16}$$

where m_x and m_y are the overall mean rates of the input and output spike trains. This equation is based on the assumption that the effects of separate x-spikes are additive, meaning low firing rates, no spike bursts, etc.

A simple expression results from Eq. (9.16) in case the x-spike train is a realization of a Poisson process; then $R_{xx}(t) = m_x$ and for $t \neq 0$:

$$h(t) = R_{xy}(t) - m_y \ . \tag{9.17}$$

For a solution of the more general case, a transformation into the spectral domain is generally used. For instance, Perkel (1970) writes the cross-covariance, defined as:

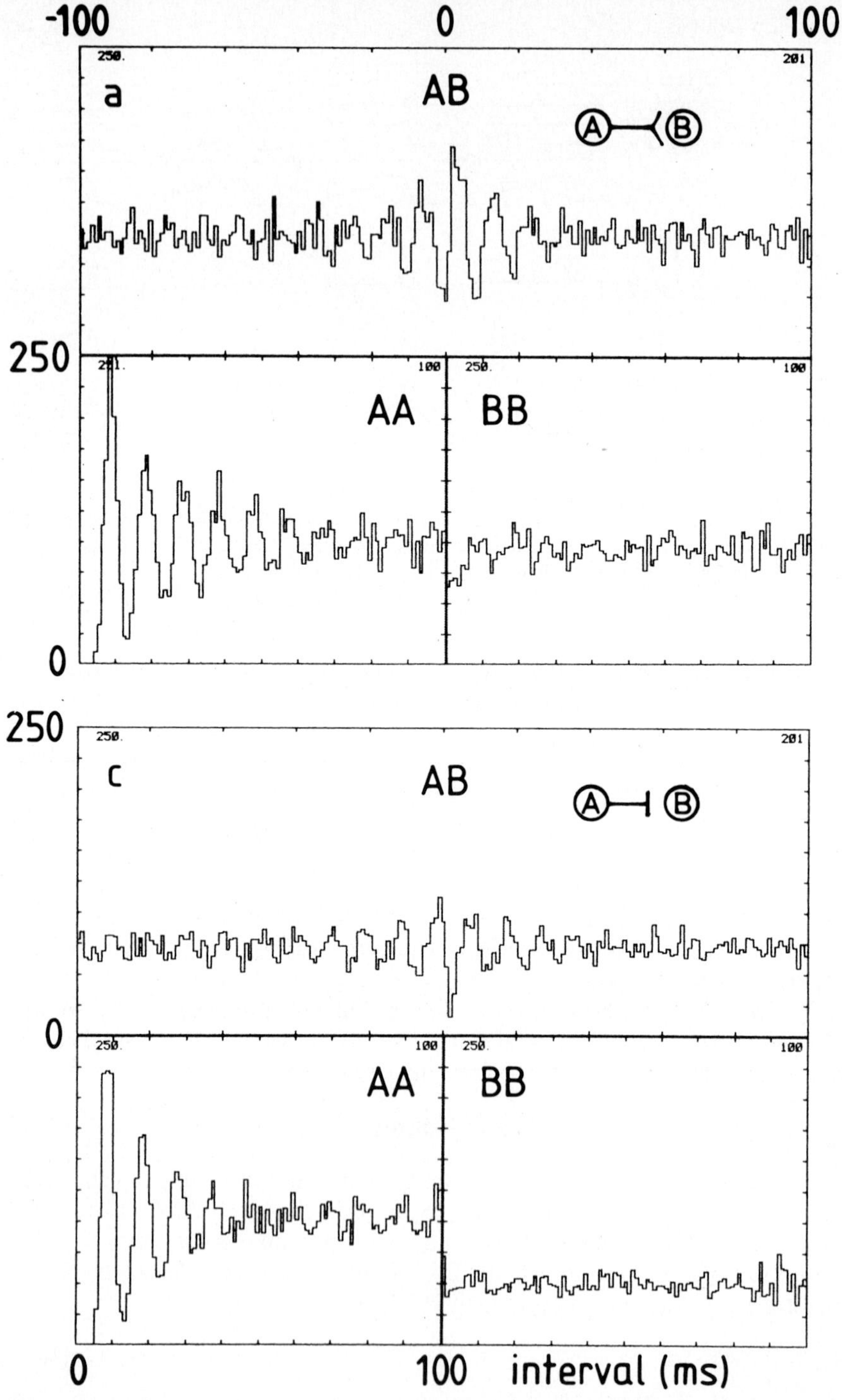

Fig. 9.3 a–d. Examples of cross-correlation histograms with superimposed autocorrelation structure of the input neuron. Shown are the auto correlogram (*AA* and *BB*) for both neurons and their cross-correlogram (*AB*). **a** and **b** show cases for direct excitation: in **a** through a

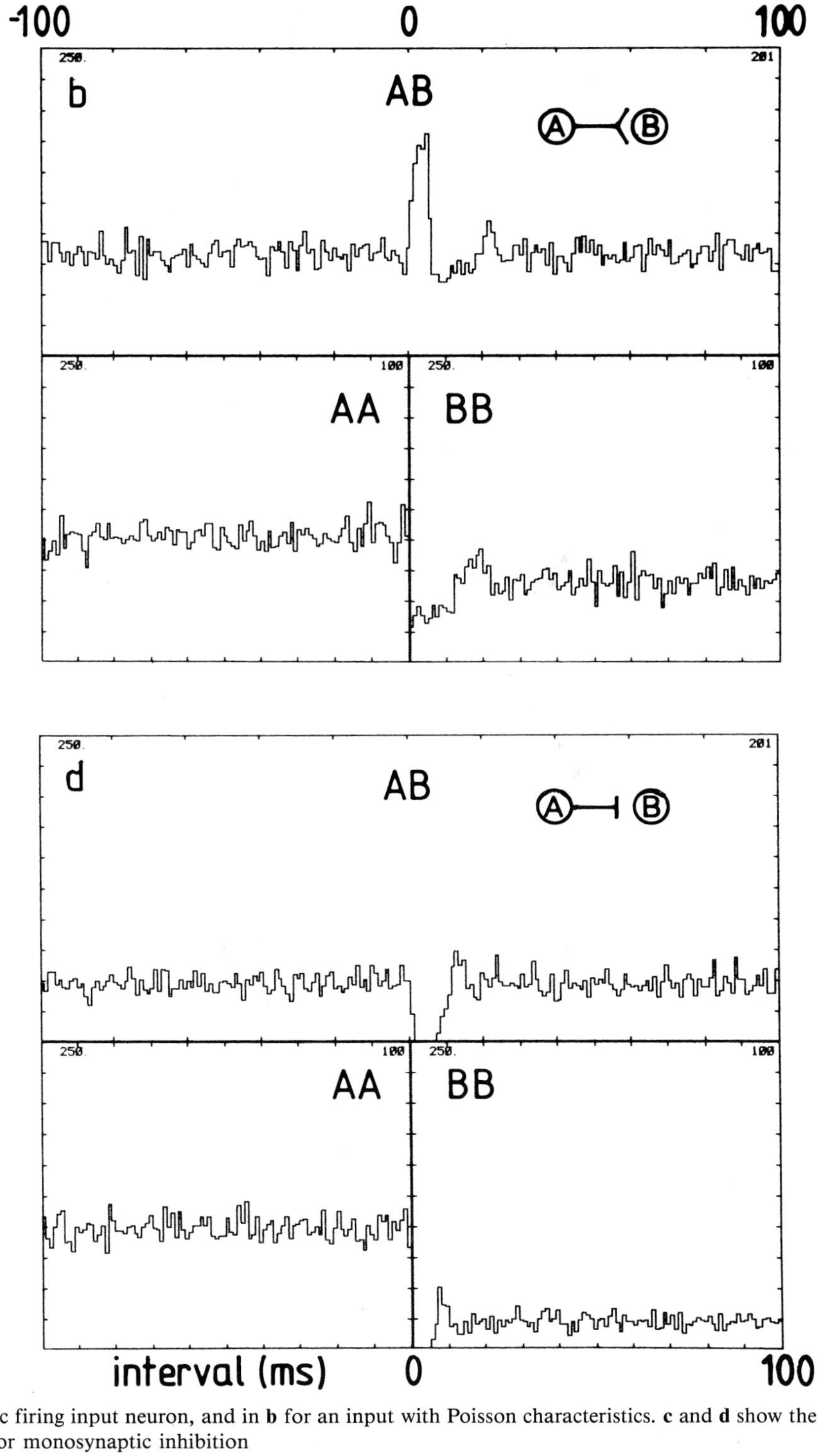

periodic firing input neuron, and in **b** for an input with Poisson characteristics. **c** and **d** show the same for monosynaptic inhibition

$$\xi_{xy}(\tau) = m_x\{R_{xy}(\tau) - m_y\} \tag{9.18}$$

as a convolution between the presynaptic autocovariance $\xi_{xx}(\tau)$ and a so-called synaptic response function $\varepsilon(\tau)$, defined by

$$\xi_{xy}(\tau) = \int \xi_{xx}(\tau - \sigma)\,\varepsilon(\sigma)\,d\sigma \tag{9.19}$$

or in the frequency domain:

$$\hat{\xi}_{xy}(\omega) = \hat{\xi}_{xx}(\omega)\,\hat{\varepsilon}(\omega)\ . \tag{9.20}$$

When the synapse is linear the synaptic response function $\hat{\varepsilon}(\omega) = \hat{h}(\omega)$; for a nonlinear synapse $\hat{\varepsilon}(\omega)$ depends on the properties of the input spike train. Perkel then splits up the synaptic response into a primary response $\hat{\sigma}(\omega)$ and a secondary postsynaptic response due to the autocorrelation structure of the postsynaptic spike train:

$$\hat{\varepsilon}(\omega) = \hat{\sigma}(\omega)\,\{1 + K\hat{\eta}_y(\omega)\}\ , \tag{9.21}$$

where K is a coupling coefficient expressing the influence of the presynaptic spike train upon the autocorrelation structure of the postsynaptic spike train, and $\hat{\eta}_y(\omega)$ is the one-sided power spectrum (e.g., Bendat and Piersol 1971) of the output spike train. From Eq. (9.18) we note (as is intuitively clear) that the autocorrelation structure of the output spike train will be mapped only on the positive frequency side of $\hat{R}_{xy}(\omega)$ and thus on the right hand side of the central peak in $R_{xy}(\tau)$. The autocorrelation structure of the input spike train will appear on both sides of the central peak. An example is shown in Fig. 9.3.

We may apply this spectral technique to our neural interaction model described in Eq. (7.32) as:

$$C_{12}(\tau) - C_{12'}(\tau) = E\,[z_1(t)\,z_{2n}(t+\tau)] \equiv C_{12n}(\tau)$$

in the following way. Let, analogously to Eq. (9.19):

$$C_{12n}(\tau) = \int_{-\infty}^{\infty} \alpha_{12}(\sigma)\,C_{11}(\tau - \sigma)\,d\sigma \tag{9.22}$$

be the expression that relates the neural correlation part of the CCH to the autocorrelation of spike train $z_1(t)$ via a coupling coefficient $\alpha_{12}(\sigma)$. In other words: the cross coincidence function of pre- and postsynaptic spike activity can be described as a filtering of the presynaptic autocoincidence function. In the frequency domain this can be written as:

$$\hat{C}_{12}(\omega) - \hat{C}_{12'}(\omega) = \hat{\alpha}_{12}(\omega)\,\hat{C}_{11}(\omega) \tag{9.23}$$

and thus:

$$\hat{\alpha}_{12}(\omega) = \frac{\hat{C}_{12}(\omega) - \hat{C}_{12'}(\omega)}{\hat{C}_{11}(\omega)} \tag{9.24}$$

gives an estimate of the neuronal coupling between spike trains $z_1(t)$ and $z_2(t)$ for the additive model. This may represent a single synapse transfer function in case of pre- and postsynaptic spike trains, or the combined effect of a number of connecting paths. When $\hat{C}_{11}(\omega)$ is flat, i.e., when the spike train $z_1(t)$ represents a Poisson process then the coupling can reliably be estimated from the difference of the simultaneous and nonsimultaneous cross-coincidence function. In general, $\hat{C}_{11}(\omega)$ will be strongly stimulus-dependent, and since $\hat{C}_{12}(\omega)-\hat{C}_{12'}(\omega)$ may have a different stimulus dependency one should expect a coupling $\hat{\alpha}_{12}(\omega)$ which will be stimulus-dependent. Note that $\hat{\alpha}_{12}(\omega)$ is the coupling term in the assumed black box model. Its stimulus dependence does not suggest that the synaptic efficacies change, it may mainly indicate that, e.g., the number or configuration of recruited synapses underwent a change.

Thus far cross-correlations have been expressed in terms of properties of the presynaptic spike train and the synaptic response function, which, as in the procedure used by Perkel (1970), may contain information about the postsynaptic spike train. Knox and Poppele (1977) describe the cross-correlation as:

$$R_{xy}(\tau) = \frac{m_x}{m_y} \int_0^{\tau} f_W(\sigma)\, R_{yy}(\tau-\sigma \mid W=\sigma)\, d\sigma \ , \tag{9.25}$$

where $f_W(\sigma)$ describes the waiting time distribution from an input at time 0 to the next succeeding output at time σ. $R_{yy}(\tau-\sigma \mid W=\sigma)$ is a conditional output autocorrelation function which takes into consideration only those output spikes that are preceded by an input spike within a time interval W. $f_W(\sigma)\, d\sigma$ is then the probability of a threshold crossing (i.e., a spike) in $W=(\sigma, \sigma+d\sigma)$ given an input at $t=0$ and no outputs in $(0, \sigma)$.

This definition of a cross-correlation is in fact a filtering by $f_W(\sigma)$ of the conditional autocorrelation of the output process. The cross-correlation function therefore carries information related not only to the immediate effects of an input on an output [through $f_W(\sigma)$] but also [through $R_{yy}(\tau \mid W)$] information related to the subsequent activity of the follower cell, as modified by that input. Note that a very large number of combinations of $f_W(\sigma)$ and $R_{yy}(\tau \mid W)$ can result in the same cross-correlation function. Therefore to estimate the excitability profile of the cell $f_W(\sigma)$ one must know both the cross-correlation function and the conditional output autocorrelation. For low firing rates the conditional autocorrelation will be approximately equal to the "normal" autocorrelation $R_{yy}(\tau)$ (because of the finite extent of the influence of the input spike) and, consequently, $f_W(\tau)$ takes the form of $h(\tau)$ as in the previous discussions.

9.3.2 Nonlinear Systems Analysis

A more general treatment of the problem analogous to the Wiener approach for continuous in- and output signals is given by Kroeker (1977) and covers also nonlinear synapses. We will follow his exposition. When the input spike train is a realization of a Poisson process with intensity λ, the Poisson-Charlier polynomial c_n of a single Poisson random variable is given by:

$$c_n(x) = \sum_{r=0}^{\infty} (-1)^{n-r} \binom{n}{r} \frac{x!}{(x-r)!} \lambda^{n-r} \tag{9.26}$$

i.e., $c_0(x) = 1$
$c_1(x) = x - \lambda$
$c_2(x) = x(x-1) - 2\lambda x + \lambda^2$
$c_3(x) = x(x-1)(x-2) - 3\lambda x(x-1) + 3\lambda^2 x - \lambda^3$.

Multivariate Poisson-Charlier polynomials of Poisson variables $x_{i1}, x_{x2}, \ldots x_{in}$ selected out of a collection $x_1, \ldots, x_s$ such that each i_ν, $\nu = 1, \ldots, n$ is chosen from 1, 2,..., s, are defined as:

$$c^n(x_{i_1}, \ldots, x_{i_n}) = \prod_{\sigma=1}^{s} c_{n_\sigma}(x_\sigma) \tag{9.27}$$

with $\sum_{\sigma=1}^{s} n_\sigma = n$.

These c^n polynomials have expectation zero and satisfy the orthogonality condition:

$$E[C^n(x_{i_1}, \ldots, x_{i_n}) C^m(x_{j_1}, \ldots, x_{j_m})] = \delta_{mn} \delta_{ij}^{(n)} \lambda_{i_1} \ldots \lambda_{i_n} , \tag{9.28}$$

where $\delta_{mn} = 1$ if $m = n$, otherwise $\delta_{mn} = 0$, and δ_{ij} has a value equal to the number of permutations $i_1, \ldots, i_n$ that match $j_1 \ldots j_n$. Now Poisson-Charlier functionals are defined as:

$$J_n(f_n) = \sum_{i_1=1}^{s} \ldots \sum_{i_n=1}^{s} f_n(i_1, \ldots, i_n) C^n(x_{i_1}, \ldots, x_{i_n}) \tag{9.29}$$

and therefore are a linear combination of all possible Poisson-Charlier polynomials of n variables.

The $J_n(f_n)$ have expectation zero and satisfy certain orthogonality conditions. The output y(t) of the nonlinear system under study may now be approximated by a sum of these Poisson-Charlier functionals:

$$y(t) = \sum_{n=0}^{\infty} J_n(f_n) \tag{9.30}$$

analogous to the Wiener series expansion for GWN input.

Using only the first few terms of such an expansion:

$$y(t) = f_0 + \sum_{j=0}^{m} f_1(j)(x_{i-j} - \lambda) + \sum_{j=0}^{m} \sum_{k=0}^{m} f_2(j,k)(x_{i-j} - \lambda)(x_{i-k} - \lambda) . \tag{9.31}$$

The system is said to be identified when the kernels $f_n(i_1, \ldots, i_n)$ can be determined. This can be done by cross-correlation of output y(t) with the polynomials $c^n(x_1, \ldots, x_n)$.

A feeling for the technique can be obtained by an application in which Poisson distributed click trains were used as stimuli, and single-unit recordings were made from the auditory midbrain of the frog. By a cross-correlation of

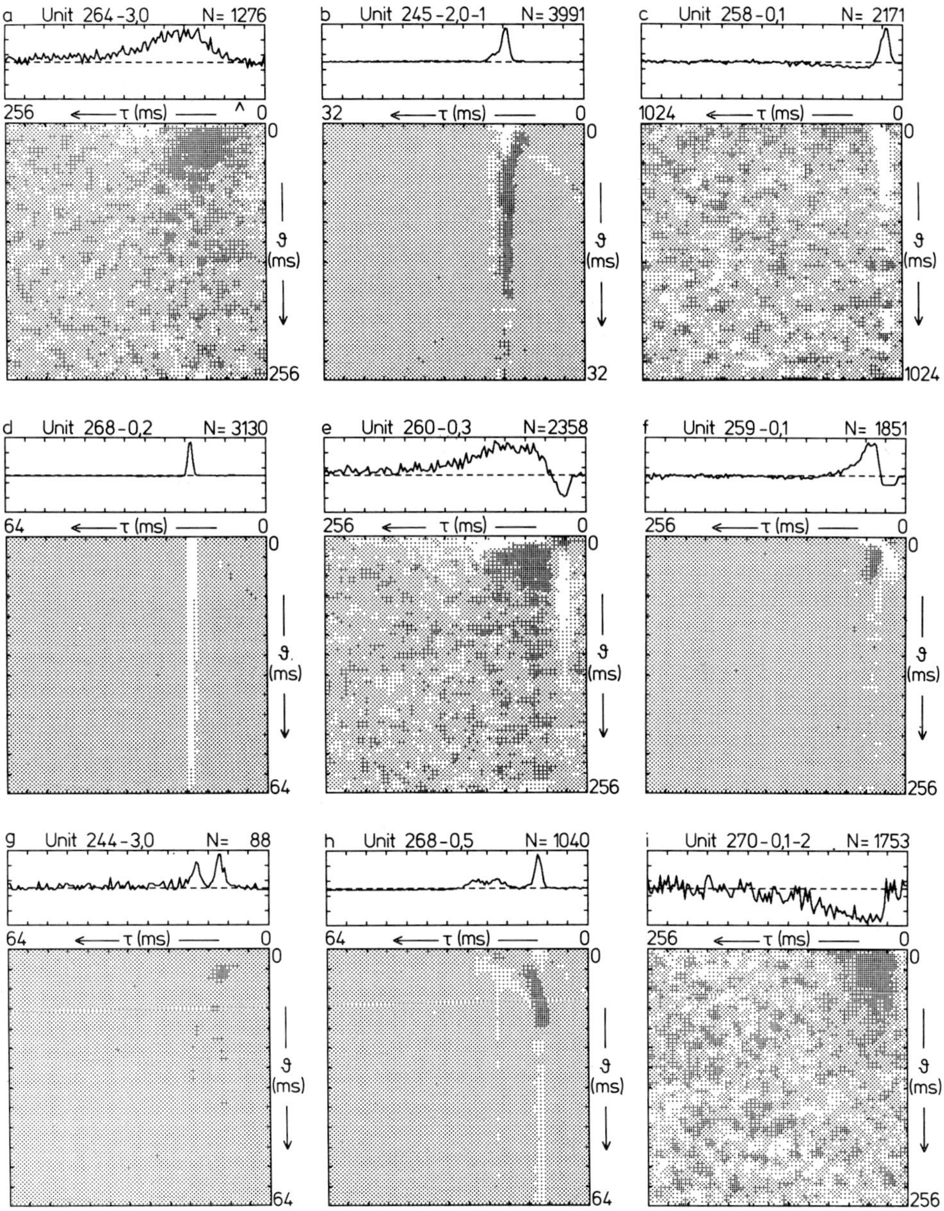

Fig. 9.4 a–i. First- and second-order Poisson kernels for some units in the auditory midbrain of the grassfrog. In the second-order kernel the amount of activation is coded along a gray scale: *black* corresponds to activation, *white* to depression. **a** shows an excitatory first-order kernel and a facilitatory second-order kernel. **c** shows an excitatory first-order kernel and a depression in the second order. **g** shows an example of a bimodal excitatory first-order kernel with a facilitatory second-order kernel, suggesting that spikes only occur in response to click pairs spaced by about 8 ms. (Epping and Eggermont 1986)

the input and output point process the first three kernels can be obtained (Epping and Eggermont 1986) as:

$$\begin{aligned} f_0 &= E[y(t)] \\ f_1(\tau) &= \lambda^{-1} \cdot E[y(t)\,x(t-\tau)] - f_0; \quad \tau \geqq 0, \ \theta > 0 \\ f_2(\tau,\theta) &= \lambda^{-2} \cdot E[y(t)\,x(t-\tau)\,x(t-\tau-\theta)] - f_1(\tau) - f_1(\tau+\theta) - f_0 \ . \end{aligned} \tag{9.32}$$

These kernels may be interpreted as follows. The kernel of order zero f_0 represents the average system response for zero input. The first-order kernel $f_1(\tau)$ represents the average system response upon a single impulse. When the system is linear, $f_1(\tau)$ is the impulse response. The second-order kernel $f_2(\tau,\theta)$ is equal to the average facilitation or depression due to a doublet of pulses θ seconds apart, regardless of the intervening pulses. When the system is at most quadratic this facilitation or depression is equal to the one measured in a two-pulse experiment.

The Poisson kernels can be obtained rapidly using first and second order cross-coincidence algorithms. Because of the discrete nature of input and output, especially the second-order kernel looks rather noisy. Therefore a bell-shaped 5×5 point smoothing window was applied to the 128×128 binned raw estimate of $f_2(\tau,\theta)$. Figure 9.4 shows a series of nine combinations of first- and second-order kernels for auditory midbrain units obtained for stimulation with $\lambda = 16$ clicks per second. The number of spikes in the y-train is indicated in the right upper corner of each part of the figure. Note that the number of output spikes is not related to the degree of noisiness of the first and second order kernels. A first-order kernel that is completely above the f_0-line indicates an excitatory action only; examples are in Fig. 9.4a and b for the simple case and in Fig. 9.4g and h for a more complex case. In the latter two cases the presence of two clicks with a distance of the time between the peaks in $f_1(\tau)$ facilitates or is necessary to generate a spike; this is corroborated by the second-order kernels. A frequently occurring type is an $f_1(\tau)$ that is above f_0 for short delays and below f_0 for larger lags. This represents excitation and post excitation suppression of activity (Fig. 9.4c and d). The reverse situation can be found in Fig. 9.4e and f. Finally a situation where the stimulation causes purely inhibition is shown in Fig. 9.4i. In this case some relief from inhibition can be expected when two clicks in that interval are present.

9.4 Identification Based upon Discrete Input and Continuous Output

For a linear synapse the mean soma potential B(t) is related to the intensity of an incoming point process $\lambda(t)$ by (Coggshall 1973):

$$\overline{B(t)} = \int h(t-\tau)\lambda(\tau)\,d\tau \ . \tag{9.33}$$

In case of a homogeneous Poisson process with intensity λ, the spectrum of the process is known to be equal to:

$$S_x(\omega) = \lambda\{1 + 2\pi\lambda\delta(\omega)\} \tag{9.34}$$

and the transfer function $\hat{h}(\omega)$ can readily be obtained. Kirkwood and Sears (1982a) used the action potentials recorded from muscle spindle afferents as the input and recorded intracellularly from α-mononeurones. Cross-correlations resulted in the epsp's produced by these muscle spindle afferent spikes – under the assumption that there were no nonlinear interactions between the spikes.

In the nonlinear situation a particular variation of the treatment by Kroeker (1977) has been presented by Krausz (1975). The output signal B(t) up to second order is expressed as:

$$\begin{aligned} B(t) = h_0 + \int h_1(\tau)\, x(t-\tau)\, d\tau + \iint h_2(\tau_1 \tau_2)\, x(t-\tau_1)\, x(t-\tau_2)\, d\tau_1\, d\tau_2 \\ - \int h_2(\tau,\tau)\, x(t-\tau)\, d\tau - \lambda \int h_2(\tau,\tau)\, d\tau \ , \end{aligned} \tag{9.35}$$

where $x(t)$ is a zero means Poisson process $x(t) = z(t) - \lambda$ with $z(t) = \sum_i \delta(t-t_i)$ and λ is the mean input rate. The kernels, again are obtained by cross-correlation:

$$\begin{aligned} h_1(\tau) &= \lambda^{-1} R_{xB}(\tau) \ - \ h_0 \\ h_2(\tau_1,\tau_2) &= (2\lambda^2)^{-1} R_{xxB}(\tau_1,\tau_2) - 1/2\, \{h_1(\tau_1) + h_1(\tau_2) + h_0\} \ . \end{aligned} \tag{9.36}$$

The second kernel can be interpreted in terms of the nonlinear effect of two inputs at the output compared to the sum of the effects of single pulses. When $x_1 = -\lambda + \delta(t-t_1)$ and $x_2 = -\lambda + \delta(t-t_2)$ and the combined input is $x_s = -\lambda + \delta(t-t_1) + \delta(t-t_2)$ it follows that:

$$B_s(t) - \{B_1(t) + B_2(t)\} = 2\, h_2(t-t_1, t-t_2) \ . \tag{9.37}$$

Thus the second kernel gives one half of the facilitation or anti-facilitation at time t measured in a two-pulse experiment, it can be interpreted as a generalized recovery function. In the simplified case that the output to all impulses is identical in shape with relatively rapid rise (as epsp's do), and one furthermore assumes that facilitation is a linear function of the past input:

$$F(t) = \int_0^\infty k(\tau)\, x(t-\tau)\, d\tau \tag{9.38}$$

then it can be derived that:

$$k(\tau) = \frac{2 h_2(\tau + L, L)}{g(L)} \ , \ \tau \geqq 0 \ , \tag{9.39}$$

where L is a fixed point (e.g., the peak) on the epsp in ms from the stimulus pulse and g(L) is the amplitude value at that point. This means that the system can be identified by its standard impulse response [the epsp, $g(\tau)$] and one fixed argument line of the two-dimensional second order kernel.

For higher order kernels a relation analogous to Eq. (9.39) is given by:

$$k_{i-1}(\tau_1, \ldots, \tau_{i-1}) = \frac{i!\, h_i(\tau_1 + L, \ldots, \tau_{i-1} + L, L)}{g(L)}, \quad i = 1,\, 2, \ldots \tag{9.40}$$

thus together with $g(\tau)$ these kernels are sufficient to characterize transmission at the synapse. In most situations it is not necessary to calculate kernels higher then third order for biological systems (Krausz and Friesen 1977), because of the linearizing effects of the intrinsic noise in the system (e.g., Marmarelis and Marmarelis 1978).

9.5 When Systems Identification Does Not Work

Systems analysis is straightforward for linear systems and somewhat more time-consuming for nonlinear systems. In neurophysiology, however, the underlying assumptions that the system has constant parameters and finite memory are generally not met. Even at the most peripheral level, e.g., in the auditory nerve, the establishment of stationarity of the spike train requires some tens of milliseconds after stimulus onset because of receptor adaptation. This adaptation, and especially that in more central structures, may depend on the level of stimulation. Thus the parameters of the system become stimulus-dependent. Studying the system with a GWN input then effectively estimates an average set of parameters for a specified intensity range of the noise. Furthermore, using GWN noise seems to linearize the system to some extent.

Another important point is that the use of GWN might show the synapse in a different way than using pulse trains or double pulses. When also other, nonobserved, inputs are present, the temporal synchrony between inputs might affect the synaptic action, a feature that may not be present as manifestly when GWN or Poisson click trains are used. The nervous system certainly is not a deterministic system, but indications are abundant that it is also not a stochastic system with constant parameters. The more we advance to more central parts of the CNS the more apparent this becomes. Two reasons can be mentioned: The first is that we observe a too small part of the nervous system and that the number of synapses on a given cell that is activated depends on the stimulus and the state of the animal (each synapse having its constant properties). The second cause is that stimulation and the context wherein this happens are indeed producing changes in the central nervous system. This is generally called learning and is a manifestation of the plasticity of the CNS. Under such conditions system identification techniques can only describe the average situation during the test and the result applies only for that stimulus condition. It is generally not possible on the basis of the identification to predict what the response of the system will be when we apply the same stimulus condition at a later time. The test has changed the system! Such situations are most likely to occur during development, the time when the system is utterly plastic. The next chapters will deal with this neuronal plasticity and electrophysiological substrates that accompany various forms of learning.

10 Plasticity – The Capacity to Change

So far we have assumed that the brain has time-invariant parameters: the connections between neurons are fixed and constant in strength. Neural systems with interconnections of this type do not change and cannot learn. However, when the neurons are spontaneously active and some form of lateral inhibition is present, they have the capacity to organize themselves in, for example, columnar fashion (e.g., von der Malsburg 1973). Such systems behave like automata: their performance develops according to innate wiring patterns. It is unlikely that such systems are found in the brain of vertebrates, and also in invertebrates examples abound of modifiability of certain reflexes (e.g., Kandel 1976). Thus, most nervous systems display some kind of plasticity; they learn (or are imprinted) during early development and they continue to learn from experience. This learning is reflected in changes in behavior as well as in changing neural firing patterns. A sensory stimulus does not provoke a fixed reaction, but rather one that depends on the context in which the stimulus occurs. If the context can be controlled, systems analysis as described in the previous chapter can still be applied and relevant parameters such as synaptic strength can still be estimated. However, one of the main goals of systems analysis, i.e., to predict the behavior of the system under arbitrary conditions, cannot be pursued. It is even highly likely that the very system identification procedures, i.e., the measurements that have to be performed, change the system. Every stimulus provides the nervous system with information about the environment that can be learned by it. If it uses this information, the system will never be the same again. Still, the accumulating information about the central nervous system in vertebrates suggests that the brain either learns very slowly or does not learn under those laboratory conditions where it is either anesthetized or resides in an immobile animal.

In this chapter a survey will be given of physiological and morphological findings that reflect plasticity in developing and adult brains. Subsequently, models and mechanisms for learning at the cellular level will be discussed. Specific discourses about plasticity in the cerebellum, the hippocampus and the neocortex will be deferred to Chapters 11, 12 and 13.

10.1 The Developing Brain

A developing brain is synonymous with a plastic brain: plasticity allows the developing nervous system to adapt to its environment. There are many signs of a changing nervous system during the first few postnatal months in animals and humans. First, I will present some physiological evidence, followed by a few examples of changes in the morphology of brain structures. This survey will not be exhaustive but merely serves to be contrasted later in this chapter with plasticity in adult brains.

10.1.1 Physiological Signs of Plasticity

In the kitten the auditory system becomes responsive during the second postnatal week, slightly ahead of the visual system. What both systems have in common during that second week is that the spontaneous activity of the neurons is very low. For instance, for the visual system Tsumoto and Suda (1984) report that cortical layer V showed spontaneous activity in about half of the neurons, while in other layers only 5–15% of the cells showed any sign of spontaneous activity. Layers IVa and the upper part of layer VI only acquired adult properties – as far as spontaneous activity is concerned – at the forth postnatal week. Layer IVc only became adult-like at the end of the fifth postnatal week. This indicates a differential maturation of the visual cortex.

The response of individual auditory nerve fibers to tonebursts is initially an "one-type" response, then the responses become rhythmic – in many ways resembling a chopper-neuron – and finally, around day 20, acquire adult-like tonic responses (Fig. 10.1). Neurons in the cochlear nucleus also acquire adult-like response properties at the 20th postnatal day. For the inferior colliculus of the kitten, the unit responses from day 2 onward also represent these aspects: on-responses, rhythmic responses, and responses with intermittent

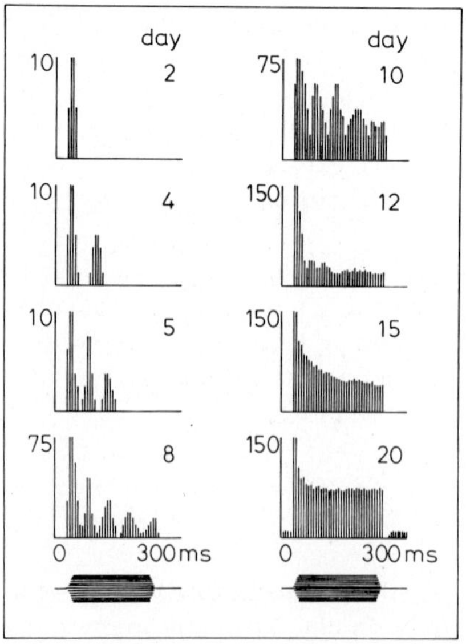

Fig. 10.1. Postnatal development of tone burst PSTH in the auditory nerve of the cat. Initial onset responses change after 4–5 days into rhythmic (chopper-type) responses. At 12–15 days sustained responses (primary like) appear. The importance of the transient character during the first 50 ms gradually decreases. (Eggermont 1985b)

silence. These types of response change gradually and disappear completely around 14 days after birth, giving way to adult-like sustained responses (see Eggermont 1985b for a review and specific references). Together with the increased responsiveness to stimuli, one observes a general decrease in the latency of firing. This has been investigated for single units in the auditory nerve and cochlear nucleus and, using evoked potentials, at most other levels in the auditory system of the kitten. The picture that emerges from these single unit studies and intracranially recorded evoked potentials shows that latency decreases in an exponential way to adult values with a time constant of around 6 days (Eggermont 1985b). When so-called auditory brainstem responses (ABR) are used to monitor the changes in latency in the auditory nervous system, a more complex picture emerges. It is reported that initially, between postnatal day 7 and 18, latency decreases linearly, thereafter the decrease is exponentially towards the adult value (Walsh et al. 1986a, b, c; Puel and Uziel 1987). It appears also possible to model these changes with two exponential functions where the longest time constant is around 12 days (Eggermont 1985c, d).

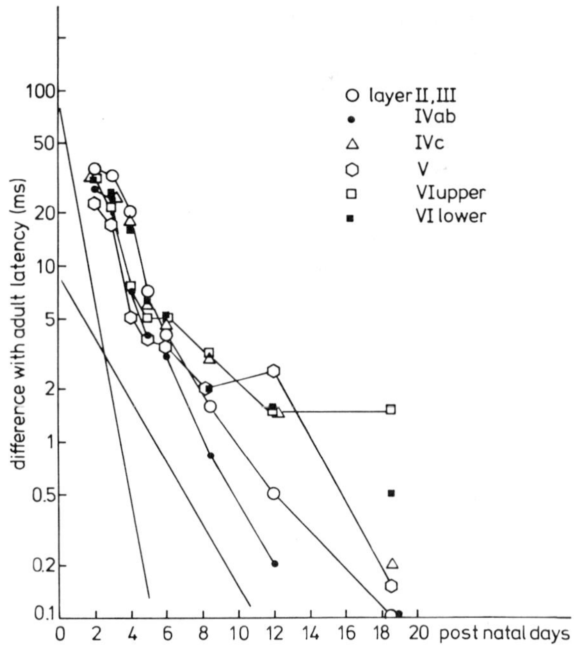

Fig. 10.2. Latency changes in single units from various layers in the visual cortex, activated by electrical stimulation of the LGN, as a function of age. Maturation starts first in layers IV and V and a few days later in the other layers. The latency changes are more rapid at first (time constant around 6 days: *steep drawn line*) and then slows down (time constants 12–25 days, *shallow drawn line* has time constant of 12 days). Adult latency values are for layers II and III: 5 ms; for layers IV abc: 2 ms; for layer V: 3 ms; for layer VI upper: 2.5 ms and for layer VI lower: 5 ms. (Tsumoto and Suda 1984)

Recordings of the N_1 component of the slow vertex potentials, supposedly generated in the auditory cortices, suggest an exponential decrease in latency with a time constant of around 25 days (Walsh et al. 1986a, b, c).

For the visual system in the kitten, electrical stimulation in the lateral geniculate nucleus and recordings from the various layers of the visual cortex (Tsumoto and Suda 1984) showed that latencies are very long (20–50 ms) at the 2nd week and rapidly become shorter in older kittens. In Fig. 10.2 the differences between the latencies found in the kitten and the adult latency are plotted as a function of age for each layer of the visual cortex. The general impression is that the maturation exhibits a fast component up to the 5th or 6th postnatal day (time constant about 6 days). Thereafter the data points for the various layers become more scattered but the slower rate of the maturation is still quite clear. Time constants in this period range from 12–25 days.

Summarizing, overall the rates of maturation found for the visual system and the auditory system are the same in the kitten. However, the auditory system, having a head start in the developmental process, completes its maturation considerably earlier than the visual system. This differential effect is also manifest for the various parts of the visual system itself. A close inspection of Fig. 10.2 shows that three cortical layers seem to be ahead in maturation by about 1 week as compared to the other layers. Especially layers IVb and V start their maturation earlier, in the 3rd postnatal week, while the other layers do not show much change in latency before the 4th postnatal week.

Cortical response latencies in the cat have acquired adult-like properties somewhere around the 20th postnatal day. It is interesting to note that recordings of ABR's and slow vertex potentials in humans suggest a much prolonged maturation period: from data collected from many laboratories it can be concluded (Eggermont 1985b, 1986, 1988) that human auditory system maturation is certainly not completed before the 5th year of life. A compilation of the data on which this conclusion is based is shown in Fig. 10.3. It appears possible to adequately describe this conglomerate with the sum of three exponentials having time constants of respectively 4 weeks, 40 weeks and 3 years. This suggests that three basically different mechanisms are operating independently and in parallel.

Analysis of visual (VEP) and somatosensory evoked potentials (SEP) in the human neonate shows essentially the same set of time constants. Again the auditory system starts to mature first followed by the somatosensory and then the visual system (Eggermont 1988).

The mechanisms responsible for the three time constants are not easy to identify. During ontogeny and early development many things change. notably the number and efficacy of synapses, and the amount and extent of myelination in the central nervous system. For the auditory system not much is known about the changes in the number of synapses in the CNS. However, there is some information for the human visual cortex (Huttenlocher et al. 1982), indicating a slow logarithmic rise in synaptic density up to 2 months of age followed by a rapid increase that doubled the density at age 4 months. Thereafter a more gradual increase followed up to 8 months, after which there is a con-

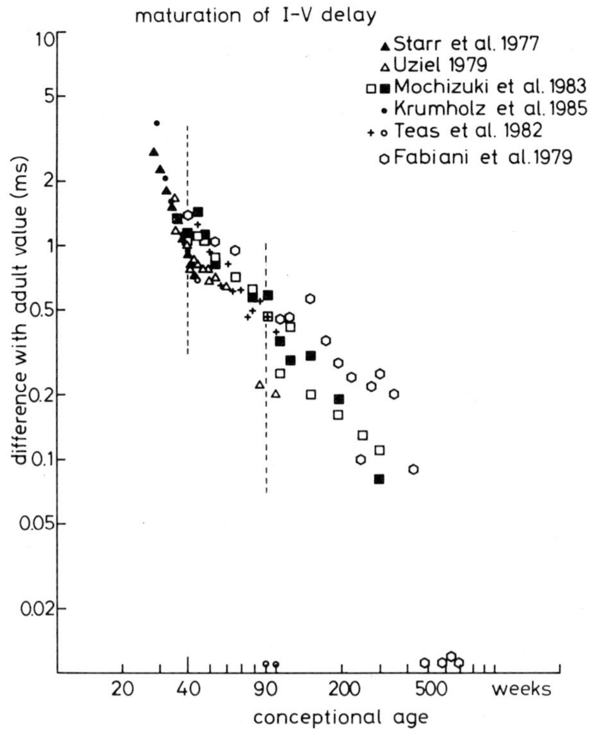

Fig. 10.3. Latency changes in the human auditory brainstem response with respect to the adult latency value. Both the latency as the conceptional age axis are logarithmic. Data shown are mean values at the various age groups. *Dashed vertical lines* indicate term birth (40 weeks) and 1 year. (Eggermont 1986)

tinuous but very slow decline in synaptic density. The onset of the rapid increase in synaptic density parallels the abrupt decrease in the latency of the flash VEP, which can be described by a decaying exponential with a time constant in the "short" range. At age 4 months latency changes in the visual system proceed according to the long time-constant process, and prior to about 2 months of age latency changes are also quite slow. Thus there is reason to assume some correlations between synaptic density changes and cortical evoked potential latency changes.

Myelination changes in the developing system have been extensively documented by Gilles et al. (1983) on the basis of a very large sample of stillborn premature fetuses. They used various qualitative criteria to describe the progress of myelination in selected parts of the CNS. Microscopically visible myelin appears in the auditory brainstem (trapezoid body and lateral lemniscus) and midbrain (lateral lemniscus) at 23–24 weeks CA. At the end of gestation about 95% of the fetuses had intensely stained myelin at all three levels of the auditory pathway, indicating a short myelination phase. This suggests that the myelination in the brainstem parallels the short time-constant process that we have identified. Tentatively then we have identified two processes that parallel the rapid latency changes in evoked potentials. For the visual cortical EP these are changes in synaptic density, and for the auditory brainstem response changes in the myelin content of the brainstem. However, also for the visual cortex myelination parallels latency changes: myelination in the optic radiations starts around term and probably requires (assuming the same rate as for the brainstem) about 2 to 3 months to completion. It is therefore possible that synaptogenesis and myelination occur simultaneously and at similar rates. Both mechanisms then, individually or combined, can explain the short time-constant latency changes in the ABR and VEP.

Besides changes in spontaneous activity and latency, pronounced effects occur also in the receptive field properties of auditory and visual neurons. In the newborn kitten, auditory nerve tuning curves are very broad and show a high threshold at the characteristic frequency. With the decrease in threshold of firing, the tuning becomes more pronounced to obtain adult sharpness at about the end of the first postnatal month. The sharpness of tuning appears to be linearly related to the improvement in threshold (Puel and Uziel 1987; Eggermont 1985b). For the visual system the receptive field properties of cortical neurons have been studied in detail. At birth a group of cortical cells already shows orientation selectivity, but other cells do not. It is not clear if this depends on the particular layer from which the recordings were obtained. What is clear, however, is that normal visual experience is required for the initially orientation-selective cells to keep this selectivity and for other cells to acquire it. A conspicuous aspect of the adult visual system are the ocular dominance columns which alternate for the left and right eye. After early monocular deprivation this normal alteration is deranged. In the cat the critical period for the rearrangement of layer IV ends some time between the 6th and 12th week. Deprivation after this period no longer affects layer IV, but still has profound effects on layers above and below layer IV. Thus binocular plasticity is a multistage process, the input layer (IV) of the visual cortex is sensitive early in life and the other layers remain sensitive until later (cf. Stryker 1982). Ocular dominance columns appear to be formed in utero in monkeys, and are present in dark-reared cats as well. This indicates that visual experience is not necessary for their formation. Upon injection of TTX (an agent which abolishes action potential generation) in the eye vitreous, no ocular dominance columns are formed, suggesting that normal spontaneous activity (possibly together with lateral inhibition) is sufficient for ocular dominance column formation.

From this short survey one may retain the impression that spontaneous activity and exposure to environmental stimuli is sufficient for normal development of cortical properties. In fact, it was shown that retinal signals never lead to changes in cortical functions when the kittens were paralyzed and/or anesthetized while exposed to visual patterns (cf. Singer 1985a), suggesting that the animal must be aware of the stimulation and be able to use it for control of its behavior, or that the anesthetic itself interferes with the development of cortical function.

10.1.2 Morphological Substrates of Plasticity

At birth the number of pyramidal cells in the cortex is well established. What one observes during development is, e.g., an increase in the arborization of the dendrites. In human infants this process definitely continues up to 2 years of age (Conel 1959). The same is found for the Purkinje cells in the cerebellum: in the newborn hardly any apical dendritic arborization is found. The formation of the dendritic tree starts after a few days of life and reaches adult-like aspects at the age of 1 year (Andersen and Eccles 1965). This increase in the density of the dendritic arborization is illustrated in Fig. 10.4.

Another aspect of maturation is the formation of dendritic spines on the pyramidal cells in the cortex. It has been shown that the growth of spines can be influenced by the environment: rearing animals in the dark showed a reduced number of spines, rearing animals in an "enriched" environment resulted in more spines. In the mouse, the development of spines more or less coincides with the early postnatal period. However, the guinea pig already has well-developed pyramidal cells at birth, densely covered with spines (Schuz 1978). Spines usually are the sites of synapses. The increase in the number of synapses in the mouse cortex closely parallels the increase in the number of

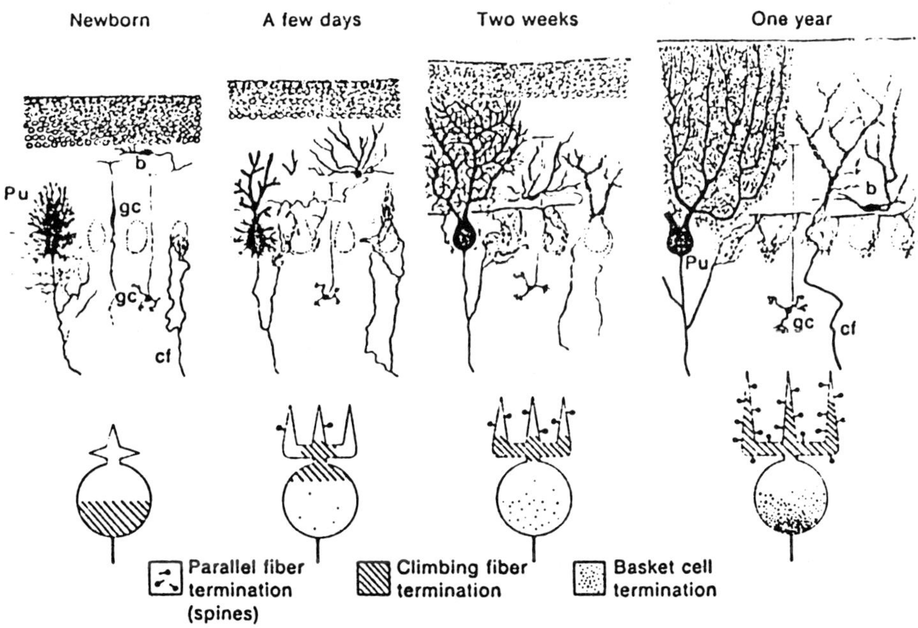

Fig. 10.4. Histogenesis in the cerebellar cortex. The most conspicuous aspect of maturation in the cerebellar cortex is the increase in size and number of the apical dendrites of the Purkinje cells. This leads to an increase in the number of synaptic contacts with the parallel fibers. Besides that the number of synaptic contacts between climbing fibers and the Purkinje cells as well as between basket cells and Purkinje cells increases. (Andersen and Eccles 1965)

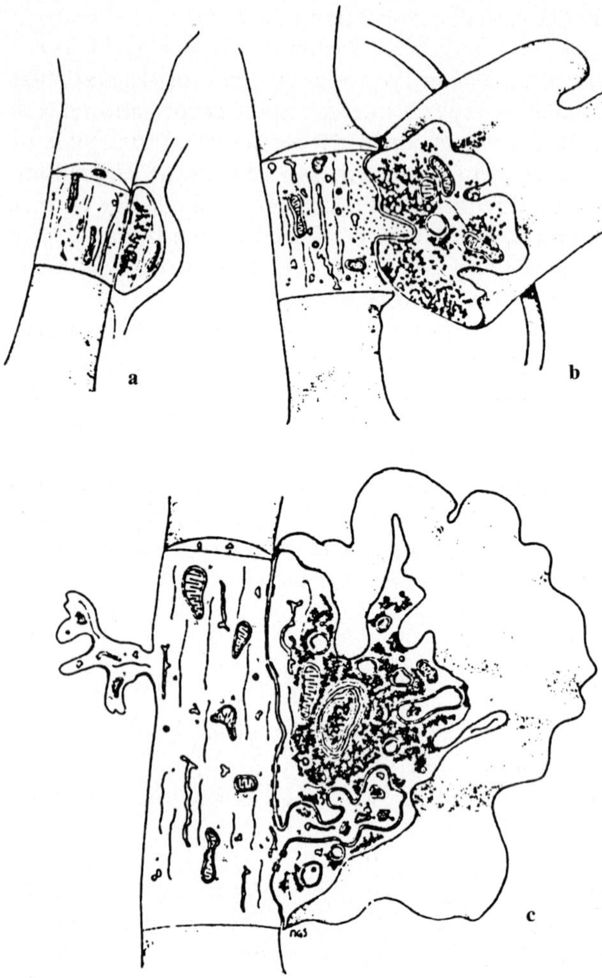

Fig. 10.5a–c. Spine and synapse formation in the hippocampus during development. (Cowan et al. 1981)

spines. However, at present it remains unclear whether spines are a prerequisite for development or whether they are formed as a result of increased neural activity during development. An illustration of the development process of spines on pyramidal cells in the hippocampus is shown in Fig. 10.5; observe that mainly asymmetrical synapses are formed on the spines.

A morphological substrate of the ocular dominance column formation is given by Stryker (1982): in adult cats the afferent Y-cell axons arborize in two or three ocular dominance columns with dense arbors in each column (but restricted to layer IV). In the kitten, single Y-axons arborize over large areas with no gaps in their terminal fields. Later in this chapter we will discuss mechanisms that may cause this columnar segregation.

Summarizing, during development the dendritic fields of pyramidal cells increase in size, the number of dendritic spines increases, and the arborization of axons becomes more specific. Since spines are the sites of synapses, the number of synapses increases dramatically during development. Schuz (1978) estimates that in the rat 6×10^{11} synapses are established in the cortex in just 26 days. The segregation of axons arborizations might be the result of localized synaptic loss, and we will see that a process of stabilization and selective removal of synapses is one of the main aspects of development.

10.2 The Adult Brain

Basically, three mechanisms have been described that can account for the plasticity of adult brains. These are the formation of new neurons, the sprouting of neurons, and the modification of synapses. Most theoretical work has been concerned with the modification of synaptic strength; we will discuss some of this work later in this chapter. First we will present a survey of physiological results that provide evidence for plasticity in adult brains.

10.2.1 Physiological Signs of Plasticity

10.2.1.1 Local Changes

In the following, the term "response plasticity" will be used to indicate that neurons respond differently to the same sensory stimulus, depending on whether this stimulus has a meaning attached to it, e.g., to signal a reward or a stimulus to avoid. Most experiments set up to demonstrate response plasticity have the character of a classical conditioning experiment (see Box K, p. 204).

This is because such experiments can distinguish changes in the response of neurons due to general arousal or performance, or to associative learning. In most cases the conditioning stimulus (CS) was acoustic, the unconditioned response frequently was a pupillary response. When using auditory stimuli in a classical conditioning experiment, one must be sure that there are no learning-related changes in the peripheral parts of the auditory system. It has been clearly established that the ventral or lemniscal pathway (ventral cochlear nucleus-olivary complex-central nucleus of the inferior colliculus-ventral part of the medial geniculate body) does not show learning-related changes. On the other hand, the dorsal pathway, projecting from the dorsal cochlear nucleus to the surround nuclei of the inferior colliculus and to the magnocellular part of the medial geniculate body, does show learning-related changes. While the lemniscal tract projects to the ventral part of the MGB in a tonotopically organized way and with target neurons having narrow tuning curves, the nonlemniscal tract projects to the magnocellular part of the MGB, which is not topographically organized and equipped with units that have very broad tuning curves. The neurons in the ventral part of the MGB respond to the physical

Box K. Classical and Operant Conditioning
In classical conditioning, use is made of two stimuli, an unconditioned stimulus (US) and a conditioned stimulus (CS). The US alone is always capable of evoking a response of the animal, the CS initially will not evoke a response. These responses can be nonspecific, such as changes in heart rate, blood pressure, pupil diameter, galvanic skin responses, or more specific motor responses. The procedure for conditioning the animal is to present the US and the CS in a close temporal relationship for a number of trials. In case a nonspecific response is observed, usually 10–15 stimulus pairings will suffice to evoke a response afterwards to the conditioned stimulus alone. The purpose of conditioning is to cause an observable response that can indicate whether the CS is perceived. After a sufficient number of times presenting the CS alone the response will gradually disappear and reconditioning has to be done.

Instrumental or operant conditioning presents the animal with various situations that it can select from; some produce reward, others punishment. The animal finds out by trial and error what the desirable or undesirable responses are, and adjusts its behavior accordingly.

parameters of the acoustic environment and their response is unaffected by learning. Neurons in the magnocellular part are not very selective for differences in physical sound parameters, but develop plasticity during learning (Weinberger et al. 1984). Single cells in AI of the auditory cortex of the paralyzed cat show response plasticity of some extend during a classical conditioning task, involving the pupillary response, in 63% of the cases. This plasticity involved either changes in background activity, in evoked discharges, or in both of these. Weinberger et al. (1984) conjectured that the plastic neurons probably received input from the magnocellular part of the MGB. This either means that there are sharply tuned neurons in the magnocellular part of the MGB or that the neurons that were presumed to be in AI actually were from AII, since it is known that the boundaries of the auditory cortex regions are very variable (Merzenich et al. 1975). In the secondary field of the auditory cortex (AII), an area which is not topographically organized and whose major inputs are from the magnocellular division of the MGB and AI, it was found that all cells recorded from (N = 22) developed physiological plasticity in either background or evoked activity during conditioning. It was emphasized that these changes were due to learning effects and not to simple arousal (Diamond and Weinberger 1984; but see also Kitzes et al. 1978 for a diverging viewpoint).

In rabbits a nicticating membrane response can be elicited using an airpuff on the cornea (US) and an auditory stimulus (CS). Kraus and Disterhoft (1982) observed the responses of 35 neurons in the association cortex of 35 rabbits that were conditioned, and compared these to 26 association cortex neurons in four rabbits that were unconditioned. In conditioned animals 51% showed changes in the CS-evoked firing rates, while only 19% of the pseudoconditioned group did. It was concluded that response plasticity is a property of some cells in the association cortex.

The same US, CS paradigm and the same response was used by Berger et al. (1976), who recorded from the hippocampus in rabbits. All conditioned animals rapidly developed a characteristic response: the unit activity showed a pronounced increase in firing rate which preceded in time and paralleled the behavioral nicticating membrane response.

Comparable findings for the visual cortex in the cat were reported by Ramos et al. (1976) during a behavioral generalization task. PSTH's were recorded for a test stimulus with properties intermediate to two training stimuli with learned behavioral responses. It appeared that 17 of the 26 cells recorded from in the visual cortex, and 23 out of 30 cells from the LGN had PSTH's which were the same regardless the behavioral action of the animal. The remaining cells (six in the cortex and seven in the LGN) showed a plastic response behavior: the later part of the response – with latencies in excess of 60 ms – clearly depended on the behavioral action of the animal. Thus, in the visual system one finds stable cells whose average firing patterns are determined by the parameters of the physical stimulus, and plastic cells whose response patterns vary, in their late components, depending on the meaning attributed to the stimulus.

In an attempt to demonstrate changes in the correlation strength between simultaneously recorded neurons during a classical conditioning task, Voronin et al. (1975) recorded from 17 units (in groups of 2–5) from the sensorimotor cortex of the rabbit. The conditioned response was a forelimb movement to an electric stimulation of the motor cortex (US) and an acoustic click or light flash served as the CS. While 7 out of 17 neurons showed an increase in response during conditioning, firing rates were in general too low to draw meaningful conclusions from the correlation studies. More success can be attributed to the study by Wyler (1985), who examined the degree of covariation of instantaneous firing rate as well as the cross-correlation between the spike trains of simultaneously recorded neurons in the precentral cortex of a monkey, that was reinforced to control the firing pattern of one neural unit (operant conditioning). A total of 125 neuron pairs was recorded from. In only 20% of the cases was there covariance between the firing rates of the units during conditioning; however, in 67% of the pairs there were significant peaks or troughs in the cross-correlation histogram within 13 ms (and in most cases within 1 ms) from the origin. Of 25 pairs with correlations in firing rate, 13 had a significant CCH. It remains difficult to attribute the effects shown to learning per se: a general improvement in performance such as an overall increase in firing rate might account for it as well (see discussions in Thompson et al. 1983; Paisley and Summerlee 1984).

10.2.1.2 Global Changes

One of the most dramatic demonstrations of adult brain plasticity was reported by Merzenich and coworkers (Merzenich et al. 1983, 1984; Kaas et al. 1983). They demonstrated that receptive fields in the somatosensory cortex of the adult monkey are alterable by lesions in the sensory periphery – such as median nerve section or digit amputation – or by increasing peripheral stimulation (cf. Merzenich and Jenkins 1983). The changes in receptive fields or digit representation upon the somatosensory cortex are subject to certain rules. Immediately after nerve section or digit amputation the cortical areas that were previously the target of the severed sensory endings are almost completely devoid of activity. However, there are small areas in these regions that now respond to stimulation of the dorsal surface of the hand. This input to the cortical areas presumably was present before the lesion, but apparently was suppressed by the normal input. After the lesion the receptive fields of the previously masked inputs increase in size, and ultimately take up most of the cortical area previously devoted to digit input. During this change of maps, the topographic ordering is strictly maintained, as in the partial overlap of receptive fields. There appears to be a distance limit to map reorganization (Fig. 10.6).

There are two types of mechanisms that can explain this central nervous system plasticity in adults. The first in that previously ineffective connections become effective in activating cortical neurons as a result of local changes in the amount of inhibition or an increase in sensitivity. Secondly, new connections may be formed through axon growth and/or dendritic extension. It is believed that a temporal coincidence in inputs or a spatiotemporally organized activity pattern constitutes the basis of a dynamic maintenance and alteration of these maps (Clark et al. 1988).

Map alterations occur as well in nondeprived adult brains and probably constitute the basis of acquisition of skill. The varying history of each animal will therefore be reflected in the receptive field mapping on the somatosensory cortex: there is a great variability in the digit representation among different monkeys. The same phenomenon is also found for the binaural interaction stripes in auditory cortex which resemble ocular dominance columns in the visual cortex.

While the tonotopic organization in auditory cortex is the result of hardwired connections and largely indicated by the organization of the anatomical projections, sound localization which is represented across isofrequency domains must be organized by experience and will be alterable throughout life (Merzenich et al. 1984b). An illustrative example of this alterability is given by Knudsen and Knudsen (1985), who showed that barn owls adjust their auditory localization with respect to a visual frame of reference (which might be erroneous when, e.g., prisms are mounted before the eyes). Such alterations probably involve changes in the strength of pre-existing synapses. Variability in brain structure is also found in organisms which have the same genetical make-up, illustrating that environmental factors and individual differences are important for shaping the brain (Changeux and Danchin 1976).

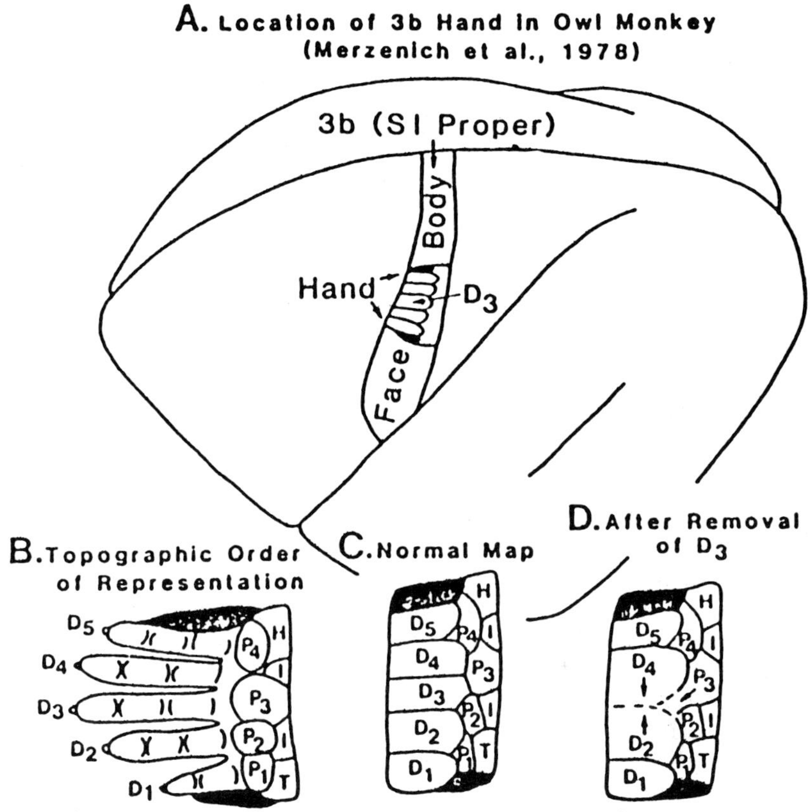

Fig. 10.6 A–D. Changing receptive fields in adult monkey after removal of one digit. (Merzenich et al. 1978)

The types of brain plasticity discussed above all reflect relatively long-term changes. Several other forms of modifiable functional connections have been termed "plastic" (e.g., Frostig et al. 1983) although they appear to be instantaneous and possibly due to the recruitment of the same cells into different assemblies, depending on the type of stimulation used. Frostig et al. observed in the auditory cortex of the cat that the number of direct interactions in local neuronal groups (cells recorded simultaneously with one electrode) is surprisingly small. In addition it was observed that certain functional connections showed a stimulus-dependent correlation. This type of plasticity was thought to be operative on a time scale of 100 ms or that order. Abeles (1982) described that 9% of the neural units in the primary and secondary auditory cortex are not exclusively sensory, but in addition show associative activity. In this respect it is noteworthy that Rebillard et al. (1977) found an enhancement of visual evoked potentials from the primary auditory cortex of the cat after an early destruction of cochlear receptors. Subsequently, they (Rebillard et al. 1980) reported visual evoked activity recorded from the primary auditory cortex of the genetically deaf cat. Thus initially the primary cortices may turn out not to be primary at all. In adult normal cortex it became clear (Abeles 1982) that the same neurons could be involved in different functions at different times, and that even in primary cortex nonsensory activity might occur after the initial sensory-bound activity (e.g., Ramos et al. 1976). In the cortex, groupings of cells are postulated (Abeles 1982) as a chain of cells which appear as a dynamic entity and which may be turned on and off according to the spatio-temporal pattern of activity in the cortical network. This concept of dynamic formation and dissolution of functional groups is similar to that of the cell assemblies postulated by Hebb (1949). A neuron may be recruited into different chains at different times and thereby appears as having "plastic" properties. Such "plasticity" can arise in nervous systems with completely constant synaptic connectivities, as demonstrated in simulated networks (Eggermont 1988; Gerstein 1988; Palm et al. 1988).

10.2.2. Morphological Changes in Adult Brains

When synaptic plasticity is defined as the capacity of synapses to modify their function, to be replaced, and to increase or decrease in number when required (Cotman and Nieto-Sampedro 1984), it is evident that there need to be structural correlates to this plasticity. In mature animals changes are expected in response to changing hormonal levels, to environmental modification or to injury. We have already seen that sensory deprivation or rearing in enriched environments decreases, or increases, the number of dendritic spines on pyramidal cells. Lesions may induce sprouting, which in turn may lead to the formation of new synapses. Tsukahara (1982) linked this formation of new connections to adaptive behavior. He considered two types of synapses in the brain – which can be on the same cell: those that sprout and could represent environmental determinants of behavior, and those that do not sprout and are assumed to be the genetic determinants of behavior. This sprouting occurs either induced by lesioning or without lesioning, and it may occur in the neonatal as well as in the adult stage. The preparation used by Tsukahara is the red nucleus in the midbrain, a motor nucleus receiving input from the pyramidal cells of the cerebral cortex and from the cerebellum through the interpository nucleus; its axons go to the spinal cord. It could be established that the excitatory inputs consist of epsp's with two different time courses: a fast-rising epsp from the cerebellar input and a small, slowly rising one from the cerebral cortex input. On the basis of model calculations, it was made clear that the cerebellar input synapses at the soma and the cerebral input at the periphery of the dendritic tree. This was later confirmed by electron microscopy. Sprouting could be demonstrated in the kitten after sectioning of the cerebral or interpository nucleus connections took place over considerable physical distances and was somatotopically organized. In the adult cat, sprouting after lesioning was limited to the denervated field of the deafferented neuron and less prominent than in young animals.

Unilateral ablation of the entorhinal cortex (Cotman and Nieto-Sampedro 1984) resulted after 1 day in the loss of 90% of the synapses in the outer two-thirds of the dendritic tree of granule cells in the hippocampus on the side of the lesion. Thereafter a massive growth response resulted in synapse formation on sprouting fibers from the contralateral entorhinal cortex, the septum, and the CA_4 area of the hippocampus. It appeared that synapse replacement is complete after 60 days and that the synapses are functional. The sprouting observed was only over very short distances: the greatest hippocampal CA_4 fibers sprout only 20–30 μm into the dentate outer molecular layer.

Recently, another morphological substrate of plasticity has received renewed attention. Nottebohm and coworkers demonstrated that new neurons can be formed in the adult bird brain. Seasonal changes were found in the volume of HVc, a forebrain nucleus: its volume is large in spring and reduces in late summer to approximately half the size. Male canaries with large song repertoires tended to have large HVc nuclei. It was demonstrated that the newly formed neurons – in a period coinciding with the learning of the new song – are in fact recruited into functional circuits since they respond to auditory stimuli (Patton and Nottebohm 1984). In addition is could be shown that the newly formed neurons in HVc were interneurons (Patton et al. 1985). This formation of new neurons is not restricted to these "song-learning" nuclei but rather is a widespread phenomenon in the bird forebrain (Nottebohm 1984). This addition of interneurons seems to be a developmental means of adding sensitivity to experiential factors. Neurogenesis has also been demonstrated in the adult rat cortex (Kaplan and Hinds 1977) and may thus be a rather widespread phenomenon.

Besides sprouting, the formation of new synapses, and the genesis of new interneurons, changes in the synapses themselves have been demonstrated by Vrensen and Cardozo (1981). During an operant conditioning task in adult rabbits, requiring a visual pattern discrimination task, it was found that the frequency of complex synaptic grids increased, that the surface area and the number of dense projections became smaller, and that there was a significant increase in the length of the postsynaptic density. No changes were found in the size of the pre- and postsynaptic terminals or in the number of synaptic vesicles. This suggests that only a subset of the structural dimensions of nerve cells in altered as a result of conditioning.

10.3 Mechanisms and Models for Modifiable Synapses

10.3.1 General Considerations

The idea of modifiable synapses has a rather long history: Young (1938, cited in Sholl 1956) suggested that repeated excitation will lead to a permanent facilitation in the transmission between neurons. McCulloch and Pitts (1943) actually were among the first to phrase the properties of what later would be called the Hebb synapse (cf. Chap. 5). According to Changeux and Heidmann

(1987), the first mention of changes in strength or number of connections in neural networks can be found in Descartes' *Traité de l'homme* (1677). In this case we have to convert Descartes' concept of a hydraulic nervous system to the present electrochemical one. Hebb (1949) formulated this as "when an axon of cell A is near enough to excite cell B and repeatedly or persistently takes part in firing it, some growth process or metabolic change takes place in one or both cells such that A's efficiency, as one of the cells firing B, is increased". In consequence "any two cells or systems of cells that are repeatedly active at the same time will tend to become associated, so that the activity in one facilitates activity in the other ... such that a reverberation in the structure might be possible ... which might frequently last for periods of time as great as half a second or a second". This last idea of reverberating activity is basic to the cooperative models, e.g., based on the Ising model, that we have discussed in Chapter 5.

Various mechanisms have been advanced to account for modifiable synapses. For instance, Szilard (1964) proposed that proteins contained in presynaptic terminals diffuse into the postsynaptic membrane whenever the pre- and postsynaptic membrane are simultaneously depolarized during a learning situation. This protein dimerizes with a protein in the postsynaptic membrane through an immunological reaction, the new dimer then enhances transmitter efficacy of the synapse by binding substances that normally destroy this transmitter. An example would be the binding of acetylcholinesterase in a cholinergic pathway.

A slightly modified hypothesis from that proposed by Hebb was advanced by Stent (1973): "When the presynaptic axon of cell A repeatedly and persistently fails to excite/inhibit the postsynaptic cell B, while cell B is firing under the influence of other presynaptic neurons, metabolic changes take place in one or both cells such that A's efficiency, as one of the cells exciting/inhibiting B, is decreased". This model relates to the selective removal of unused synapses and may explain changes during the developmental stage of the brain. Whatever the formulation regarding the mechanism of synapse modifiability by learning, di Prisco (1984), in a review of Hebb synaptic plasticity, was forced to conclude that there was not much physiological evidence for the existence of such synapses. Then it would seem that modifiable synapses remain only a useful and necessary element in the work of brain theorists. However, plastic changes occurring in the time lapse of some minutes to half an hour can hardly be accounted for by sprouting or the formation of new cells. In a comprehensive survey of developmental plasticity, Singer (1985 a, b) comes to the conclusion that modifiability of existing synapses definitely has its place in this process. We will extract a few points from these papers. It is assumed that the activation of the postsynaptic neuron is essential for a long-term modification of excitatory transmission, and that such a modification depends on the correlation between the pre- and postsynaptic activity, i.e., a Hebb-like synapse. Recently, strong experimental indications for such a mechanism in the hippocampus and associated with long-term potentiation have been obtained by Gustafsson et al. (1987). By recalling that behavior is an important aspect in developmental as well as adult plasticity – visual signals only influence the development of cortical visual function when the animal plays attention to these signals *and* uses them for the control of behavior – Singer stated that the process mediating cortical plasticity has a threshold which is reached only when retinal signals are coincident with additional facilitatory input. This facilitatory input then should be provided by the behavior of the animal. The threshold process was supposed to operate on the postsynaptic side of those cortical neurons in which both retinal and nonretinal inputs converge. As a substrate for such a threshold process, the action of voltage-dependent Ca^{2+} channels was suggested.

Such channels were also mentioned to play a role in the selective stabilization theory for developing synapses (Changeux and Danchin 1976). In this theory, the genetic program directs the proper interaction between main categories of neurons; however, during development within a given category a significant redundancy of connectivity exists. The early activity of the circuits, spontaneous or evoked, would increase the specificity of the system by reducing that transient redundancy. Three synaptic states were distinguished: labile (L), stable (S), or regressed (R). A growth process can then be characterized by the emergence of L-states, which then may become stabilized (L→S) or regress (L→R). Both processes were originally thought to be irreversible, but in a later version of the theory (Changeux et al. 1984) the labile-stable transition was considered to be reversible. Transitions of the L, S, and R states refer to the localization of receptor proteins in the postsynaptic membrane. In the L-state, receptors can diffuse through the postsynaptic membrane, and become fixed in the S-state. If insufficient numbers of receptors agglomerate at a specific site, the nerve terminal at that location regresses. This synaptic stabilization takes place during critical periods. For the adult brain the postulated mechanism is slightly different (Changeux et al. 1984): it is assumed that the postsynaptic receptor exists in at least two states which are in irreversible equilibrium. One state, A, is susceptible to activation by agonists, the other, D, is desensitized or refractory to activation. The amplitude of the postsynaptic response is then proportional to the number of receptors in the A-state, and the A-to-D ratio simply determines the efficiency of the synapse. By postulating that the concentration of Ca^{2+} effects the A-to-D ratio, and that the timing relationship between the activities of the neighboring synapses may create changes in synaptic efficiency, the optional use in a learning task in possible. Considering that the A–D transition is the limiting factor and is slow, the duration of the long-lasting changes in synaptic efficiency might be several minutes. A reversible compartmental model like this one could well be treated according to a master equation formalism (cf. Chap. 4, Box E) along the lines outlined in an auditory adaptation model by Eggermont (1985a).

10.3.2 Formal Models for Modifiable Synapses

Brindley (1967, 1969) has given an exhaustive classification of modifiable synapses. According to Brindley, all modifiable synapses fall into three classes.

The best-known class is that where the simultaneous presence for pre- and postsynaptic activity increases the synaptic weight: a Hebb synapse. It is assumed that two inputs are initially needed to cause postsynaptic activity but after a sufficient number of pairings one input has acquired a synaptic weight strong enough to fire the postsynaptic cell alone. A second class consists of a modification of the Hebb synapse: one of the input neurons is unmodifiable and always capable of firing the cell alone. Marr (1969) has termed such synapses Brindley synapses. Several other modifications of the Hebb synapse have been introduced, most of them containing also a normalization condition (see, e.g., Chap. 5). In most theories about organization of the brain, Hebb-like or Brindley synapses seemed to be all that is needed to obtain consistent results. When these modifiable elements are incorporated between "logical neurons", i.e., neurons that do not show any delay in their action and do not depend on previous activity, simple networks can be devised that feature classical conditioning or operant conditioning. Still, the performance required of these networks is rather modest compared to what behavior paradigms require from experimental animals or human subjects.

The Hebb learning rule needs modification according to Sutton and Barto (1981), who gave the start to what they call a modern theory of adaptive networks. In a recent paper (Barto 1985), this theory is extended by introducing "goal-seeking" elements based on Klopf's (1982) "hedonistic neurons". Adaptive or learning networks are, in fact, signal-transforming systems, whereby the parameters of the system are slowly changing under the influence of the signal. In the following we will consider a set of input signals

$$\vec{x} = \{\xi_i\} \ , \quad i = 1, \ldots, n$$

and a set of output signals

$$\vec{z} = \{\zeta_j\} \ , \quad j = 1, \ldots, m$$

as well as a "key input", a teacher signal or reinforcement signal η_j for neuron j. We will discuss various adaptive elements and learning rules.

a) The adaptive linear element (Widrow 1962) with analog elements obeys:

$$\zeta_j = \sum_{i=1}^{n} w_{ij}\,\xi_i + w_0 \tag{10.1}$$

with learning rule:

$$\Delta w_{ij}(t) = \alpha(t)\{\eta_j(t) - \zeta_j(t)\}\xi_i(t) \tag{10.2}$$

and $0 < \alpha(t) < 1$, $\eta_j(t)$ in this case represents the desired output. If the input patterns that are applied in the learning phase are linearly independent, then $w_{ij}(t)$ converges to a unique limit. In order to avoid oscillations around that limit value, one slowly diminishes the learning ability of the elements:

$$\alpha(t) = \left\{ \sum_{p=1}^{t} (\overrightarrow{x(p)})^2 \right\}^{-1} . \tag{10.3}$$

Table 10.1. Truth tables for three types of learning

	Perceptron learning			Steinbuch learning			Willshaw learning	
	η_j			η_j			η_j	
	0	1		0	1		0	1
ξ_i 0	*	$w_{ij}\uparrow$	0	*	*	0	*	*
ξ_i 1	$w_{ij}\downarrow$	*	1	$w_{ij}\downarrow$	$w_{ij}\uparrow$	1	*	$w_{ij}\uparrow$

b) The perceptron (Rosenblatt 1962) is basically similar to the adaptive linear element, however, $\xi_i(t)$, $\eta_i(t)$, and ζ_i are now binary elements. The output $\zeta_j = 0$ if:

$$\sum_{i=1}^{n} w_{ij} < \delta_j$$

and $\zeta_j = 1$ if:

$$\sum_{i=1}^{n} w_{ij} > \delta_j$$

with δ_j the discrimination threshold of neuron j. The learning rule is:

$$\Delta w_{ij}(t) = \alpha\{\eta_j(t) - \zeta_j(t)\}\xi_i(t) \qquad (10.4)$$

c) The learning matrix (Steinbuch 1961; Willshaw et al. 1969) with $\xi_i(t)$ and $\eta_i(t)$ as binary signals but with $\zeta_i(t)$ as a continuous variable. The results of the various combinations of input and teacher signals can best the expressed in truth tables (Table 10.1, see also Palm 1982a).

There are various other learning rules, based on aspects of human learning. We will follow the overview given by Barto (1985). For this purpose we introduce the function $\chi_j(t)$ to indicate what the neuron usually does for a given input pattern:

$$\chi_j(t) = \sum_j w_{ij}(t)\xi_i(t) \ . \qquad (10.5)$$

d) The correlative or Hebb rule with:

$$\Delta w_{ij}(t) = \alpha(t)\xi_i(t)\zeta_j(t) \ . \qquad (10.6)$$

This rule adjusts the synaptic weight according to the correlation between pre- and postsynaptic signals. Adaptive elements using variants of this rule are suited to a type of learning that resembles classical conditioning. The Hebb rule can also be applied to supervised learning in which a key – or teacher – signal η_j is present. In that case the learning rule is:

$$\Delta w_{ij}(t) = \alpha(t)\xi_i(t)\eta_j(t) \ . \qquad (10.7)$$

A correlation matrix with such a learning rule results in a perfect recall of orthogonal (learned) patterns.

e) The Widrow-Hoff rule with:

$$\Delta w_{ij}(t) = \alpha(t)\{\eta_j(t) - \chi_j(t)\}\xi_i(t) \ . \tag{10.8}$$

$\eta_j(t)$ is the value of a specialized training signal giving the desired response of the unit. This rule is useful in supervised learning tasks. According to Kohonen (1984), this rule can be interpreted as an iterative algorithm for the solution of a set of linear equations. A solution of the learning rule exists if the input patterns used are linearly independent. If the input patterns are not linearly independent, then solutions exists for a monotonic decreasing $\alpha(t)$, e.g., $\alpha(t) = ct^{-1}$. In this case the $w_{ij}(t)$ converge under the rule of minimizing $\sum_j (\eta_j - \chi_j)^2$ for the whole net of adaptive elements. In this form the Widrow-Hoff rule is an iterative algorithm for forming the Moore-Penrose pseudo-inverse of a linear operator. It is noted that a Hebbian learning rule for perfect recall requires a learning set to be orthogonal, while the Widrow-Hoff rule only requires the elements of the learning set to be linearly independent. It appears that the Widrow-Hoff rule is identical to the Rescorla and Wagner (1972) model for classical conditioning. They stated that "organisms can only learn when events violate their expectations", in other words, learning occurs only when the information content of the message is high.

f) Uttley's rule, is a variant of the Hebb rule and has appropriately been termed also the anti-Hebb rule (Palm 1982a):

$$\Delta w_{ij}(t) = -\alpha(t)\xi_i(t)\eta j(t) \ .$$

Applying such a rule has a stabilizing influence on the adaptive network.

g) The Klopf rule:

$$\Delta w_{ij}(t) = \alpha(t)\{\varrho_j(t)\zeta_j(t) - E[\zeta_j(t)/\chi_j(t)]\}\xi_i(t) \ , \quad \varrho_j(t) = 1 \tag{10.9}$$

and for $\varrho_j(t) = -1$, $\alpha(t)$ is multiplied by a factor smaller than 1. The element using this rule has been introduced by Barto (1985) and called the "associative reward-penalty element". This element adjusts its synaptic weights on basis of an additional reinforcement signal $\varrho_i(t)$. The use of the expectation $E[\zeta_j(t)/\chi_j(t)]$ makes the learning process to converge properly; it causes the $w_{ij}(t)$ to change according to the discrepancy between an estimate of what the element should have done in the presence of stimulus pattern $\{\xi_i(t)\}$ and what it usually does in the presence of this stimulus pattern. A plot of the expectation has the form of a logistic function, very similar to Fig. 5.1. The philosophy behind the Klopf neuron or "hedonistic" neuron is that the single neuron is selfish, i.e., a goal-seeking system in its own right, its goal being to obtain excitation and to avoid inhibition (Klopf 1982).

See also Palm (1982a) for a survey of learning rules and some algebra with them.

10.3.3 Stability Considerations for Modifiable Elements

The Hebb learning rule introduces a positive feedback, the synaptic weights can only increase. Such a system, when linear, is inherently unstable and results in unlimited growth of excitatory synapses. The system can be kept stable through a nonlinear saturation or through normalization conditions. An example of a normalization is that proposed by von der Malsburg (1973), which keeps the total connectivity to one element constant. Edelman and Reeke (1982) give an example of saturation by making the learning constant α dependent on the current synaptic strength: α decreases when the correlation between input and output is positive and stays constant for zero or negative correlation. Some learning rules, like the Widrow-Hoff rule, have an inbuilt stability through an internal feedback. The same is found in the Klopf neuron. Another example of such a feedback-modifiable element is given by Oja (1982), which provides a stabilization of the Hebb-like properties proportional to the squared output of the neuron.

Other possibilities to stabilize the Hebb synapse are through inhibition. Easton and Gordon (1984) consider the cases with feedforward inhibition as well as with recurrent inhibition. It appears that the inherently unstable Hebb excitatory synapse, for a constant input signal, can only be stabilized by inhibitory feedback when also the inhibitory synapse is modifiable.

10.3.4 Associative Recall and Learning

Associative recall is defined (Kohonen 1984) as a mapping in which a group of input vectors (representing neural activity in a group of cells) is transformed into given output vectors. The input and output vectors can be combined in an association matrix:

$$M = \sum_k |y_k\rangle\langle x_k| \ . \tag{10.10}$$

Recall is then represented as the product of M with a key pattern $|x_i\rangle$:

$$|y_i\rangle = M|x_i\rangle = \sum_k |y_k\rangle\langle x_k|x_i\rangle \ , \tag{10.11}$$

indicating that the recalled pattern $|y_k\rangle$ is a linear combination of stored patterns $|y_i\rangle$ with relative intensities equal to $\langle x_k|x_i\rangle$. When the key patterns $|x_k\rangle$ are independent, recall is perfect. Kohonen suggests that processes such as lateral inhibition and the action of the Laplacian operator (taking the second spatial derivative, see Chap. 3) are effectively making independent representations of primary sensory input patterns. Pattern completion basically uses the same correlation mechanism. Whenever an incomplete key is presented, then that pattern is recalled for which the best correlation is obtained. Kohonen's procedures for associative learning use average firing rates and Euclidian distance qualifiers. Other approaches are possible, for instance Palm (1982a, b) uses sparse coding matrices and Hamming distances. Both methods lead to different memory capacity and error tolerance.

Pre- and postsynaptic spike trains can be considered as realizations of stochastic processes in synapses and nerve cell membranes. It is therefore advantageous to describe the learning process in terms of probabilities. For most cortical cells we may assume that there are 1000 to 10000 inputs and only one output. The probability that input x_i and output y are jointly occurring at times t and u is then described by:

$$P_{xiy}(t,u) = P_{xi}(t) P_y(u) + C_{xiy}(t,u) \ . \tag{10.12}$$

For all practical purposes we can regard them as independent in a functional sense, i.e., the correlation between x_i and Y is small. Kohonen (1984) assumes that the synaptic strength $w_i(t)$ changes in proportion to $P_{xiy}(t,u)$:

$$dE[w_i]/dt = \alpha P_{xiy}(t,u) = \alpha P_{xi}(t) P_y(u) \tag{10.13}$$

i.e., the expectation of the synaptic strength changes in proportion to the input and output activity. In my view, it would be better to let the small but consistent covariance $C_{xiy}(t,u)$ determine the changes in synaptic strength:

$$dE[w_i]/dt = \alpha C_{xiy}(t,u) \ . \tag{10.14}$$

This avoids unlimited growth of w_i as will be the case using the mechanism proposed in Eq. (10.13). The factor α is a learning parameter that describes the plasticity of the synaptic connection. It is assumed that α changes slowly with time, being largest in the early stages of ontogeny and development, that is in the critical period.

11 Learning – The Cerebellum

Voluntary movements require activity of the motor system, goal-directed movements also require the sensory system. The interplay between the sensory and motor systems is called sensorimotor coordination. For the initiation of limb movements one distinguishes four levels in the motor hierarchy, which are all acting on all lower levels (Fig. 11.1). The lowest level is the spinal cord, the next higher level is the brainstem, followed by the motor cortex and the premotor and association areas. When saccadic eye movements are included, one has to add the superior colliculus, which acts through the motor nuclei in the brainstem and in turn receives input, amongst others, from the premotor areas (Sparks 1986).

Besides the above-mentioned areas there are two other parts of the brain that are important for the overall motor function: the cerebellum and the basal

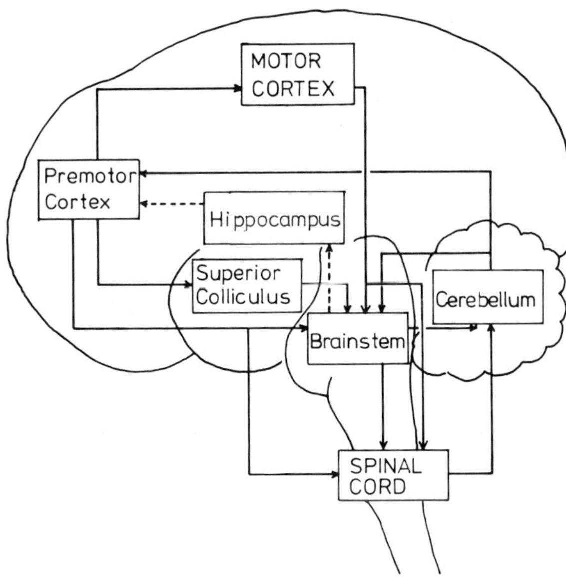

Fig. 11.1. Hierarchies in sensorimotor control. Direct control of activity at the spinal cord level is present from the brainstem, the premotor cortex and the motor cortex. Indirect control, through the brainstem, is provided by the superior colliculus, the motor cortex, and the cerebellum (which also acts through the premotor cortex). The hippocampus is also considered to play a role in the control of voluntary movements

Box L. Anatomy, Physiology and Pharmacology of the Cerebellum
The cerebellum (Fig. L.1) is a highly regular structure with a nearly crystalline order: one can distinguish a kind of unit cell that repeats itself every 10 μm or so. This unit is marked by the Purkinje cells of which there are 15×10^6 in the cerebellum of man. Purkinje cells are characterized by a fan-shaped ($250 \times 250 \times 6$ μm) dendritic tree, oriented along the longitudinal axis of the animal. The dendritic trees of neighboring Purkinje cells do not intermingle, and their axons go to the cerebellar nuclei and exert there an inhibitory action. There are two afferent systems to the Purkinje cells: the mossy fiber system and the climbing fiber system. The mossy fibers convey input from the vestibular system, the reticular system, the cerebral cortex via the pons, and from various other receptor organs in, e.g., muscles, joints, and skin. In humans the number of mossy fibers is estimated at 5×10^7, they exclusively synapse on 5×10^{10} granule cells, which are found in the granular layer just below the Purkinje cell layer. The innervation is such that one granule cell is connected on average with four mossy fibers. This implies an enormous divergence of the mossy fiber input: every mossy fiber contacts some 4×10^3 granule cells. Granule cells are excitatory, their axons are called parallel fibers and they run perpendicular to the Purkinje cell dendritic trees, with which they make synaptic contacts. It is estimated that one parallel fiber contacts about 300 Purkinje cells along its 3-mm course. Each Purkinje cell in turn is contacted by approximately 2×10^5 parallel fibers. In contrast, every Purkinje cell is connected by only one climbing fiber and every climbing fiber synapses with only one Purkinje cell. The climbing fibers originate from the inferior olive, which receives afferents from proprioceptive end-organs, as well as from the cerebral cortex and the red nucleus. Both the mossy fiber and the climbing fiber systems are excitatory.

This agglomeration of cells and axons constitutes the skeleton cerebellar cortex. In the frog this cerebellar cortex in addition contains a small type of inhibitory interneuron: the stellate cells, which receive parallel fiber contacts as well as climbing fiber collaterals. They are situated in the molecular layer with their dendritic trees in between the Purkinje cells, on which they have a strong inhibitory influence. In the alligator cerebellum yet another type of inhibitory interneurons is found in addition to the stellate cells. These are the Golgi cells which have a cylindrically structured dendritic tree which overlaps with the flat dendritic fields of about nine Purkinje cells. Every Golgi cell inhibits about 10^5 granule cells. The Golgi cells are innervated by the parallel fibers and by climbing fiber collaterals. In the bird cerebellum a third type of inhibitory interneuron is added, the basket cells. They also have rather flat dendritic trees which extend in the 2–4-μm spaces between the

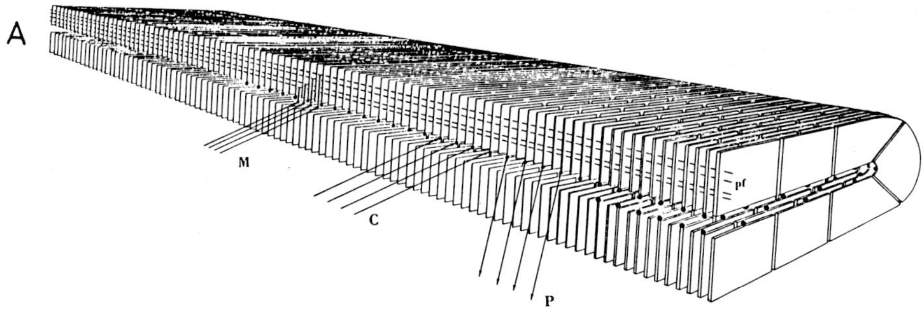

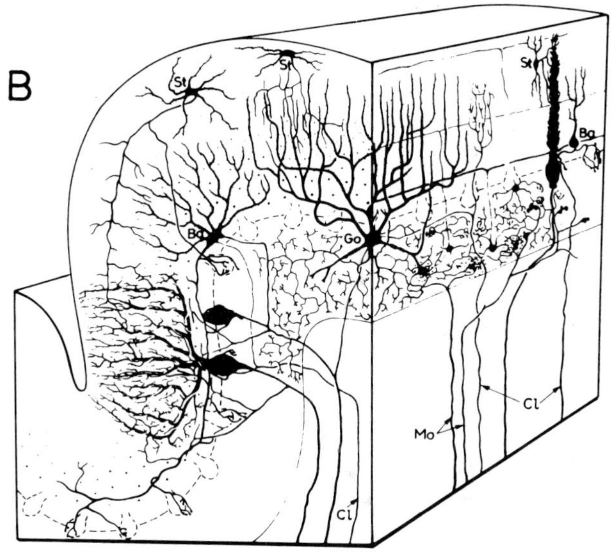

Fig. L.1 A–C. Views of the cerebellum. **A** The nearly perfect order of the Purkinje cells and extensive apical dendrites is shown as an arrangement of parallel plates; the parallel fibers run perpendicular to the dendritic sheets. (Braitenberg 1973). **B** More detailed view of a few dendritic plates showing Purkinje cells, interneurons, and the various fiber systems. (Eccles et al. 1967). **C** Wiring diagram of the cerebellum. (Eccles et al. 1967)

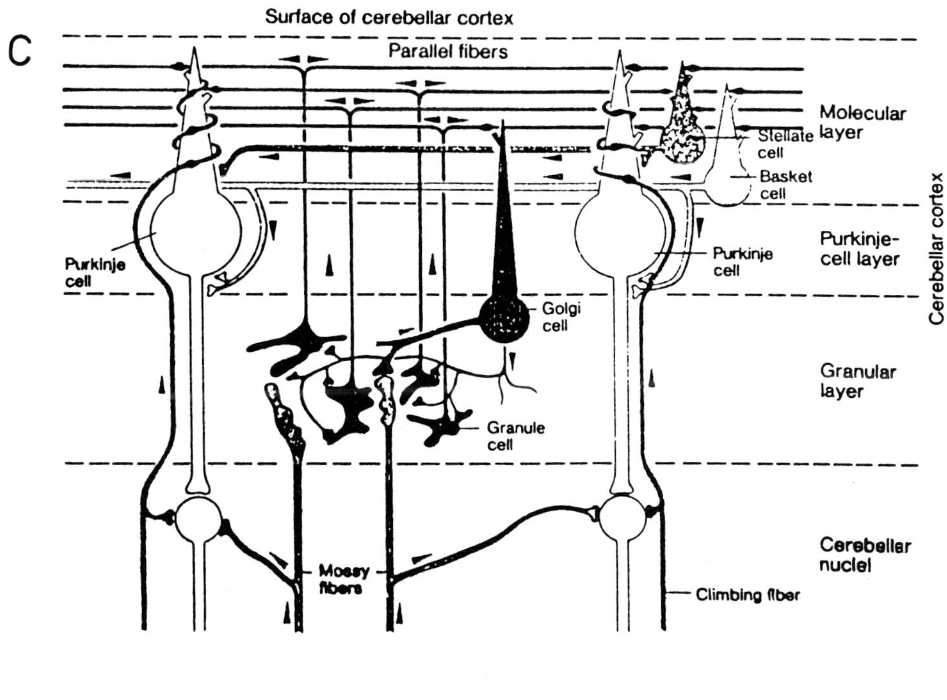

Box L (continued)

Purkinje cell dendritic trees. The number of basket cells is approximately the same as the number of Purkinje cells and their axons make strong inhibitory contacts with them excluding, however, the most nearby ones. Like the other interneurons, basket cells receive input from both the parallel fibers and climbing fibers. In mammals the cerebellum becomes still more elaborated but no further fundamental changes are observed.

Thus, with the exception of the granule cells, all cell types in the cerebellum have an inhibitory action. This is reflected in the neurotransmitters found in the cerebellum. GABA mediates the inhibitory action of the Purkinje, basket and Golgi cells. Taurine has been claimed as the mediator of stellate cell inhibition. In addition, motilin, a 22 aminoacid peptide, has been claimed to act for a certain group of Purkinje cells. The excitatory action of the climbing fibers is mediated by glutamate and for the mossy fibers transmitters may include acetylcholine, substance P, or somatostatin. Noradrenalin and serotonin may act as neuromodulators. For more extensive information consult Eccles et al. (1967), Szentagothai (1968), and Ito (1984a, b).

ganglia. While the knowledge about the basal ganglia effects is limited, a wealth of information is available about the motor coordination action of the cerebellum. The cerebellum is considered to adjust the actions of both the brainstem motor nuclei and the motor cortex by comparing signals from the proprioceptive system accompanying the intended motor response with the actual result of the motor action. This results in an update and a control of the movement in case it deviates from its intended course.

The hippocampus is also considered to play some role in sensorimotor integration. Bland (1986) assumes that the so-called "theta subsystems function in the capacity of providing voluntary motor systems with continually update feedback on their own performance relating to changing environmental conditions".

Systems such as the cerebellum and the hippocampus are in themselves as such not necessary to generate movements, they rather perform fine tuning. Complete destruction of the cerebellum, for instance, produces no sensory impairment and leaves muscle strength intact. However, lesions of the cerebellum interfere with the coordination of limb movements, affect eye movements, and impair balance. The cerebellum consists of three main regions: the vestibulocerebellum, which coordinates eye movements and controls balance, the spinocerebellum, which coordinates limb movements, and the cerebro-cerebellum, which is thought to be involved in the initiation, planning, and timing of movements. The cerebro-cerebellum is supposed to elaborate upon input from the "sensory" association cortex and report back to the premotor cortex. The spino-cerebellum takes part in two feedback loops: one where input from the sensory systems and the motor cortex is compared and returned to the motor cortex, the other where this information is used to adjust movements through the action of the brainstem motor nuclei.

In this chapter I will discuss theories of global cerebellar action that may broadly be classified into two categories: those that consider the cerebellum as a site for learning of motor tasks and those that interpret cerebellar action as a modifying and coordinating influence. In addition, evidence against or in favor of plasticity in the cerebellum will be presented.

11.1 The Cerebellum as a Timing Device

When the cerebellar cortex is compared to the cerebral cortex, one immediately is struck by the regularity of the former. The Purkinje cells are arranged in almost geometric order, all their dendritic branches oriented in parallel to the longitudinal axis of the animal and the beams of parallel fibers perpendicular to them. This suggested Braitenberg and Onesto (1960, cited in Braitenberg 1973) that the cerebellar cortex is important to preserve distances in the transverse direction in which the parallel fibers run. The hypothesis was put forward that the parallel fibers act as a delay line that is tapped by the Purkinje cells. The differences in arrival time between neighboring Purkinje cells of parallel fiber activity would be in the order of 0.1 ms, ergo, the cerebellum acts as a clock in the millisecond range. Such a clock is handy for critical timing of, e.g., rapid eye movements. Delays, furthermore are essential components of devices which perform auto- and cross-correlations of time sequences. Such operations are of fundamental importance in the detection of signals in noise and for the initiation of movements (Braitenberg 1973).

The timing hypothesis was investigated experimentally in the frog by Freeman (1969) through simultaneous recording from pairs or triplets of Purkinje cells lying along a common parallel fiber beam. First order cross interval histograms (see Chap. 7) were used to detect correlated firing in 43 of such Purkinje cell pairs during spontaneous activity, during electric stimulation of the cerebellar surface, and during natural stimulation (angular rotation). Only one out of the 43 pairs showed correlated firing during spontaneous activity, and in this particular place the recording sites were only 50 μm apart. During stimulation, either electrically or naturally, all pairs showed correlated firing. The time delay revealed by the peak in the cross-interval histogram was plotted against the distance of the cells recorded from: this resulted in a straight line, the slope of which coincided almost exactly with the conduction velocity of the parallel fibers ($v = 0.2\ \mathrm{ms}^{-1}$). Six pairs of Purkinje cells lying on different parallel fiber beams were studied as well; none of these pairs showed correlated activity under spontaneous or stimulated conditions.

Bell and Grimm (1969), also using two independent electrodes, recorded from 28 Purkinje cell pairs that were not necessarily in the same parallel fiber beam. Activity could consist of so-called "complex spikes" (climbing fiber responses) and "simple spikes" (parallel fiber responses). With one exception, the complex spikes of two Purkinje cells showed no correlation, while the simple spikes of neighboring Purkinje cells showed a peak in the cross-correlation histogram (CCH) around zero. Bell and Kawasaki (1972) studied the climbing fiber responses in 81 longitudinal pairs and 40 transverse pairs of Purkinje cells. The climbing fiber responses from a pair or triplet of Purkinje cells often occurred at about the same time, and a remarkable uniformity in the shapes of CCH's was found in different cell pairs, thereby contrasting the earlier results of Bell and Grimm. Broad symmetrical peaks (width = 50 ms) were in almost all cases centered at the point of zero delay. Symmetrical valleys were always present and secondary peaks at around $\tau = 130$ ms were common in both directions. The amplitude of the correlation peak decreased more or less exponentially in the longitudinal direction, reaching an asymptote at approximately 1.5 mm. In the transverse direction large correlations were still present at distances between cells in a pair of up to 2 mm. Fifteen pairs with simple spike activity were recorded as well. In three out of 15 pairs, the cross-correlation histograms had large symmetrical, bimodal peaks, in two out of 15 there were small monomodal peaks, and the remaining ten pairs had marginal or flat histograms. In five climbing fiber response pairs the influence of anesthesia was investigated: no difference could be found when compared with the awake state.

Spontaneous climbing fiber responses were recorded from pairs of Purkinje cells in immature rats (Dupont and Crepel 1979), and it was found that the firings were correlated from day 3 on. On postnatal day 7–9, the strength of the correlations was actually higher than in the adult. This may be related to the fact that in early development, climbing fibers contact more than one Purkinje cell.

Ebner and Bloedel (1981 b) studied cross-correlations in the simple spike firing rates of Purkinje cells in the decerebrated unanesthetized cat, both for spontaneous activity and under natural stimulus conditions. In 69 pairs, 17 triplets and 2 quadruplets natural proprioceptive stimuli (stretching the gastrocnemius-soleus muscle) were used, and in 18 pairs cutaneous stimuli (stimulation of the footpads) were used. Short latency, short duration correlations were found when both Purkinje cells were recorded on the same electrode; these peaks usually were not present during spontaneous activity. No short latency correlations were found if the sagittal distance was larger than 100 μm. In six pairs, cross correlation histograms with peaks of 1–2 ms width and lag times of around 2 ms were found when the cell pair was recorded on one electrode. Long duration correlations were found in 25 pairs under spontaneous activity. In 14 out of 25 pairs there was a positive correlation, in nine pairs a negative one, and in the remaining two pairs a combined negative-positive correlation was found. For long duration correlations no obvious relationship with the spatial orientation was noted. It appeared that this long duration (up to 500 ms) correlation was present in about one-quarter of the pairs studied, and could be changed by peripheral stimuli.

Application of conditional interspike interval and conditional cross-interval techniques (Tam et al. 1988) to the simple spike discharge of neighboring

Purkinje cells revealed an even more intricate temporal relationship than the standard cross-correlograms. Thus, the conclusion suggested on the basis of the morphology that the cerebellum is a timing device seems to be corroborated by electrophysiological data. If cells in the cerebellar nuclei could perform a kind of correlation analysis on the spike trains from Purkinje cells, then the timing information present in the cerebellum could be used.

This idea is also put forward by Pellionisz and Llinas (1979), who envision Purkinje cells connected by parallel fibers, and assume that Purkinje cells can take the kth-order time derivative ($k = 0, 1, 2, 3, \ldots$) from the parallel fiber activity. Some Purkinje cells take the first-order derivative, some the second, and so on. By summing the activity of a group of Purkinje cells converging on a single nuclear cell, this cell can perform a look-ahead operation. Using the time derivatives up to order K one can expand $M(t+\delta)$ in a Taylor series of $M(t)$. Thus on the basis of $M(t)$, the present summed firing rate, and derivative operations up to the Kth-order, one can predict the activity δ time units ahead, $M(t+\delta)$. Actual recordings from nuclear cells appear to suggest that components representing zero-, first- and second-order derivatives are indeed present in its output. The functional advantage of such a look-ahead module is obvious, especially in regard of goal-oriented, ballistic movements. When the spontaneous rate of the cells is taken as a reference and if that activity is sufficiently high, negative as well as positive differentials can be computed.

11.2 The Cerebellum as a Perceptron

11.2.1 Theoretical Considerations

The classical perceptron (Rosenblatt 1962) consists of an array of sensory cells that convey excitatory and/or inhibitory signals to an array of an equal number of association cells; each association cell typically receives convergent input from a large number of sensory cells. Association cells project with adjustable weights onto response cells. All the cells are logical switching devices: they fire when a threshold level is surpassed and remain silent when the input stays below that level. The synaptic weights between the association cells and the response cells can be adjusted, allowing the perceptron some form of learning. Note that in the modern terminology (e.g., Rummelhart and McCleland 1987) the perceptron is a “connectionist model”. The learning in the perceptron is of the classical conditioning type (see Chap. 10) and appeared to be inadequate in the classical form to learn complex patterns. This could be remedied to some extent by replacing the association cells with an “expansion encoder”. The idea is that if N sensory cells connect divergently and convergently with 100^N association cells, then there are more than 100^N possible ways to map the 2^N possible input patterns onto these 100^N cells (Marr 1969). The restriction is, however, that in order to curtail the enormous redundancy in the output pattern, only a very small fraction of the input fibers is active at the same time (sparse coding). Such a recoder can function as an associative memory (Palm 1982a; Kohonen 1984).

This model, including the idea of the expansion coder, has been used independently as a basis for the cerebellum theories of Marr (1969) and Albus (1971). The recording takes place at the mossy-fiber granule-cell junctions. Since there are approximately 1000 times as many granule cells as mossy fibers and each cell is on average innervated by four to five mossy fibers, the recoding is extensive. The output of the granule cells is sampled by the Golgi cells which each feedback onto 10^5 granule cells. Since the Golgi cells are also innervated by mossy fibers, this constitutes in addition a feedforward inhibition of the granule cells. Albus (1971) considers this an automatic gain control, resulting in and responsible for the fairly constant firing level of the parallel fibers. In the analogy with the perceptron the output of the granule, association, cells to the Purkinje cells should be modifiable. Marr (1969) assumed that those parallel-fiber Purkinje-cell synapses are facilitated by the conjunction of presynaptic parallel fiber and postsynaptic (induced by the climbing fibers) activity and that no other synapses in the cerebellum are modifiable. Albus is not so restrictive and assumes that all parallel fiber synapses with Purkinje cells, basket cells, and stellate cells are adjustable since the stellate as well as the basket cells are also contacted by climbing fiber collaterals. Because after climbing fiber activity there is a 15–30 ms pause (the "inactivation response") in activity of the Purkinje cells, it was assumed that the climbing fiber activity weakens any parallel fiber synapse that tends to cause firing during the inactivation response. Furthermore, synaptic weakening stabilizes the system. Thus the granule cell recoder output excites through modifiable parallel fiber synapses the Purkinje cells as well as the stellate and basket cells, which in turn inhibit the Purkinje cells. By weakening the synapses to the interneurons a kind of disinhibition occurs, leading to a stabilization of the Purkinje cell firing rates.

In terms of classical conditioning learning, the climbing fiber activity is the unconditioned stimulus, and the parallel fiber (or mossy fiber) activity is the conditioning stimulus. One could then conceive that the climbing fiber input carries control patterns to be stored in the cerebellar network. Later, the activity of the mossy fiber system alone would then be sufficient to activate this control pattern. Marr supposed that during learning the cerebrum organizes the movement, which causes the appropriate olivary cells to fire in a particular sequence. In turn the Purkinje cells learn, from the climbing fibers, the context (provided by the mossy fibers) within which their corresponding elementary movements are required. The next time that such a context occurs, a specific mossy fiber activity pattern stimulates the Purkinje cell, which in turn evokes the elemental movement. Thus the cerebellum in this view is a learning device which comes into action every time a prelearned context is presented.

Fujita (1982) has reformulated this perceptron model in terms suitable for linear systems analysis. He postulates that a mossy fiber-granule cell-Golgi cell input network functions as a phase lead-lag comparator. The output of this network is gathered through variable synaptic connections to form a Purkinje cell output. The error signal that is to be minimized is provided through the climbing fibers and regulates the strength of the parallel fiber Purkinje cell synapses. The result is that the output of the Purkinje cell converges into the desired response.

In these models inputs and outputs are expressed as firing rates, and the underlying assumption is that rate coding is the way information is processed. Sejnowski (1977) reformulated the Marr-Albus theory in the framework of covariances in firing activity. The learning algorithm then becomes: "changes in synaptic strength are proportional to the covariance of the neural activity in climbing fibers and mossy fibers". For constant average rates such covariances can be positive or negative, leading to strengthening or weakening of the parallel fiber Purkinje cell synapses. In this version of the model the temporal modulation of the co-firing activity in the afferent inputs to the cerebellum becomes important.

11.2.2 Experimental Evidence

Experimental evidence for the effect of climbing fiber activity on the subsequent output of Purkinje cells is provided, for example, by Ebner and Bloedel (1981 a), who studied the effect of climbing fiber activation on the simple spike response in 206 spontaneously firing Purkinje cells in decerebrated unanesthetized cats. By using the autocorrelation of the spike patterns they could distinguish two basic types of spontaneous discharge patterns. A type I autocorrelation histogram, with a very prominent positive peak occurring after a short lag (≈ 10 ms) and decaying over a time course of 100–200 ms to baseline level, was found in 121 cells. A type II autocorrelation histogram, featuring a somewhat longer dead time with the peak occurring around 15 ms, decaying within 10 ms, and occasionally showing secondary peaks, showed up in 85 cells. By using a time-dependent autocorrelation function (see Chap. 7, definition analogous to the time-dependent cross-correlation function) they could show that electrical stimulation of the inferior olive modified the simple spike pattern. The long duration type I autocorrelation appeared to be concentrated in the simple spike activity immediately following the response to climbing fiber input. Thus the dependency among the type I simple spike activity is at least partly the result of the action of a climbing fiber on cerebellar cortical neurons. Electrical stimulation of the inferior olive was in fact capable of changing a type II pattern into a type I pattern. In addition, Ebner and Bloedel (1981 c) showed that a natural peripheral stimulus, preceded from 7–20 ms by climbing fiber activity, could induce an augmented simple spike response. For many cells a response to flexion of the forepaw was only observed if this flexion occurred at short intervals following a climbing fiber input.

Ito et al. (1982) showed that conjunctive stimulation of the vestibular nerve at $20\,\mathrm{s}^{-1}$ and of the inferior olive at $4\,\mathrm{s}^{-1}$ for 25 s effectively depressed the excitation of Purkinje cells by subsequent vestibular nerve stimulation without any sign of an associated change in spontaneous activity. This depression disappeared after 10 min, but was followed by a second slow depression phase lasting about 1 h. This effect was not observed when Purkinje cells were excited from the contralateral vestibular nerve. This suggests that the depression occurs specifically at the parallel fiber Purkinje cell synapses involved in conjunctive stimulation. An iontophoretic application of glutamate to the Purkinje cells in combination with $4\,s^{-1}$ stimulation of the inferior olive depressed subsequent glutamate sensitivity also for about 10 min, again followed by a second slow depression lasting 1 h. This indicates that the putative neurotransmitter for the parallel fiber Purkinje cell synapses is involved in the depression.

These and other studies by Ito (1984a, b) have consistently revealed that the parallel fiber Purkinje cell synapses undergo a long-term change in transmission efficacy subsequent to conjunctive stimulation of the parallel fibers and the climbing fibers. The change observed has been a depression, no evidence has been obtained for facilitation. Attempts to reproduce this long-term depression in vitro have so far been unsuccessful, however. Ito proposes that this long-term depression is the physiological counterpart of the synaptic weakening postulated theoretically by Albus (1971). Paulin (1985), however, argues against evidence that this supports the Marr-Albus theory. It appears that the timing of the effective conjunction of the parallel and climbing fiber stimulation is not critical. Parallel fiber spikes anywhere between 20 ms prior to 150 ms following climbing fiber stimulation are all equally effective in introducing the long-term depression. Ito excludes a classical Hebbian mechanism for this depression since repetitive stimulation of parallel fibers does not introduce such depression, despite the fact that these stimuli produce postsynaptic firing of the Purkinje cells. A more likely mechanism was suggested in that inflow of Ca^{2+} ions into Purkinje cell dendrites as introduced by climbing fiber stimulation interferes in some way with the glutamate receptors in the parallel fiber Purkinje cell synapse (cf. Singer 1985a; and Chap. 10). Ca^{2+} inflow may induce a desensitization of receptor molecules or it may cause a change in the shape of the dendritic spines. An analogy to the long-term potentiation in the hippocampus, attributed to the action of NMDA on postsynaptic receptors, suggests itself. Ito points to the coincidence that the three different types of synaptic plasticity so far discovered in the cerebellar cortex, the hippocampus, and the red nucleus are all glutaminergic. A plastic process that may be related to cerebellar activity is the vestibular compensation which occurs after unilateral labyrinth or vestibular nerve destruction. The recovery process (at least in humans) does not consist of a repair of damaged circuits: there is no nerve regeneration or labyrinth repair. Thus compensation occurs through a reorganization of the remaining neural structures and the formation of new circuits through deafferentation, hypersensitivity, and collateral sprouting. Flohr (1983) proposed that the establishment of new synapses results from a controlled goal-seeking process induced by the systems error. The consequences are first that the compensation process does not directly result from the lesion itself but is induced and maintained by the systems error: an increase in the g-forces acting on the otoliths during the initial phase of compensation greatly accelerates the compensation process. The second consequence is that there must be a concrete mechanism that is activated by this performance-assessing subsystem which induces the desired neural changes. This, Flohr suggests, could be the role of an ACTH-like neuropeptide system and/or a monoaminergic system since amphetamine significantly accelerates the compensation process while its antagonists (chlorpromazine, epinephrine, dopamine) inhibit compensation. This allows a modification of the Hebb learning rule: pre- and postsynaptic activity, in connection with a neuromodulator signal, induces synaptic modification.

11.3 The Cerebellum as a Space-Time Metric

Pellionisz and Llinas (1979, 1980, 1982, 1985) developed a geometric theory for the cerebellum from the premise that its major function is regulating, by setting motor commands into the context supplied by the rest of the nervous system. Thus in this view the cerebellum is not assumed to *learn* motor acts as proposed in the perceptron theories, but *performs* sensorimotor acts. Localization of prey and orientation toward it (in order to intercept it) requires an internal correspondence of space and timing information regarding the target with the events occurring in the external world. A sensorimotor theory therefore should formulate this internal relation between space and time information. In classical physics, space and time information can be treated separately because of the very large velocity of light that conveys the timing. In the central nervous system this is impossible because (a) simultaneous external events cannot be represented simultaneously in the central nervous system as a result of conduction velocity differences in the neurons, and (b) the simultaneous onset of firing in a group of neurons does not result in a simultaneity of external actions (motor actions). Both reasons arise because conduction velocities along nerve tracts are of the same order or magnitude as the velocity of the observed target.

It is proposed in this theory that the brain functions as a tensor (see Box M). The arguments are that first the activity of an array of neurons can be expressed as a vector of, say, the firing rates of the individual neurons. Secondly, because movements can be represented by vectorial expressions they have magnitude and direction in the chosen frame of reference. Thirdly, vector expressions related to the same physical entity but in different frames of reference are tensorially related. Physical entities that can be defined independently of frames of reference will be called invariants. Examples of invariants are, e.g., targets in the external world.

The coordination of fast and accurate muscle movements (e.g., saccadic eye movements) requires some "space-timing". To know the way in which this space-time information is processed in the CNS is essential to understand how saccades are made. In the CNS a certain neuronal activity pattern $(f_1, f_2, \ldots, f_n)$ of a set of n neurons is assumed to represent internally ("efference copy") the same invariant that is represented externally in Cartesian coordinates and separate time coordinates. The Newtonian reference $M(x, y, z, t)$ and the CNS reference $F(f_1, f_2, \ldots, f_n)$ are completely similar

Box M. Vectors and Tensors in Oblique Frames of Reference

In an oblique two-dimensional frame of reference a point $U(\alpha,\beta)$ can be represented in basically two ways. One way consists of projecting perpendicularly to the coordinate axis α and β, resulting in the components U_α and U_β which are called covariant components. The other way is through the parallelogram method resulting in components U^α and U^β (see. Fig. M.1). The components transform into each other through a metric tensor **g**:

$$U^i = g^{ij}\,U_j \tag{M.1}$$

$$U_i = g_{ij}\,U^j \tag{M.2}$$

g^{ij} represents the components of the contravariant metric tensor, g_{ij} those of the covariant metric tensor. The matrix of tensor **g** for the example given in Fig. M.1 reads:

$$g = \begin{pmatrix} \frac{1}{\sin^2\omega} & \frac{-\cos\omega}{\sin^2\omega} \\ \frac{-\cos\omega}{\sin^2\omega} & \frac{1}{\sin^2\omega} \end{pmatrix} . \tag{M.3}$$

This is an orthogonal matrix with determinant equal to 1. The matrix of the contravariant metric tensor is equal to the inverse of the covariant metric tensor and vice versa. For orthogonal matrices $g^{-1} = g^T$, thus the matrices for the covariant and contravariant metric tensor are equal in this case. Note that for $\omega = 90°$ the matrix reduces to its main diagonal form with 1's as entries. In more general cases than the one illustrated in Fig. M.1 the matrices corresponding to the tensors are defined by:

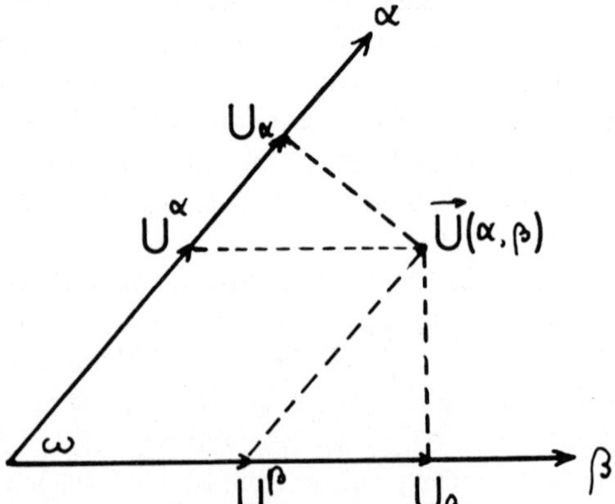

Fig. M.1. Covariant and contravariant vector components in an oblique frame of reference. (After Pellionisz and Llinas 1979)

Box M (continued)

$$g^{ij} = \sum_a \frac{\delta x^a}{\delta x_l} \frac{\delta x^a}{\delta x_j} \tag{M.4}$$

and

$$g_{ij} = \sum_a \frac{\delta x_l}{\delta x^a} \frac{\delta x_j}{\delta x^a} \tag{M.5}$$

Any metric tensor has a set of principal axis (orthogonal) on which the covariant and contravariant components are equal. These are found from the eigenvalue equation:

$$g_{ij}\, U^j = \lambda\, U^i \ . \tag{M.6}$$

The principal axes are formed by the eigenvectors of matrix g_{ij} (see Box G). Therefore a metric tensor can geometrically be represented by a hyperellipsoid. For a general reference see Kron (1959).

representations of the invariant, and are therefore related through a tensor. A profound difference in these two representations is that the Cartesian reference frame is an orthogonal one and the CNS one is not. In a nonorthogonal reference frame two types of vector representations exist, respectively with *covariant* and *contravariant* components (see Box M). In Cartesian coordinates, vectors are both covariant and contravariant in their components. Covariant components are formed by orthogonal projections of the vector onto the frame of reference. They are therefore independent but do not add up to reconstruct the invariant by vector addition. Contravariant components are formed by projections parallel to other coordinate axes (by a parallelogram construction), thus they are not independent but they add up vectorially to the invariant. It is postulated in the theory of Pellionisz and Llinas that sensory systems, extracting independent features, use a covariant representation of an invariant. In this way the receptive field of a neuron is a subspace of a higher dimensional input hypespace. Motor systems use a contravariant representation since they have to reconstruct the invariant. The relation between the covariant sensory representation and the contravariant motor representation of an invariant can be represented independently of any frame of reference through a *metric tensor* in an n-dimensional hyperspace. Sensorimotor coordination then is the transformation of the sensory system representation.

So far time has not been introduced; the sensorimotor transformation is considered instantaneous. In order to account for the nonuniformity in conduction delays, the U_j must represent position not only in space but also in time. In other words, the metric tensor must be a space-time metric tensor. If the space metric in itself is correct, but delays are occurring in the system, then a spatial distortion of the resulting motor patterns will still result. To compen-

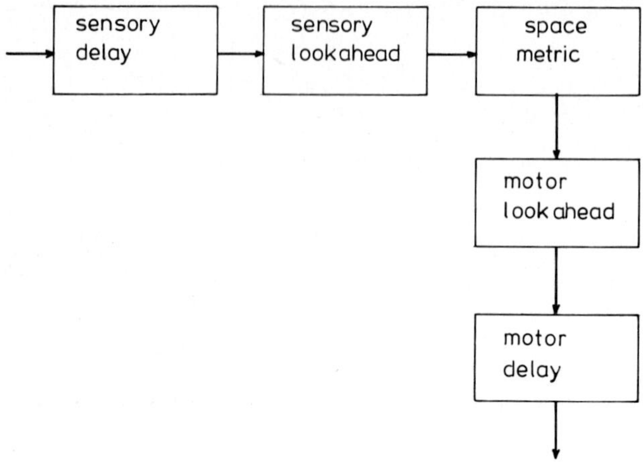

Fig. 11.2. Space-time transformations in the sensorimotor system. In order to compensate for the slow speed of information transfer in the nervous system, prediction units (*look ahead*) are needed, both in the sensory and the motor parts of the brain

sate for this delay one needs a look-ahead module to adjust the covariant components. As we have seen in Section 1 of this chapter, such a procedure can be based upon a group of Purkinje cells or other neural units that are able to take time derivatives. Thus after compensation of delays in the sensory system, the space metric can perform the contravariant transformation. In fact, another look-ahead module is needed to compensate for delays occurring in the motor system. This procedure of space-time transformations with the timing as an explicity operation is illustrated in Fig. 11.2. Basically this approach treats space and time as separable entities. Since there is no a priori reason to separate the timing adjustment from the space metric, the latter might incorporate the timing factor as well. Let $\Omega = (t-\tau_1, t-\tau_2, \ldots, t-\tau_n)$ be the covariant timing vector, and $T = (t+\tau^1, t+\tau^2, \ldots, t+\tau^m)$ be the contravariant timing vector. Then:

$$U_{i\Omega}\, g^{ij\Omega T} = U^{jT} \tag{11.1}$$

and the metric is now a tensor of rank four. The matrix element $g^{ij\Omega T}$ thus comprises a delay $\tau^{ij} = \tau_i + \tau^j$. By this construction the vectorial expressions for the sensory coordinates $U^{i(t)}$ and motor coordinates $U_{j(t)}$ refer to the same space-time point.

In most sensorimotor coordination systems, e.g., the cerebellum of the cat or frog, the look-ahead module solution requires far less neurons to form the interconnection matrix than the space-time metric solution. For low-dimensional systems, such as the vestibuloocular reflex, the implementation of a space-time metric in the available neural network is feasible.

The metric tensor theory discussed so far implies that there always is an adequate internal representation of the external world invariants and that a motor action automatically follows perception. This may be true for reflexes, but in

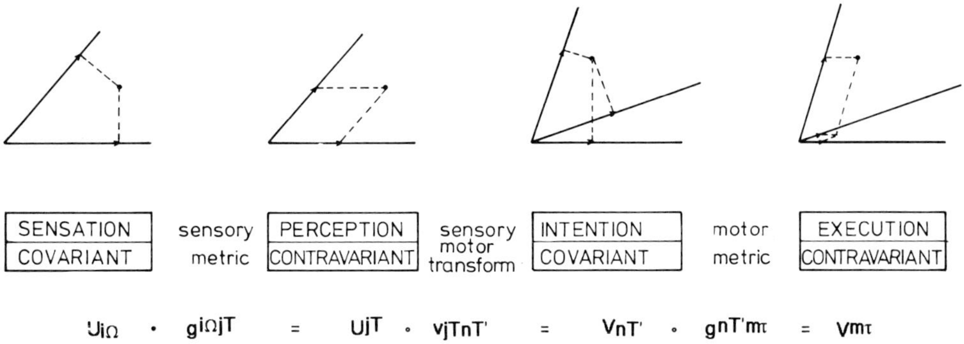

Fig. 11.3. Geometric and tensor representation of the sequence: sensation, perception, intention, and execution

other cases a decision has to be made before a motor action is executed. Thus the sensory perception requires the building-up of a sensory metric which – independently of the motor system – transforms the covariant sensory vector components into contravariant sensory vector components in a sensory, internal, coordinate frame. On the basis of this contravariant representation a decision is made whether it is feasible to execute the motor action, i.e., create a motor intention vector. The motor coordination action will then be a geometric transformation of this covariant intention vector into an executable contravariant displacement vector. Between the sensory part and the motor part of the system we still require a sensorimotor transformation (see Fig. 11.3). This transformation can be interpreted as a covariant embedding operation, basically consisting of correlating the contravariant perception vector and the covariant proprioception vector.

In lower vertebrates (e.g., frogs) the sensory space-time metric could be residing entirely in the optic tectum, where the presence of different sensory modalities can be interpreted in geometrical terms as representing different axes in a unified sensory hyperspace. In higher vertebrates the sensory cortex will take part in this sensory metric. The motor system contravariant metric is thought to be implemented by the cerebellar connectivity matrix. A substrate of the covariant embedding is thought to be residing in the telencephalon in case of a frog or in the sensorimotor cortex in higher vertebrates. By a correlation procedure between the covariant target reception vector and the contravariant target perception vector a distance is calculated. Whenever the distance is outside the reach of the animal an inhibitory signal is issued and no action is taken.

Both the sensory metric and the motor metric emerge during development. The theory proposes that the motor metric derives from a process in which the eigenvectors of a covariant metric of the motor system – which relates a contravariant motor action to a covariant proprioception – can be established by reverberation of neural activity, resulting from the reentry of covariant proprioception information to motor effects as if they were contravariant. This reverberation will rapidly reach a steady state of covariant-contravariant eigenvectors. Such a covariant-contravariant pair, where the coefficients differ by only a constant factor (the eigenvalue) is an eigenvector in a given frame of reference. Once the first eigenvector (with the largest eigenvalue) is established it can be filtered out and the procedure can start all over again in order to find the second eigenvector and so on. The theory assumes that the covariant proprioception information enters the cerbellum via mossy fibers that sent collaterals to the cerebellar nuclei. This covariant "imprinting" process requires a convergence of the identical eigenvector on both the row and column elements of the connectivity matrix in order to establish their product. In addition, ionic mechanisms that are capable of triggering chemical changes at the postsynaptic element may modify the intrinsic responsiveness in a manner proportional to this product. The mossy-fiber parallel-fiber Purkinje-cell system is in principle capable of generating such a modification, since both these pathways carry the eigenvectors at the required steady state of the oscillation. In this respect the learning or imprinting process does not differ in its substrate from the perceptron theories. However, since the mossy fiber and climbing fiber systems also converge on the nuclear cells, these could as well be a candidate. Moreover, this would fit in with the knowledge that ablation of the cerebellar cortex does not impair vestibular adaptation.

The cerebellum is capable of learning – or of modification – in the adult phase as well. Such learning requires modification of existing geometries. When only small adjustments are needed, the process of updating the eigenvalues by iteration can update the metric. This assumes that the eigenvectors have been precisely established and need no change. If the eigenvectors themselves are improper, the system needs to regress and redo the entire oscillatory reverberative process. Such a situation might, e.g., occur after vestibular nerve section. One then observes in the early phase a violent shaking and oscillatory behavior that could be interpreted as the required reverberative process.

The theory also suggests that this organizational process starts with the establishment of a covariant motor metric (a proprioceptive calibration). Thereafter the generation of the sensory metric can take place: a sensory function cannot evolve without an active motor participation.

This theory thus appears to be rather complete and can be used to describe and predict (relatively simple) motor actions. It could form the basis for building a robot featuring an intrinsic sensorimotor integration. Whether the cerebellum really acts as a space-time metric will not be resolved before an experimental method is established to read out the connectivity matrix and the way it is formed. Optical recording could form one of the methods needed in this kind of endeavor (see Chap. 6).

12 Learning – The Hippocampus

Memory and learning are intricately connected. When something is learned it is more or less permanently stored in the brain, from which it can later on be recalled. The movements that are coordinated by the cerebellum are memorized during the learning or rehearsal processes described in the previous chapter. Therefore in the cerebellum the memory resides in the connectivity matrix or, alternatively, in the metric tensor describing these connectivities. The memory of the cerebellum is of the *reflexive* or *procedural* type, i.e., it has an automatic quality and neither its information nor its recollection is dependent on awareness, consciousness or other cognitive processes. Usually the learning process for reflexive memory requires many trials, and this type of memory relates to perceptual and motor skills or the learning of procedures and rules. It is therefore also called *procedural* memory.

In this chapter I will discuss forms of *declarative* memory. Declarative memory depends on conscious reflection for acquisition and recollection. Although such memory can be established in single-trial learning, I will argue that the mechanisms by which declarative memory is established and by which items from such memory are recollected must be similar to those we described for the cerebellum.

Memory is distributed throughout the entire brain. It has not been possible to pinpoint any region of the central nervous system in which a memory is stored, nor to isolate single neurons or groups of neurons, the activity of which corresponds to cognitive functions. Presumably storage and retrieval of memory requires only relatively small changes in the activities of the individual elements of large neural groups. Of main interest will be the larger changes in the simultaneous or near-simultaneous activities of many neurons: this exemplifies our idea that memory is a correlative property of the brain.

By establishing relations between the severity of memory loss and the extent of damage to the brain, structures important for declarative memory appear to be the dorsomedial thalamus and the hippocampus (Winocur 1984). The thalamus seems to be involved in the early sensory stages of processing, ultimately leading to the formation of memory traces. The memory trace – to be visualized as a chain of strongly connected neuronal elements, e.g., a synfire chain (Abeles 1982) – can occupy large parts of the brain and probably has a sufficient redundancy in actual implementation as to be quite resistant to local damage. Since the memory is distributed throughhout the brain it cannot be addressed in a "labeled-line" fashion, e.g., by specifying the neurons that take part in it. It is much more likely to be "content-addressable" or "associative"(Anderson 1970, 1972; Marr 1971; Cooper 1981; Kohonen 1984). In this view, memory resides less in neurons than in the relationships between neurons. Following Hebb (1949), a given memory trace is formed in an assembly of neurons by way of their interconnections. A given cell or group of cells may be part of many such networks and therefore of many memory traces. Hinton and Anderson (1981) consider such a memory as a large matrix (first proposed by Steinbuch 1961), just like the connectivity matrix of the cerebellum but not so well localized in a circumscribed part of the brain (see also Palm 1982a). Items to be stored are described as vectors, which upon presentation, alter the matrix in some way. If a certain item is repeatedly presented the matrix will change in a predictable and persistent fashion (cf. Chap. 5.3). Hence a vector (input-item) can be recognized by the matrix as either "old" or "new". Old items, represented by firm connections, are recalled: new items can be associated with already stored old items, if they are sufficiently similar, or else they may lead to new changes.

This theoretical concept will form the basis for our discussion of declarative memory, and in this chapter I will review theories along these lines and outline the possible neural substrates in the brain. However, it is wise to take the attitude that at present there is no practical method at hand for experimentally studying the functional associations within and between the cell groups of, e.g., associative cortex (Fuster 1984). Such a study should be done by monitoring cellular activity in several domains simultaneously, either by optical methods or with multielectrode arrays (see Chap. 6). The analysis of the vast amounts of data associated with the changing activity patterns as a function of (repeated) stimulus presentation and the behavior of the animal is at the moment technically becoming possible (see Chap. 7), but conceptually still in its infancy. Without a skilful application of these recording and analysis techniques, it will be very difficult to establish a neural substrate of memory, which, as we will see, is individualized as well. At present, most attempts are based on lesion studies, field potential analysis, and single-unit studies. Since, however, theory predicts that it will be the correlations between neural activities that matter, one should make these correlations the subject of study both experimentally and theoretically. This, evidently, means going beyond single unit and towards multiunit experiments, however immature they may be at present (see discussions in von Seelen et al. 1988).

12.1 Types of Memory

We have already encountered a subdivision of memory on the basis of what is stored. Reflexive memory stores rules, "know how"; declarative memory stores data, "know what". There is some evidence that reflexive and declarative memories may involve different neuronal circuits: reflexive or procedural memory may require "climbing fiber" – mediated context input, while declarative memory cells can do without. Some structures identified as having reflexive memory properties are the amygdala and the medial dentate and lateral interpositus nucleus of the cerebellum. Declarative memory is attributed to the temporal lobe, associative structures of the limbic system, and the diencephalon. Eccles (1983) considers the neocortex as the location of the data banks that subserve declarative memory. In his view the hippocampal input to the neocortex (see Box N for details) is part of a reentrant loop that periodically (1 – 3 years) refreshes the "data banks". Failure of this replay leads to forgetting.

Memory can also be subdivided according to the time period over which it can store data. I will follow the conceptual scheme used by Ervin and Anders (1970) to give a brief survey. The first memory, with retention times less than one second, is called the *sensory memory*: it can briefly hold sensory data for scanning, attention, and further processing. The contents of the sensory memory are lost by fading and by transfer to more permanent memory. Visual memory seems to be able to store for less than 250 ms. Before the stored information is transferred to more permanent memory it is recoded. This recoding can take two forms: verbal or nonverbal. Verbally coded information is transferred to a so-called *primary memory* which has a rather small storage capacity. When new information arrives, old information disappears, i.e., the memory is of the "first in, first out" type. Primary memory has rapid recollection (of the order of 100 ms) and retains its information – at least in the absence of interference by new data – for a few seconds. Simply rehearsing the facts can hold them in memory in the absence of new input. Ervin and Anders (1970) claim that nonverbally coded information may be directly transferred to the *secondary memory*, as is eventually the contents of primary memory. The secondary memory has a large storage capacity, is more permanent – it can store information for minutes to years – but has a longer access time (0.7 – 1.4 s). Information is forgotten by interference with other data. Finally, the contents of secondary memory may be transferred to *tertiary memory* which has a nearly lifetime storage durability, a very large storage capacity, and a much faster access than secondary memory.

A somewhat distinct and overlapping subdivision of memory into "short-term" and "long-term" memory exists as well. *Short-term memory* definitely comprises primary memory and part of secondary memory. Usually the division between short- and long-term memory is set at a retention time of 5 min. *Long-term memory* then covers most of secondary and all of tertiary memory. One can interpret this as short-term memory being comprised of mainly verbally coded memory. Arbib et al. (1976) consider short-term memory to contain information that represents our model of the current state of the environment. Only discrepancies between short-term memory and a new state are signaled and used to update short-term memory. This view is comparable to the learning model of Rescorla and Wagner (1972), discussed in the contect of the Widrow-Hoff rule in Chapter 10, which states that only those facts are learned, which violate our expectations.

Long-term memory has been subdivided in various ways, more or less analogous to reflexive versus declarative memory. Olton (1983) refers to *working* memory and *episodic* memory to describe memory that stores information about the context and current items of rules. *Reference* memory and *semantic* memory store general rules and procedures, information about logical relationships, but ignore the context. All these subdivisions were useful in classifying clinical amnesia syndromes; for our purpose, however, not all these distinctions will be needed.

12.2 Brain Structures Involved in Memory

The discoveries of the brain structures that are important for memory nearly all came from lesion experiments. Thalamic damage, for instance, appeared to affect the early stages of processing, those where information is registered and encoded. It is likely therefore that the thalamus and especially the dorsomedial region thereof is involved in sensory and possibly also primary memory. Memory loss is reported in animals where the hippocampus was electrically stimulated and probably damaged. When the interval between learning and testing was short, no effect was noted; however, for long intervals between learning and testing the effect was obvious. This indicates that animals with damaged hippocampus have problems with long-term memory. It is assumed (e.g., Seifert 1983; Squire and Butters 1984) that the medial temporal lobe contributes to the process whereby important informational elements and their relationships are identified and stored. The medial temporal lobe is crucial in specifying the elements which are to be incorporated into the existing body of knowledge as part of a consolidation process (Winocur 1984). Its function is also supposed (Halgren 1984) to contribute to the re-experience of events that have occurred in the recent past which then results in a subjective experience of familiarity. It is assumed that the direct neural substrate for the initial experience and the reexperience of events resides in the associative cortex.

The hippocampal formation appears to be critically important in laying down a memory trace. Lesions in the hippocampus produce a devastating anterograde amnesia (inability to store new information) and a less marked retrograde amnesia that may extend to 1–3 years prior to the lesion. This suggests (Routtenberg 1984) that the hippocampus itself stores information for a considerable period of time, well in the range of the long-term memory. For this impairment to be observable, however, a near-total bilateral removal of the hippocampus is required. Complete unilateral removal has little or no effect on memory formation or retention. This elucidates the importance of the commissural pathway between both hippocampi.

It became clear that patients with extensive bilateral hippocampal damage still can store new rules; their procedural memory seems not to be affected by the lesion. Such patients cannot store new data, indicating that declarative memory is impaired (see, e.g., Squire 1986). Routtenberg (1984) considers the hippocampus as a cognitive system that mediates selective processes determining which subset of memories, out of many potentially relevant memories stored in the cortex, will guide the behavior at any given time. Braitenberg and Schuz (1983) consider the hippocampus as associational cortex with a re-en-

Box N. The Hippocampus

The hippocampus (Fig. N.1) has a very striking design and is a contiguous part of the medial temporal lobe and the cerebral cortex. The hippocampus consists of a series of long strip-like cortical layers, adjoining and folding along its longitudinal axis. Cells from one cortical strip send their axons to cross the border to the next strip. The main input to the hippocampus is from the entorhinal cortex, which in turn receives efferents from the frontal-, parietal-, occipital-, and temporal association

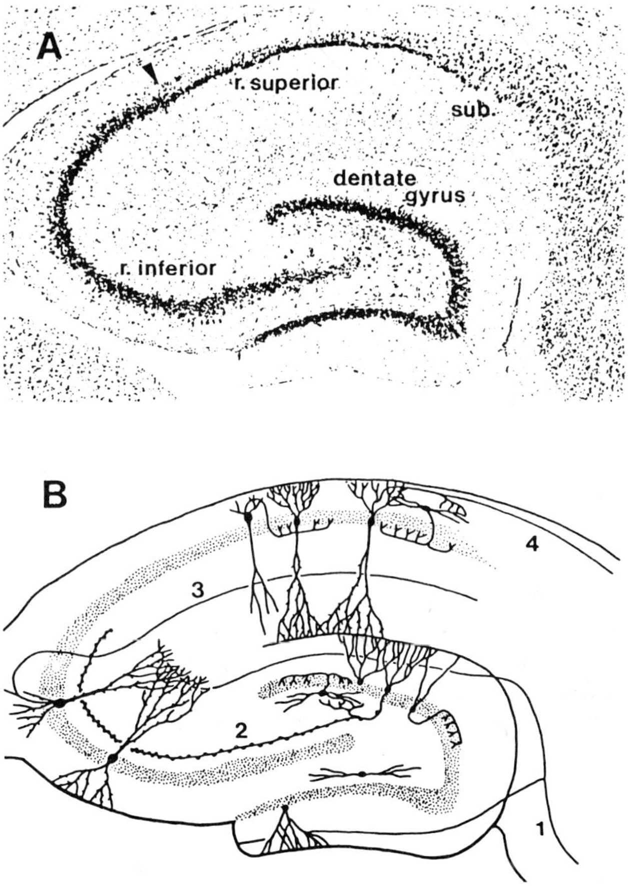

Fig. N.1 A, B. Schematic view (**B**) of the hippocampal cortex (**A**). Input arrives through the perforanth path (*1*), goes via the granule cells and mossy fibers (*2*) to the pyramidal cells in the CA_3 region, from which the Schaffer collaterals (*3*) contact the apical dendrites of the pyramidal cells in the CA_1 region, whose output connects through the commissural fibers (*4*) with the contralateral hippocampus, and the subicular cortex. (Rose 1983)

Box N (continued)

cortex, by way of a fiber system, called the perforant path. It enters the dentate area in parallel lines, in such a way that successive regions of the entorhinal cortex are connected with the granule cells in the dentate region in a series of the parallel strip-like layers. The other main afferent source to the dentate area are the commissural fibers from the contralateral hippocampus. From the dentate granule cells a major afferent fiber system, the mossy fibers, connects with the pyramidal cells in the CA_3 region. Mossy fibers contact many pyramidal cells by forming en passant excitatory synapses on the dendritic spines on the proximal parts of the pyramidal cell apical dendrites. It appears that about 100 synapses (out of 10000) on a single pyramidal cell have to be activated simultaneoulsy to discharge the cell. The axons of CA_3 pyramids, the Schaffer collaterals, are the main afferent input to the CA_1 pyramidal cells, again together with commissural fibers. Both input systems to the CA_1 form excitatory synapses on the apical dendrites, the commissural input also synapses on the basal dendrites. The CA_1 pyramids send axons to the subicular cortex, which projects directly to the limbic cortex, which in turn projects to the association cortex. Thus within the hippocampus there is a profound unidirectionality from entorhinal cortex to subicular cortex. Braitenberg and Schuz (1983) link this unidirectionality to the preservation of the spatiotemporal structure of the input. However, the connections of the hippocampal regions with the association cortex offer the possibility of multiple feedback loops between the association cortex and the hippocampus. This excitatory loop is kept under control by a poweful inhibition that can be recorded extracellularly from the granule cells and pyramidal cells after each excitation. Note also the low threshold of "epileptic" activity of hippocampal neurons. The inhibitory inputs are probably originating from the basket cells which are activated by collaterals of the pyramidal cells in CA_3 and CA_1. This thus forms a recurrent pathway, since the basket cells synapse on the soma and the initial axon or on straight regions of primary and secondary dendrites of the pyramidal cells. It appears that the epsp's as well as epsp's and ipsp's sum perfectly linear for all synapses. The synapses in the excitatory loop appear to be highly plastic (for more detail see Andersen (1975, 1983) and Halgren (1984)).

A number of transmitters have been identified and described for the hippocampus (Haas 1983). A powerful pathway associated with theta-activity and originating from the septum uses acetylcholine (ACh) as a transmitter (e.g., Bland 1986). Histamine (HA) has been attributed to supramammillary input, noradrenaline to input from the locus coeruleus, dopamine to input from the brainstem, and serotonin (5-HT) to input from the median raphe. Besides these, glutamate acts as an excitatory

Box N (continued)

transmitter in the Schaffer-commissural fibers (Wieraszko 1983) and GABA as an inhibitory transmitter decreasing the excitability of the granule cells, as does 5-HT whose action is quite similar (Crunelli et al. 1983). A variety of neuropeptides modulates this transmitter action (Gahwiler 1983).

trant unidirectionality (see Box N) and speculate that it deals with the sequential aspects of association. In addition the also present, very rich, diffuse set of nonoriented input fibers from the brainstem, locus coeruleus, etc. suggests a less specific arousal, attentional type of input that can modify this association process. The unidirectionality of the entorhinal-hippocampal loop could serve a reverberation in an assembly. The hippocampus then could be involved in the igniting of cell assemblies. A somewhat comparable view is that the transition from large irregular activity to theta-activity which accompanies self-generated motion through space (Bland 1986) could serve as a "start recording" signal (Doty 1984). In other words: as the animal starts to move, it opens a memory register for the sequence of stimuli to be encountered. This could well be seen as the ignition of a particular assembly of neurons. In this way the hippocampal, or more generally, the medial temporal lobe's function, could be to contribute to the re-experience of events that have occurred in the recent past.

Of course, more brain areas are involved in memory. It has already been mentioned that the association cortex probably functions as a data bank, but we will leave neocortical aspects of memory to the next chapter. For a comprehensive general review of memory see the recent book by Squire (1987).

12.3 A Mechanism of Memory Formation in the Hippocampus

Memory formation can occur through the process of changing synaptic strength following stimulation of the perforant pathway. McNaughton (1983) considers four processes that form a mechanism for such changes. Three of these mechanisms are well known to occur at neuromuscular synapses as well as at other peripheral synapses. These mechanisms are all related to the number of transmitter quanta released from the presynaptic terminal. The underlying processes all seem to decay in a more or less exponential way and thus each process can be characterized by one time constant. The processes are respectively *facilitation* with a decay time constant of the order of 0.1 s, *augmentation* with time constant around 5 s, and *potentiation* with a time constant ranging from 20–240 second depending on the conditions. One can distinguish short-term and long-term potentiation based on the value of the decay time constant.

The fourth process, called *enhancement*, does not involve an increase in the transmitter release probability, occurs instantaneously, and seems to be critically dependent on the correlated activity of a fairly large number of afferent fibers innervating a granule or a pyramidal cell. Such enhancement has been demonstrated in the medial and lateral perforant pathways of the fascia dentata, in the mossy fiber terminations of the granule cell axons in CA_3, and in the Schaffer collateral and commissural terminals of CA_3 pyramidal cells in CA_1.

Routtenberg (1984) distinguishes two components to the process of synaptic plasticity: an *intrinsic mechanism*, which forms (via biochemical and structural changes in synaptic strength) the circuit underlying the memory, and an *extrinsic mechanism*, which can enhance or prevent this change in synaptic strength. It has been demonstrated that electrical stimulation of the granule cells in the dentate gyrus has a quite strong and specific erasure effect on remembering a particular location in the environment. Opioid peptides seem to be potential transmitters in this system, since naloxone, an opioid peptide-blocking agent, completely blocks the erasure effect of the electrical stimulation. Naloxone applied alone, i.e., without electrical stimulation, has a facilitating effect and does not block memory. This suggests that the mossy fiber system is not intrinsic to the memory formation process. The fact that blocking the granule cell system actually enhances memory storage suggests that the granule cell system is an extrinsic system that in its normal condition dampens memory formation. Routtenberg (1984) then assumes that there is a direct entorhinal input to the CA_3 pyramids which must be considered intrinsic. Another intrinsic system is most likely the commissural input from the contralateral CA_3. Thus the data and facts in the environment coded by neocortical systems enter the CA_3 pyramidal cells via the entorhinal and commissural pathways. Figure 12.1 gives an overview of the intrinsic and extrinsic connections to the CA_3 cells. As extrinsic inputs, besides the granule cell axons, are considered the cholinergic input from the septum which is driven by activation from the brainstem and the limbic system. The impact of this input may be directed at specifying the significance of the data. Lesions in the medial septum have only a minor influence on already formed memories.

The mossy fiber opioid input overlaps with the cholinergic input on the apical dendrites. There is evidence that opioids can modulate, specifically decrease, the release of ACh from presynaptic terminals and it is assumed (Routtenberg 1984) that this also happens at the CA_3 pyramidal cells.

The assumption that CA_3 pyramidal cells actually are the sites of memory storage or are crucial in the storage process is corroborated by the fact that kainic acid selectively destroys CA_3 cells and impairs memory. The result is a profound anterograde amnesia and a less profound retrograde loss.

During memory formation it is proposed that inputs from both the entorhinal and the cholinergic system are required in a time-locked way. A high correlation of septal and entorhinal activity will be needed. It appears that both entorhinal input alone and septal input alone are not able to generate CA_3 spikes. Since septal activation leads to theta activity, memory formation will be enhanced during such activity.

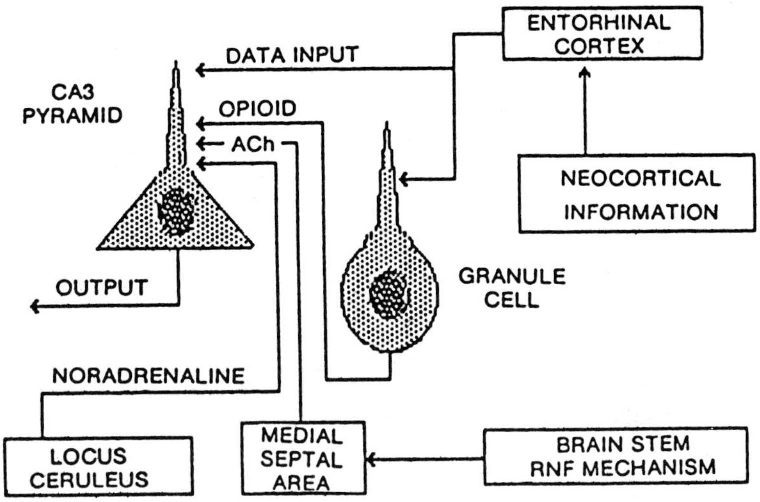

Fig. 12.1. Schematic hippocampal circuit. Intrinsic input comprises that from the entorhinal cortex and the input from the contralateral CA_3, indicated as *data input*. The extrinsic inputs are from the locus ceruleus, the medial septal area, and the granule cells. (Routtenberg 1984)

The role of the granule cells in this memory model of Routtenberg is that of modulating, or "editing" the contents of the memory, since opioids control the release of ACh onto the CA_3 pyramidal cells. In more detail, Routtenberg suggests that the consequence of cholinergic input would be to change the K^+ conductance of the CA_3 pyramidal cell. Inputs from the entorhinal cortex would be influenced by this alteration of CA_3 pyramidal cell membrane properties because the depolarization would be enhanced. This co-occurrence of presynaptic entorhinal activity and alterations in postsynaptic membrane conductance may be an essential step in a bidirectional synaptic communication, leading to morphological alteration of those synapses where increased presynaptic data input and elevated postsynaptic activity occur. This is essential the Hebb model for synaptic plasticity.

Essential in Routtenberg's model is the split in the perforant pathway. One path goes directly to the CA_3 pyramids to form an intrinsic input, the other path arrives via the denate granule cells and constitutes a delayed modulation system. In his various reviews of hippocampal circuitry Andersen (1975, 1983) does not mention the direct entorhinal-CA_3 pathway, nor does Eccles (1983). Both stress the unidirectionality of the path via the dentate gyrus, together with the stabilizing inhibitory action of the basket cells. Chronister and White (1975) review evidence for an entorhinal-CA_3 pathway.

Recent investigations into the pharmacological mechanism of long-term potentiation (LTP) in the hippocampus have shown that LTP is dependent upon simultaneous pre- and postsynaptic activity and, in addition, dependent upon the activation of a specific class of excitatory amino acid receptor: N-methyl-D-aspartate (NMDA). This NMDA activation depends on glutamate binding concurrent with membrane depolarization (Cotman et al. 1988). Upon activation, the NMDA receptor channel allows Ca^{2+} influx into the cell, thus resulting in a long-lasting enhancement of other excitatory input to the cell (LTP).

12.4 Formal Models of Memory Formation

Models of brain function should reflect that the brain is best adapted to interacting, complex, distributed, and parallel operations. Models of memory should in addition reflect that memory is associative. A memory model that is conceptually simple and takes into account most of these assumptions has been described by Anderson (1970, 1972) and modified and elaborated upon by Anderson et al. (1977) and Cooper (1981). The model assumptions are basically that first of all *neurons are independent* during memory tasks, and secondly that *memory traces* corresponding to large patterns of individual neuron activity, *interact* at the synaptic level such that different traces overlap in storage.

The first assumption is based on the observation that neurons usually are correlated in their firing properties when the brain is in a sleep stage or under anesthesia, but that in the waking stage this long-term correlation disappears (e.g., Noda and Adey 1970, and Chap. 8). Whether this means that neurons are also independent on a much shorter time scale is not clear; however, this assumption makes the treatment of the model much simpler. When the activity of individual nerve cells is characterized by their average or instantaneous firing rates, large-scale activity of a neural net can be represented by a vector with independent components (cf. Chap. 5).

Memory basically can be divided into recognition and association. I will discuss a minimal association model first. Assume two large groups of neurons, **a** and **b**, each of N neurons. Every neuron in **a** projects to every neuron in **b** through excitatory connections (this can be relaxed so that only a fraction of neurons projects to **b** in such a way that the convergence is equal to the divergence, cf. Anderson 1972). Neurons are assumed to be simple linear integrators (as in the models discussed in Chap. 4), which seems to apply to, e.g., hippocampal neurons. Neuron a_j in **a** is connected to neuron b_i in **b** with synaptic strength w_{ij} such that the activity of neuron b_i in **b** is given by:

$$g(i) = \sum_j w_{ij} f(j) \ . \tag{12.1}$$

where $f(j)$ is the firing rate of neuron a_j and $g(i)$ is the firing rate of neuron b_i. The set of all $f(j)$ is represented by a vector $|f\rangle$, the set of all $g(i)$ by $|g\rangle$ [for the use of the bra(c)ket convention of representing vectors see Box H]. Let w_{ij} initially be zero for all i, j. The essential learning assumption is that in order to associate $|f\rangle$ in **a** with $|g\rangle$ in **b**, w_{ij} should change according to the correlation of $|f\rangle$ and $|g\rangle$. If W is the matrix $\{w_{ij}\}$ then:

$$W = |g\rangle \langle f| \tag{12.2}$$

and the activity in **b** is given by:

$$W\,|f\rangle = |g\rangle \langle f|f\rangle = |g\rangle \ . \tag{12.3}$$

Let us assume that there are K sets of associations $|f_1\rangle$, $|g_1\rangle;\ldots;|f_k\rangle, |g_k\rangle$ each generating a matrix W_k. Under the assumption of linear nonsaturating synapses the connectivity matrix W that results after applying these K sets of associations is given by:

$$W = \sum_k W_k \ . \tag{12.4}$$

When input pattern $|f_j\rangle$ is applied, the output pattern in **b** will be:

$$\begin{aligned} W\,|f_k\rangle &= \sum_k W_k\,|f_j\rangle \\ &= W_j\,|f_j\rangle + \sum_{k \neq j} W_k\,|f_j\rangle \\ &= |g_j\rangle + \sum_{k \neq j} |g_k\rangle \langle f_k|f_j\rangle \ . \end{aligned} \tag{12.5}$$

Thus the result is the correct association $|g_j\rangle$ together with a noise term. If the input patterns are selected orthogonal, i.e., $\langle f_i|f_j\rangle = 0$ for all i = j, then the noise term becomes zero and perfect association results. Orthogonality of input patterns requires that the neurons in **a** behave independently. If the neurons are not independent, then the association will not be perfect. When now an arbitrary input signal $|f_x\rangle$ is applied, the memory mechanism can be instructed, e.g., through the incorporation of a nonlinear mechanism, to produce that association for which $\langle f_x|g_j\rangle$ is maximal.

Recognition of previously learned inputs basically is self-association (Anderson 1972) because $|f\rangle = W|f\rangle$, i.e., it makes use of the diagonal elements of W only. Thus generally the connectivity matrix W can be described as:

$$W = R + A \ , \tag{12.6}$$

i.e., as a diagonal part, R, serving recognition and an off-diagonal part, A, serving association (Cooper 1981). During (repeated) presentation of an association pair, element w_{ij} of W is changed according to the Hebb rule:

$$dw_{ij} = c\,g(i)\,f(j) = \sum_k w_{ik}\,f(k)\,f(j) \ , \tag{12.7}$$

and therefore the change in the connectivity depends on the already memorized input events as well as om the current one. The set $\{w_{ij}\}$ thus can be considered to form an internal representation of events in the outside world. The representation in a certain brain of some event in the outside world is not necessarily the same as that in another brain simply because of the differences between the respective learning histories. Therefore if we want different animals to agree on their description of the external world, we have to require that their representations of the world are the same, that is their learning histories should be the same.

12.5 A Model for the Role of the Hippocampus in Memory

A more explicit model for memory and the role of the hippocampus in it has been described on the assumption (Marr 1971) that the simple structure of the hippocampal formation probably results in a simple function and as a consequence should exhibit a simple memory. Since the pyramidal cells in the hippocampus lack climbing fibers, Marr postulates that this memory only associates and does not classify or place events in a context. Thus, unlike the cerebellum, presenting the context alone would not generate a memory or recall a memory. We have seen that contextual input is provided to the hippocampus by the entorhinal pathway; it may, however, not be used to induce recall. A simple memory is basically an associative memory as described above. Marr assumes that it consists of two neuronal groups **a** and **b**, that are fully interconnected with each other. The feedback loop from **b** to **a** is assumed to involve the hippocampus. Set **a** is considered a group of codon formation cells with input synapses of the Brindley type (Chap. 9) and output to set **b** through Hebb synapses. Both sets **a** and **b** contain inhibitory interneurons for threshold setting. The role of the hippocampus is to perform a diagnostic operation: by a more or less random process a set of classes unique to each input is set up. Each class is represented by a separate cell, although a given cell may represent more than one class. In this way a representation is formed by the set of cells which represents a given input. This representation in the hippocampus serves the storage of events, the recognition of subevents, and the completion (by association) of events. Marr assumes that the memory in the hippocampus elaborates on an input of 1.25×10^6 fibers divided into 25 blocks of 50000 fibers each. The blocks are conceived to represent the parallel strips of cells in the hippocampal cortex. An input event is assumed to activate approximately 2500 fibers – 100 in each block – thus forming the representation of the event. Cells in the next level of operation (e.g., CA_3) are assumed to have on the order of 10^4 afferent collateral synapses. After 50000 events have been learned, recovery of a specific event E, according to this scheme, will already have a high probability of recall from stimulation of about 30 of the fibers per block that were active during the learning of event E. A prerequisite of this theory is that (Marr 1971) the pyramidal cells in the hippocampus should have modifiable synapses of the Brindley type, and that the excitatory fibers from the entorhinal cortex should terminate on the pyramidal cells of CA_1 and CA_3 through such synapses. Collateral synapses between CA-pyramids should be modifiable in the Hebb sense. In addition, the "memory" cells should have modifiable inhibitory afferent synapses, serving at least two functions. The first is to control the synaptic modification conditions on the memory cell dendrites during the learning of events, and secondly to control the cell's threshold during the recall of previously learned events.

Thus while the architecture of the hippocampus and its connectivity suggest a very useful scheme for its function, I cannot refrain from quoting Arbib et al. (1976) on this matter: “In a system as far from the periphery as the hippocampus, the significance of its input and output firing pattern is virtually unknown..., as a result the anatomy has proved something of a Rorschach test for neural modelers”.

13 Learning – The Neocortex

Perception in its broadest sense comprises a generalization from particular stimuli to classes of stimuli, including stimuli that were never encountered before. This generalization has among others to deal with stimuli that are translated and rotated over the receptor surface, are scaled in size and intensity, or are distorted. Categorization of classes of stimuli as being similar is an important aspect of perception. This chapter once more will discuss learning and memory; however, by now we know that we cannot really discuss these items in isolation for certain parts of the brain. As we have seen, the act of perception requires additional nervous tissue than that available in the neocortex and its sensory afferents. Our memory capacities and those of the primates in general have nevertheless greatly benefited from the increase in size of the neocortex. Numerous theories about brain function concentrate on the action of the neocortex, notably the visual cortex (e.g., Rose and Dobson 1985). Before discussing some of these (brain) theories in more detail, I will present an overview of certain aspects of neocortical organization: the way it differs from or resembles that of the cerebellum and hippocampus (Chaps. 11 and 12) and the question whether any difference in structure can be related to a difference in function. Subsequently, certain developmental aspects will be summarized, since I take the point of view that during development, primitive perceptual memory units are formed and laid down in the global and local organization of the cortex. Models for such self-organizing capacity of the cortex are abundant and available for some time. However, only after von der Malsburg's (1973) demonstration that very simple organizational rules can result in a realistic columnar organization, has some progress been made. Aspects of genetic and experiential determinants in this process emerge automatically to culminate in instruction-selection theories of brain function. At the moment computer simulations form the only "tests" of such theories, but some of their predictions can, and have already, been tested using electrophysiological techniques (Edelman 1987).

13.1 The Neocortex: Pinnacle or Way-Station?

Braitenberg's (1974) interpretation of the connectivity pattern among cortical cells resulted in the statement that "the cortex is to a large extent a thinking machine working on its own output"; later (Braitenberg 1982) extended with "and works in a reflexive mode". This can be interpreted in a way that the cortex can function on its own, that is, it elaborates predominantly upon its own generated data. The cortex is, of course, our most impressive memory store containing all records of relevant, or once relevant, features of the environment and is capable of storing new records. Above all, the cortex is a necessary instrument in our conscious actions. Add to this that the neocortex is the most obvious and most developed part of the human brain and one arrives easily at the conclusion that the neocortex is the pinnacle of the brain, the center that coordinates and creates all our actions. On the other hand, one can consider the neocortical network as only a link in the stream of activities from the thalamus and other afferent structures to the effectors. In other words: the neocortex is nothing but a way-station between the thalamus and the higher brainstem (Penfield 1975). In this view the neocortex is not the final place of recognition, decision making, and planning, although none of these functions is possible without the neocortical link. Creutzfeldt (1978) considers the neocortex only as part of the circuits in the forebrain in which the thalamocortical connections are the essential ones. Penfield puts it this way: "the brainstem and spinal cord provide us with inborn reflexes, the diencephalon (or higher brainstem) is the target area for pain, but information about the other senses continues in a detour to the gray matter in the cerebral cortex from which it returns directly to the target nucleus of cells in the higher brainstem." The visual cortex then is a way-station between the eye and higher brainstem.

Large parts of cerebral cortex can be removed without abolishing consciousness, but injury or interference with the function of the higher brainstem, even in small areas, abolished consciousness completely. Penfield also reflects upon the mind-brain problem in this sense: "there is no place in the cerebral cortex where electrical stimulation will cause a patient to believe or to decide (a mind action) . . . the mind seems to act independently of the brain in the same sense that a programmer acts independently of his computer." However much the mind may depend upon the action of the cortex for certain purposes, it seems that there is a brain mechanism that switches off the power that energizes the mind each time it falls asleep and switches it on when it awakes. The mind can only communicate through its brain mechanisms that control the effectors. In other words, the (higher) brainstem controls the perceptual mechanism. Harth and Unnikrishnam (1985) consider this the consequence of the many reciprocal pathways between neocortex and thalamus. The reticular formation, through a direct action on the neocortex, functions as a center for arousal, controlling the transition between sleep and wakefulness. In addition, the reticular formation, together with the neocortex, controls neural transmission in the dorsal thalamus by inhibiting the perigeniculate and intrageniculate interneurons, thereby disinhibiting the relay cells in the thalamus (see also Crick 1984). This reticular input to the dorsal thalamus may cause a feature-specific modification of the sensory input to the neocortex. Perception in this view (Harth and Unnikrishnam 1985) consists of an interaction between sensory input and templates which are internal to the perceiver, modifiable by experience, and allowing generalization. The presence of templates or memories makes the sensory messages acceptable for classification. The reciprocal pathways rehearse the stimulus and modify it until it fits into a certain class, i.e., until it is associated with a particular memory. Thus the neocortex is an important part of reverberant pathways in the same sense as the hippocampus: although essential for declarative memory, the hippocampus is probably not the site of this memory.

Sensory cortex in mammals and especially in primates seems to elaborate on the neural activity relayed from the periphery in two ways. The *first* way comprises a *recording* of the message, usually accomplished by a great divergence of the specific afferents on their target cells in layer IV of the cortex. It has been suggested (Barlow 1979) that in the visual system this divergence can compensate for the coarse spatial sampling in the retina. The granular cell layer of the striate cortex with its extremely dense packing – there are 30–100 times as many granule cells as terminals from thalamic afferents – can be considered as an interpolation device to obtain a smooth cortical representation of a coarse sampled retinal image. This could, for instance, be accomplished if the synaptic connection strenghts between one specific afferent to the various granule cells would be distributed in, e.g., a sin x/x way (corresponding to a spatial low pass filter or smoothing operation). In the memory systems we discussed in the two previous chapters such a recoding stage was indeed present: the granule cells in the cerebellum and in the dentate gyrus, the number of which is considerably larger than the number of afferents to them. Marr (1969, 1970, 1971) has considered these stages instrumental in the recoding of a sensory message carried by a relatively large number of active thalamocortical fibers (in the case of the neocortex) into a message carried by a small amount of active granule cells, in order to make optimal use of the storage capacity of the associative memory formed by the projection cells of these granule cells. The *second* way of elaboration upon the sensory message is related to the ability of the cortex to detect *symmetries* and *patterns* and to *fill in* where neede to allow for continuity. Correlated neural activity of neighboring cells may be instrumental in this respect; it is shaped and determined by, yet at the same time in itself affects, the intracortical activity pattern.

The increased sensory specialization is especially clear in the visual cortex. Receptive fields of visual cortical cells are distinctly different from those in the retina and LGN. One might consider the single cell's receptive fields as a memory trace for a certain stimulus feature, created during ontogeny. Each pyramidal cell then represents a certain aspect of the input conveyed to it through the parallel fibers in layer I in conjunction with a certain pattern of activity from neighboring pyramidal cells on its basal dendrite. The input in layer I may result either from outside: a direct projection from the more diffusely organized parts of the thalamus such as the magnocellular part of the medial geniculate body, the C-lamina of the LGN, or the ventromedial nucleus, or from within the cortex: the parallel branching of the Martinotti cell axons. Both inputs thus can be viewed as memory elements which, due to their double dendritic tree, "learn" a situation which comprises an input feature provided by the layer I input, associated with certain other features represented by other pyramidal cells. To put it in the perspective of consciousness: "percep-

Box O. The Neocortex
The histological appearance of the cerebral cortex is characterized by six layers, some dense with cell bodies, some only containing fibers (Fig. O.1). The major cell type is the *pyramidal cell.* Pyramidal cells have apical as well as basal dendritic extensions covered with spines. Their axons leave the cell body at the basal part and generally proceed perpendicular to the layers; in addition they form horizontal collaterals somewhat below the basal dendrites. While the cell bodies of the pyramidal cells may be found in nearly all layers, their apical dendrites always extend into layer I. The second cell type is that of the *stellate cells*, which come in two types. One type is covered with spines, the other type has a smooth appearance. Unlike the axons of the pyramidal cells, those of the stellate cells do not leave the cortical hemisphere and are generally short. Stellate cells are therefore considered as local circuit neurons or interneurons. The spiny stellate cells are abundant in layer IV and receive sensory input from the thalamic efferents. They in turn provide excitatory input to pyramidal cells in layers III and V (Fig. O.2). Smooth stellate cells and basket cells are assumed to be inhibitory, their axons terminate on pyramidal cell bodies. A third class of cells is formed by the *Martinotti cells* which have axons that run vertically upwards to end in layer I and form parallel fibers that make thousands or more synaptic contacts with the apical dendrites of pyramidal cells (Fig. O.3). Pyramidal cells make up one-half to tree-quarters of all cortical cells, depending on which part of the cortex and what animal species is taken under consideration.

The cell types described so far are organized in *modules*, columnar units of remarkably similar build-up and about 200 – 300 μm in diameter in the cat (in the mouse these are smaller, and in monkey and human larger than in the cat). The modules extend throughout the entire depth of the cortex, are innervated by cortico-cortical fibers and have a predominantly vertical internal connectivity. In a column spiny stellate cells can establish repeated contacts, so-called cartridge synapses, with a single or a few closely spaced pyramidal cells. Their collaterals can be considered as equivalent to the climbing fibers in the cerebellum (Fig. O.2). The specific thalamic afferents end exclusively on the excitatory interneurons in layer IV. The inhibitory interneurons receive input from collaterals of the pyramidal cells and are so of the recurrent feedback type (Szentagothai 1978 a, b, 1979). A cortical module typically consists of about 4000 neurons, half of it being pyramidal cells the other half stellate cells and other interneurons. The axons of the pyramidal cells project for about 10% to the modules in the contralateral hemisphere through the callosal fiber tract. It is estimated that the human cortex contains approximately 3×10^6 modules (Eccles 1981). In this respect it

Box O (continued)

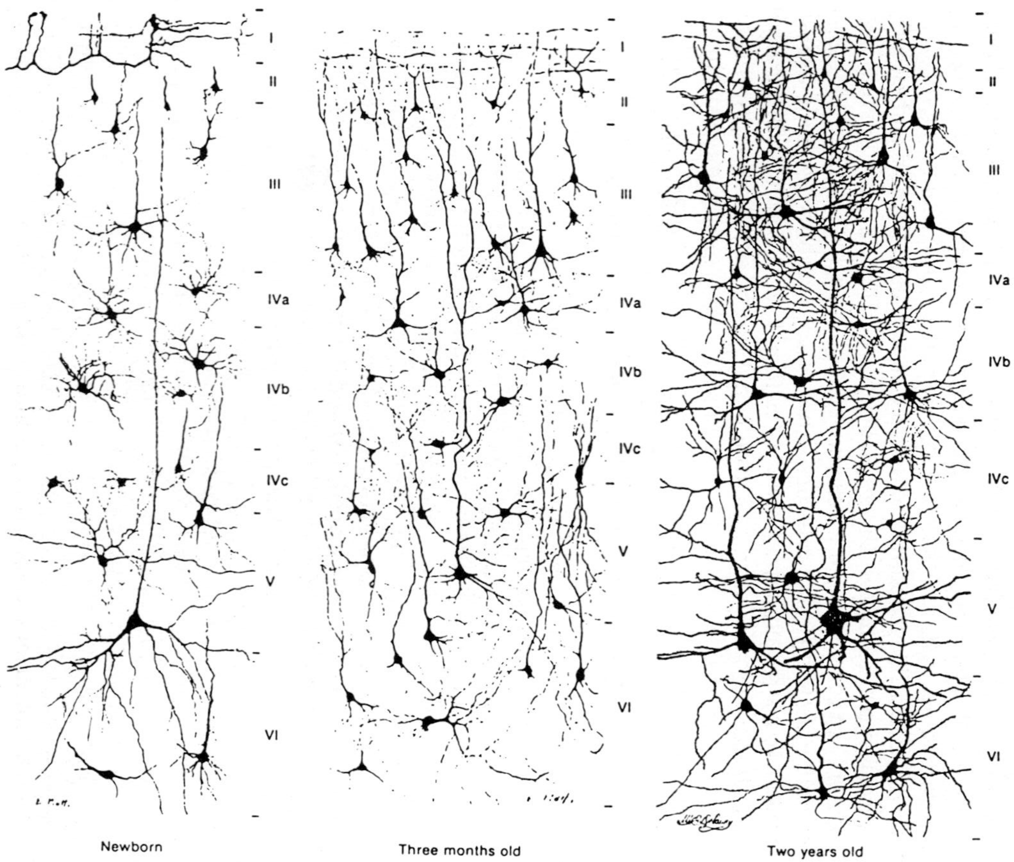

Fig. O.1. Layer structure in the developing visual cortex. (Conel 1959). The development of the cortex is characterized by the extensive growth of the dendrites and thereby in the increase in the number of synaptic connections

Box O (continued)

is interesting to consider Braitenberg's (1978a) conjecture that if one wants all cell groups in the neocortex to be informed about the status of all other cell groups then one has to organize the N cells into $\sqrt{N}$ cell groups. Assuming 10^{10} pyramidal cells in the cerebral cortex, this would result in 10^5 modules of 10^5 neurons each. The estimates of Eccles (1981), resulting in more modules of smaller size, imply that not every module is informed about the state of all other modules.

An interesting observation about the organization of cortical modules is that apical dendrites of nearby pyramidal cells tend to bundle. Bundles consist of 6–30 dendrites and are formed at intervals of 50–100 µm. It is conjectured that the resulting ephaptic connections can synchronize the activity of the pyramidal cells in a given cortical microcolumn (e.g., Roney et al. 1979).

At present it is fairly well established that the inhibitory transmitter in the cortex is GABA. The orientation specificity of cells in the striate cortex can be reduced by GABA blockage, e.g., with bicuculine. In addition to GABA there is evidence for acetylcholine (ACh) and norepinephrine (NE) as neurotransmitters. NE is found associated with input from the locus coeruleus and has an inhibitory action, ACh is found in ter-

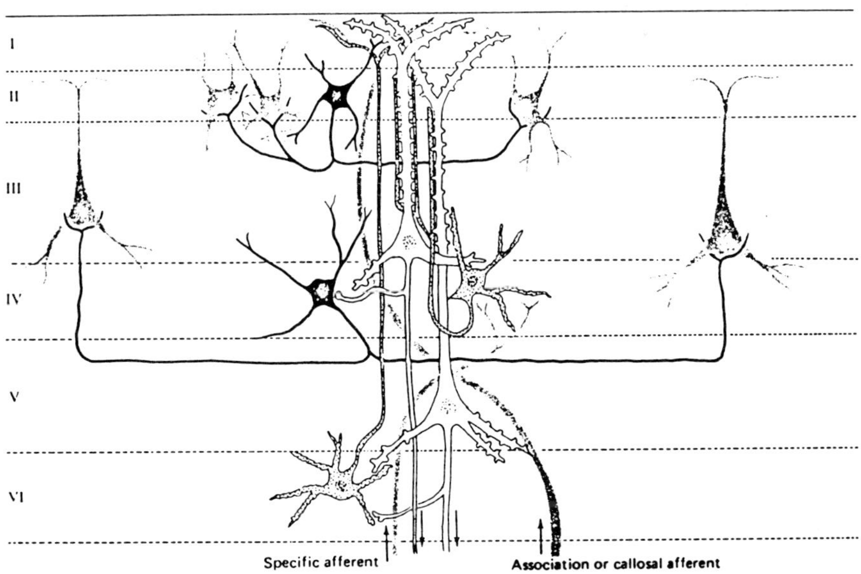

Fig. O.2. Neuron types in the neocortex. One distinguishes pyramidal cells, that receive afferent input via the stellate cells (*shaded*). In addition inhibitory interneurons (*black*), in this case basket cells are present. (Szentagothai 1979)

Box O (continued)

minations of the efferents from the basal forebrain nuclei. The transmitter for the specific thalamic afferents is presently unknown, but aspartate and glutamate have been associated with particular classes of cortical cells (Toyama et al. 1981 a, b; Gilbert 1983).

General references about cortical structure and function may be found in Sholl (1956), Braitenberg (1974) and Shepherd (1979).

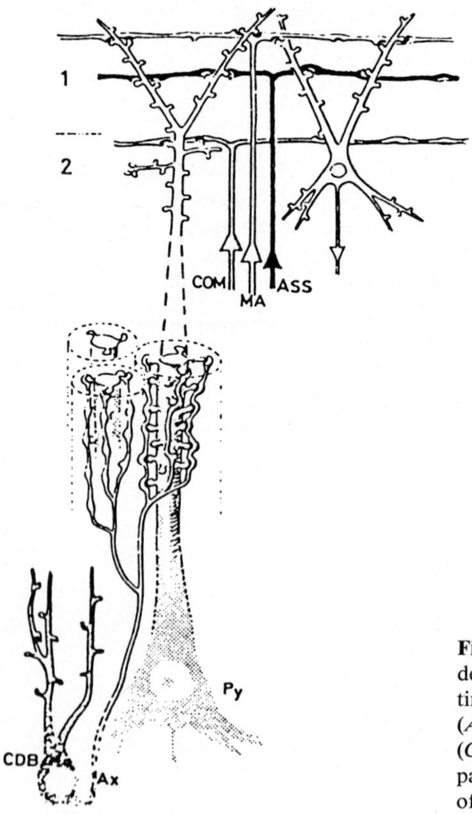

Fig. O.3. The pyramidal cell's apical dendrites receive input from the Martinotti (*MA*) cells, the association (*ASS*) fibers and the commissural (*COM*) fibers. These fibers form a vast parallel array connecting large numbers of pyramidal cells. (Szentagothai 1979)

tion (external input through layer I) always occurs on a background of imagination (collateral input) and so does memory" (Braitenberg 1974).

Eccles (1978) noted the remarkable similarities between the dual input system to the pyramidal cells in layers II and IV of the neocortex and that to the Purkinje cells of the cerebellum. The climbing fiber input from the spiny stellate cells may have an instructional role to the parallel input. If a particular pyramidal cell represents a particular stimulus feature, then an assembly of such "memory cells" may constitute the image of the stimulus in its coherent activity pattern (see Box P for an overview of assembly related concepts). Such an assembly need not be confined to the striate cortex but will encompass, in the case of visual stimuli, most visual cortical areas. Damage to areas 18 and 19 usually does not produce blindness but interferes in a destructive way with the ability to interpret what is seen: the image is present but the perception is lacking.

Hebb (1949) has pointed out that there are two extreme views of such assembly action. One was called the switchboard theory: the cortex is an elaborate kind of telephone exchange with precise connections; the other was called the field theory, which regards the cortex as an aggregate of cells forming a statistically homogeneous medium. A recent example of a switchboard theory is presented by Ballard (1986); examples of field theories were given in Chapters 5.4 and 5.5. Hebb's own assembly model was somewhat intermediate in assuming that precise connections existed but with modifiable synapses that could be changed by experience. An elaboration of such an assembly theory was presented by John (1972) in what he called a "statistical configuration theory". In this theory, learning and memory are envisioned as the establishment of a representational system of a large number of neurons in different parts of the brain; the activity of these neurons has been affected in a coordinated way by the spatiotemporal characteristics of the stimuli presented during the learning task. This was assumed to initiate a common mode of activity in various brain regions specific for that stimulus. Information about an event is represented by the average behavior of such a responsive neural ensemble. Another event can be represented by the same ensemble but with a different coherent pattern. New responses result from the building-up of new temporal patterns in such an ensemble. I find it somewhat peculiar that intricate spatiotemporal patterns are required to lay down a memory while the output of the recall procedure is assumed to be the average activity of that ensemble. How is such average activity distinguished for one coherent pattern from another, unless there are very specific read-out connections that have to be established for the representation of each new event? Such connections furthermore are specifically sensitive to the temporal pattern rather than the overall activity, if the former is supposed to be the important characteristic. I will take up this point again in the last part of this chapter.

13.2 Development Aspects of Neocortical Organizations

Maturation of the brain takes place through execution of a genetic program as well as through influences of the environment. For understanding and modeling of the self-organizing capacities of the cortex it is important to know which part of that process can take place in the absence of sensory stimulation. Although we like to generalize knowledge about one part of the brain to that of another, it is probably wise to restrict ourselves for the time being to what is known about the development of ocular dominance and orientation columns in the visual cortex, since this has been most thoroughly investigated. Most of the research in the development of the visual cortex has been carried out in monkeys and cats (Swindale 1982; Wiesel 1982, for reviews). Since both animal groups show differences in their development, it will be necessary to relate the findings in both systems. In the following we will follow the viewpoints from the cited review papers.

The monkey is visually alert at or very soon after birth. Single cells in the visual cortex show orientation preferences and the preferred orientations form columnar sequences that are nearly as well organized as those in adults. Ocular dominance columns are present as well, but they are incompletely formed. In the kitten the visual cortex is less mature and eyes do not open until in the second postnatal week. Ocular dominance columns are poorly formed and many more cells than in adults lack orientation selectivity. Experiments with deoxyglucose have failed to show orientation columns in 3-week-old kittens. However, at 4 to 5 weeks of age orientation responses of single cells become adult-like. From the findings in the monkey it is clear that cells can develop adult response properties without visual experience; however, a certain fraction of the cells will never develop orientation selectivity when visual experience is prevented. Ocular dominance columns in the monkey continue to form after birth, even when the animal is light-deprived. This may have its origin in the spontaneous activity of retinal ganglion cells, since blocking this spontaneous activity with an injection of tetrodotoxin in both eyes completely abolishes the formation of ocular dominance columns.

It appears that impulse activity plays an important role in shaping the pattern of connections in the developing brain. Assuming a rather crude genetically determined topographic organization, impulse activity can refine these ordered projections before vision is present (Fawcett and O'Leary 1985). Neurotransmitters play an important role as well, norepinephrine appears to be crucial for the plastic organization of ocular dominance columns (Singer 1983). Later, when visual stimuli can influence the organizational process of the cortex, a feedback signal from the eye muscles to the brain is required for the further development of orientation selectivity and ocular dominance. Such eye movements probably reflect an attentional aspect or show that the animal uses visual stimuli to control its behavior. Thus some neural activity signaling "meaning" must reach the brain together with the visual stimuli in order to induce changes in activity (Singer 1983).

Experience of visual stimuli is necessary to establish or further develop ocular dominance columns. Two types of theories are commonly invoked to explain the competitive interactions between afferents of both eyes. The first type roughly states that visual experience instructs the acquisition of new response properties, the second assumes that all response properties are already present at birth and that visual stimulation selects and consolidates those that appear to be important. The presence of innate selectivity in visually inexperienced cats that improves after visual stimulation certainly points to selection as a major mechanism in the development. Rauschecker and Singer

(1981) argue that the difference between both theories may just be a different expression of basically the same mechanism. This mechanism is characterized by the following three properties: (1) excitatory synaptic connections are enhanced each time presynaptic afferents and postsynaptic target cells are simultaneously active; (2) synaptic efficiency decreases when the postsynaptic element is active and the presynaptic fiber is silent; (3) synaptic efficiency decreases slowly in time irrespective of presynaptic activity when postsynaptic activity is absent. These rules, which are elaborations on the Hebb learning rule, lead to competitive interactions between afferents converging on the same cell. In all three rules the postsynaptic activity appears to be critical, indicating that innate tuning properties are critical for later development. What then is the effect of visual experience? In the opinion of Rauschecker and Singer, visual experience leads either to (1) adaptation of response properties that are already present, or (2) to the persistence of responses when they are matched to present stimuli or to degeneration of original responses when visual stimuli do not match. The first mechanism is of the instructional type, the second is of the selection type. However, both originate from the same postulated learning rules. We will take up the selection versus instruction issue again in the last section of this chapter.

Any theory about brain function should primarily account for the previsual organization in the visual cortex and its fine adjustment due to visual experience. In the next section I will present a survey of some of these theories that attempt to explain the changes in the cortical function during development.

13.3 The Neocortex as a Self-Organizing System

Self-organizing systems can quite crudely be described as systems that obtain a larger degree of order as time progresses. Examples are abound in physics, where, e.g., magnetic order can arise in initially randomly oriented spin assemblies (see the discussion of the Ising-model in Chap. 5), in chemistry, such as the Belousov-Zhabotinsky reaction, and in biology, such as morphogenesis (cf. Haken 1983). The developing brain also acquires an increase in the degree of order, exemplified by the formation of cortical columns. In many ways, the problem faced when describing such self-organizing processes resembles that of how to obtain a jellyfish out of jelly.

13.3.1 General Considerations

Pask (1960) presented an abstract model of a self-organizing system that might be used when discussing theories of columnar organization in the brain. The model comprises four rules: (1) the network is defined through the connectivities between pairs of points: initially the distribution of connection strengths is assumed homogeneous, (2) there is a currency which on the average is preserved; this means that the self-organizing system is closed or in a steady state with respect to that currency, (3) there is a set of currency seeking servomechanisms (compare Klopf's hedonistic neurons, Chap. 10), and (4) there is a set of rules that determines changes in connectivity under the influence of activity in the network. Pathways, structures of network connectivity, or organized networks arise necessarily if the servomechanisms are active and they continue to exist only if these mechanisms are used.

In this model, the state of the system is coupled to parameters of the environment in such a way that the parameters of the system can be modified. Thus a self-organizing system does not differ fundamentally from a learning system. The model described above assumes – through the assumption of initial homogeneous distribution of connection strengths – that all organization results from an interaction with the environment. Applied to the brain it would assume that initially it is a tabula rasa: the theory is clearly of the instructional type.

A model based on the same idea has been described for the organizational process in the visual cortex (Perez et al. 1975). Starting from an unspecified initial condition as present in the newborn, the final mature organization was assumed to depend only on certain adaptive growth rules. These rules include a conservation of synaptic strength, a Hebb-like learning rule, and a general activity-limiting process. Visual stimuli had to be present to instruct the network but intracortical interactions were not explicitly taken into account (although they probably are responsible for the activity-limiting process). Von der Malsburg (1973) had earlier formulated a comparable model, in which nearest-neighbor neurons mutually excite each other, and excitatory cells excite inhibitory cells. Inhibitory cells, on the other hand, would exert their influence only on the next-to-nearest excitatory cells. Grossberg (1976), in a review of both models, concluded that the general activity-limiting process as assumed by Perez et al. and the next-to-nearest neighbor interaction postulated by von der Malsburg essentially served the same function. In addition, he showed that a center-surround retinal organization, capable of adaptation, may serve this function as well. Experimental evidence in the mean time indicated that intracortical inhibition is mandatory for orientation selectivity. This rules out a mechanism for orientation selectivity that is based on excitatory afferent organization alone. Thus von der Malsburg's (1973) model is the only one physiologically tenable, and has the following essential properties:

1. there are plastic synapses that connect sensory afferents to cortical cells,
2. learning occurs at these excitatory synapses,
3. synaptic strength and total neural activity are conserved, e.g., through lateral inhibition.

Such models show all aspects of learning and imprinting for visual patterns that can then later be recognized. Mathematical formulations for such models have been given by Legéndy (1978), by Amari and coworkers (Amari 1977, 1980, 1982, 1983; Takeuchi and Amari 1979), and by Kohonen (1982), all of which stressed the importance of lateral inhibition in organizing the network. Nass and Cooper (1975) showed that, on the basis of a learning rule for synaptic strength and repetitive presentation of a certain stimulus pattern, an initially homogeneous group of nerve cells can become committed to a certain pattern. Inclusion of excitation as well as lateral inhibition in the interaction pattern between cortical cells can greatly enhance the resulting specificity.

The models discussed so far do not take into account that a certain amount of organization is already present in the brain before any visual stimulus has been processed. This may not be a problem for the somatosensory system and to a lesser degree for the auditory system, since both systems are clearly already activated in utero. However, since the visual system is not, this does present a problem: the models described above cannot explain the presence of orientation and ocular dominance columns at birth. Von der Malsburg and Willshaw (1981) proposed some organizational rules for this pre-vision process of self-organization. These rules apply to two levels of organization, the global level and the local level. The *global level* needs the specification of boundary conditions: to what part of the brain should the growing nerve fiber migrate? In addition, some initial specification of orientation sensitivity of the fiber may be required. On the *local level* synaptic learning rules and conservation of synaptic strength for each cell has to be prescribed. In an elaboration on this general idea (von der Malsburg and Cowan 1982), the developmental process was distinguished into a pre-visual period and a visual period. During the pre-visual period, intracortical connections organize themselves by restricting their activity to an appropriate family of patterns. This is followed by a one-to-one mapping that couples each stimulus orientation (although as yet no stimulus has been perceived) to one of the cortical patterns that are spontaneously generated. Prior to eye opening, the cortex is assumed to be active and to generate periodic standing waves of activity. The coupling may take place in two steps. In the first step, still before eye opening, two sets of afferent pilot fibers, which possibly originate from W-cells in the retina and carry only horizontal and vertical orientation preferences, serve to fix the mapping in two points. In the second step, after eye opening, a more precise mapping of orientations into cortical columnar patterns develops under the influence of visual stimuli and synaptic plasticity obeying Hebb learning rules, and proceeds according to von der Malsburg's 1973 model. This second step takes place in the time span covering the critical period, but it may extend beyond it. Essentially then the mechanism for the development of iso-orientation columns in the visual cortex is twofold (Cowan and von der Malsburg 1985): intracortical interactions spontaneously generate strip-like aggregates of coupled cells and peripheral afferents carrying genetically determined orientation preferences then become associated with such aggregates.

13.3.2 Theoretical Descriptions

A more formal way of describing the self-organization of nervous structures is presented in von der Malsburg (1983) as follows. Assume a one-dimensional chain of retinal neurons that projects to a one-dimensional chain of, e.g., tectal

Box P. Neural Assemblies

The concept "neural assembly" is one of the most commonly used in contemporary brain theory (e.g., Palm and Aertsen 1986), although not always by that name. It has been introduced by Hebb (1949) as "a three-dimensional lattice with no regular structure and with connections possible from one intersection to any other through modifiable synapses". This is therefore an example of a very general, fully interconnected, plastic neural network, reminiscent to the ones used in the mathematical theories of Little (1974) and Little and Shaw (1978) and as discussed in Chapter 5.

Other formulations which come close to that of Hebb are "a set of neurons, each of which receives excitation from and gives excitation to some other members of the same" (Braitenberg 1978a), and "sets of collective units" (von der Malsburg 1986). These formulations become either more restrictive, as in the first one, or very general, as in the latter. In Braitenberg's view, the entire nervous system, with the exclusion of motorneurons and sensory neurons not involved in feedback loops, is a cell assembly, of which all others are subassemblies (for which the same definition holds). Palm defines a cell assembly "as a set of neurons that fire together because the activity of some of them is sufficient to excite the whole set".

Von der Malsburg's (1986) concept of assemblies as "sets of coactive units" must be understood in terms of the following description (von der Malsburg 1981): "a cell assembly is a set of neurons cross-connected such that the whole set is brought to become simultaneously active upon activation of appropriate subsets which have to be sufficiently similar to the assembly to single it out from overlapping others". Thus the elaborations on Hebb's concept by German scientists all converge to an active concept of an assembly: given suitable input, the assembly can be ignited and then acts as a logical unit by going through a spatio-temporal activity pattern characteristic for that assembly. Contrasting views exist as well: Shaw et al. (1982) adopt the formulation of John (1972) defining "assemblies of neurons so that the firing response (to a single presentation) of an assembly averaged across the localized group of neurons in it would be the same as the PSTH of a single neuron in the assembly". This definition takes care of the statistical fluctuations and vulnerability of coding in single neurons and replaces the two valued neural unit response by that of the assemblies' average activity. If one considers single neurons in a sensory area of the brain to correspond to individual features or properties of objects in sensory space, then in Shaw's view the assembly corresponds to the same features or properties, only in a more reliable way. In an elaboration (Shaw et al. 1985) such assemblies are characterized by three response states: the assembly's activity is lower

Box P (continued)

than, equal to, or higher than average. Small localized neuron groups having such properties are called "trions" and are assumed to be the building blocks of larger networks (see Chap. 5 for details).

In Braitenberg's (1978a) view the cell assemblies correspond to the "objects in sensory space" themselves. However, because assemblies habe an ignition threshold, the representation of sensory objects is an all-or-nothing process. Therefore his assemblies are formal neurons in the sense of McCulloch and Pitts (1943). The ignition character of an assembly is also evident in the concept of the "synfire-chain" (Abeles 1982): "the activity of the neurons that transmit information is organized along a chain of sets of neurons. Each link in the chain is made of a set of neurons that fire in exact synchrony whenever the chain becomes active". The concept also includes the necessarily hierarchical character of the organization, and is related to the concept of repertoires (Edelman 1978) defined such that "the main unit of function and selection in the higher brain is a group of cells connected in a variety of ways. Groups of cells build repertoires."

neurons. Let each of these chains consist of N cells, then the connections can be specified in an N×N connectivity matrix. In order to avoid boundary effects, both chains are considered cyclically bound. The synaptic weights, or entries of the matrix, are indicated by real numbers w_{tr} representing the connection strength between retinal cell r and tectal cell t. The synaptic weights are assumed to change according to:

$$dW/dt = \alpha + W \cdot C(W) - W \cdot B[\alpha + W \cdot C(W)] \ , \tag{13.1}$$

where · denotes element-wise multiplication, not an ordinary matrix multiplication. Written in components this equation reads:

$$dw_{tr}/dt = \alpha + w_{tr} \cdot C_{tr}(W) - w_{tr} \cdot B_{tr}[\alpha + W \cdot C(W)] \ . \tag{13.2}$$

In these equations α is a positive number describing the spontaneous formation of new synapses. C is a convolution operator with a bell-shaped kernel (e.g., a two-dimensional Gaussian) in the t-r space. C(W) is therefore a smoothed version of matrix W. The components of C, C_{tr}, can be interpreted as rate constants for the autocatalytic growth of the synaptic weights w_{tr}. B is an operator which at position (t, r) forms the mean of all elements in row t and column r of its argument matrix. The term involving B expresses the competition between one fiber with all others coming from the same retinal cell (column r of the matrix) and those which go to the same tectal cell (a row of the matrix). In this model the positive feedback between the effectiveness of connections and the local synchrony of signals is the basis of the organizational process (through the convolution operator C). The connectivity in the neural net is controlled by the action of an inhibitory system which suppresses global synchrony (through the action of operator B). Since B is a linear operator, we can rearrange Eq. (13.1) to give:

$$dW/dt = -\alpha(W-1)+W\cdot\{C(W)-B[W\cdot C(W)]\} \tag{13.3}$$

where "1" is a matrix completely filled with one's and representing a homogeneous distribution of synaptic contacts. The term $-\alpha(W-1)$ thus represents an exponential decay with rate α of W towards this homogeneous state. Individual synaptic weights w_{tr} can grow when $C_{tr}(W)$ is larger than the average as expressed in $B[W\cdot C(W)]$. As C_{tr} describes a weighted sum over the neighbors of w_{tr}, the growth term represents the cooperative effects of self-organization: weights will only grow in regions of already high w-values. The resulting process favors a diagonal chain of large weights in W, a prerequisite for a correct model of a retinotopic organization.

Rewriting Eq. (13.3) in terms of deviation from the homogeneous solution, $V = W-1$ leads to:

$$dV/dt = L(V)+Q(V)+K(V) \ , \tag{13.4}$$

in which L(V), Q(V), and K(V) are linear, quadratic, and cubic terms respectively. The linear terms are

$$L(V) = -\alpha V+C(V)-B(V)-B[C(V)] \ . \tag{13.5}$$

A linear analysis only requires to solve the equation:

$$dV/dt = L(V) \ , \tag{13.6}$$

in which L is a linear operator. The general solution of Eq. (13.6) can be expressed in terms of eigenpatterns or modes of connectivity E which are solutions of:

$$L(E) = \lambda E \ , \tag{13.7}$$

in which λ is the eigenvalue belonging to eigenpattern E. A special solution is:

$$V(t) = e^{\lambda t} E \ , \tag{13.8}$$

which will grow or decay exponentially, depending on whether λ is positive or negative. The number of modes of L(V) is equal to N^2; each one can be indicated by E^{kl} where k, l are indices running from 0 to $N-1$. The effect of the convolution operator C with eigenvalues γ^{kl} is a particular selection of modes satisfying:

$$C(E^{kl}) = g^{kl} E^{kl} \ . \tag{13.9}$$

The eigenvalues of operator L are found to be:

$$\lambda^{kl} = \begin{cases} -\alpha-1 \ ; & k=l=0 \\ -\alpha-(\gamma^{kl}-1)/2 \ ; & k=0,\ l=0 \text{ or } k=0,\ l=0 \\ -\alpha+\gamma^{kl} \ ; & \text{otherwise} \ . \end{cases} \tag{13.10}$$

If we require that C is a symmetric matrix, then its eigenvalues γ^{kl} are real positive valued and the largest eigenvalue is $\gamma^{11} = \gamma$. Obviously the largest eigenvalue of L is then $\lambda^{11} = -\alpha + \gamma$, and the smallest eigenvalue equals $-\alpha - 1$. The eigenvalues thus cover the range from competition only ($\lambda_{min} = -\alpha - 1$) to cooperation only ($\lambda_{max} = -\alpha + \gamma$). It is clear that the value α, being the rate of unspecific synapse formation, to a large extent determines which mode of retino-tectal connectivity will survive, since only modes with positive eigenvalues will result in pattern formation. The entire process of organization in this linear approximation depends on how α changes during development. It is conceivable that α initially is large but gradually becomes smaller and even zero. Thus when stable modes are formed and α decreases no big changes can take place anymore. Incorporation of the quadratic and cubic terms from Eq. (13.6) will allow interactions between the various modes.

Local enforcement and global inhibition (and their opposites) are also crucial in Swindale's (1980) model for the formation of ocular dominance stripes. The key assumptions are that synapses from cells of opposite eye types inhibit each others growth locally (at distances less than 200 μm) and reinforce each others growth when situated more than 200 μm apart. Conversely, synapses from cells of the same eye types interact, such that further growth of that type is encouraged for distances less than 200 μm and discouraged for distances larger than 200 μm. These effect are denoted by w_{RR} and w_{LR}. The kernels describing these actions have a "Mexican-hat" shape, e.g.,

$$w(r) = A \exp(-r^2/d_1) - B \exp(-r^2/d_2) \ , \tag{13.11}$$

with $d_1 < d_2$. The growth rates of right and left eye synapses can be expressed as:

$$\begin{aligned} S_R &= n_R * w_{RR} + n_L * w_{LR} \\ S_L &= n_L * w_{LL} + n_R * w_{RL} \ , \end{aligned} \tag{13.12}$$

where * stands for convolution, and n_R and n_L are the synaptic densities which change as:

$$\partial n_R/\partial t = S_R n_R (N - n_R)$$

and

$$\partial n_L/\partial t = S_L n_L (N - n_L) \ , \tag{13.13}$$

with N the uniform starting density of both right and left eye synapses. When we define $a = n_R - n_L$, we can write:

$$\partial a/\partial t = (a * w + K)(N - a)(N + a) \ , \tag{13.14}$$

with $w = w_{RR} + w_{LL} - (w_{RL} + w_{LR})$ and $K = N * (w_{RR} - w_{LL})$. For small a, n_R is about the same as n_L, i.e., initially when the densities of both eye synapses are approximately equal we can simplify this to:

$$\partial a/\partial t = N^2 (a * w + K) \ . \tag{13.15}$$

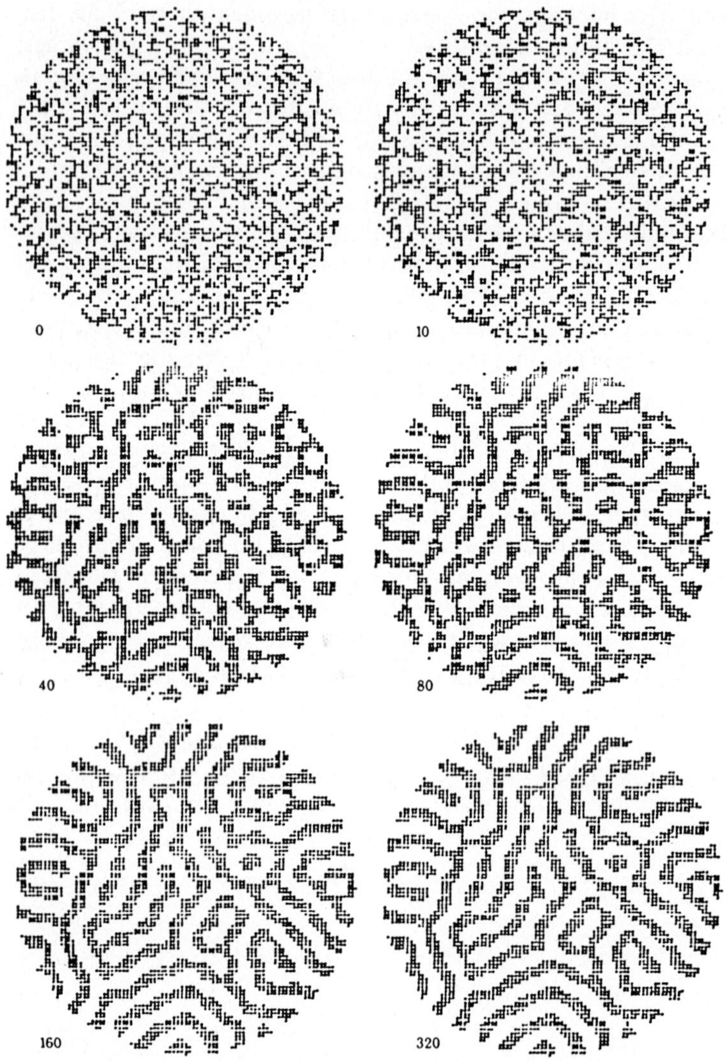

Fig. 13.1. Ocular dominance pattern formation by local enforcement and global inhibition, the numbers indicate the number of iterations. (From Swindale 1980)

In addition, we may assume that initially the effects of like eye synapses on each other are the same, therefore $K = 0$. Denoting the spatial Fourier transforms of $a(t)$ and $w(t)$ by A and W respectively allows us to rewrite Eq. (13.15) into a linear differential equation:

$$dA/dt = N^2 A W \tag{13.16}$$

with solution:

$$A = A_0 \exp (N^2 W t) \ . \tag{13.17}$$

When initially all modes are present (i.e., for t = 0), the mode with the largest W will dominate for t sufficiently large. The result is pattern formation with a predominant spatial frequency; the action of this algorithm is illustrated in Fig. 13.1. Legéndy and Salcman (1985) have pointed to the relation between the width of cortical columns and the range of intracortical inhibition.

13.3.3 Relation Between Structural Organization and Assembly Formation

The formation of stripes in Swindale's model may well serve to illustrate the assembly concept. Especially in the initial phase of the stripe formation process, seemingly unrelated regions tend to become organized. Is this organization resembling assembly formation? When two regions with a higher left-eye-synapse density are joined, do they form an assembly or just a larger region? Are the assemblies (as described in Box P) more than a mass of simultaneously active cells? Does an assembly have an internal structure? When two assemblies are joined to make up a larger assembly, is it then possible to uniquely decompose the larger assembly into its two subassemblies? In other words: are there hierarchical rules in assembly formation (cf. von der Malsburg 1981, 1986)?

The answer, according to von der Malsburg, resides in the temporal correlations between the elements that make up an assembly. Time is divided into two scales: a psychological time scale (time unit some tens of seconds) and a faster, physiological, time scale (units are milliseconds). It is assumed that the mean activity in the assembly fluctuates on the psychological time scale, while the activity around the mean fluctuates on the fast time scale. The weight of an excitatory, plastic, synapse in this view depends on two variables, representing these two different time scales:

$$w_{ij} = a_{ij} s_{ij} \ , \tag{13.18}$$

where a_{ij} is a fast synaptic modulation of a more permanent synaptic strength s_{ij}. The absence of correlation on the fast time scale will reduce a_{ij} such that w_{ij} goes to zero, while leaving s_{ij} unchanged. Strong correlations will in addition tend to increase s_{ij}. Thus units can bind to each other by correlation in their fast activity fluctuations; groups of units (assemblies) can be bound by synchronizing their fast activity fluctuations to form blocks. Several blocks can coexist independently if their activity is mutually desynchronized. This allows for a hierarchy in the formation of assemblies and leaves the contributing units identifiable. Furthermore, units can be recruited into different assemblies by mainly changing m, which can occur on a time scale of a few milliseconds. In this context we recall that memory re-establishes correlates that were previously active. If these correlations were formed with a certain refinement in their spectro-temporal relationships then one could expect that the memory recall also has this refinement.

13.4 Does the Brain Learn by Selection?

In this section I will summarize a selection theory of brain function (Edelman 1978, 1987) based upon ideas of Jerne (1967), Edelman (1975) and Young (1975). This theory comprises many of the aspects that we have dealt with before, notably in Chapter 10, but they are sufficiently important to be presented in a unified framework. Jerne (1967) was among the first to draw attention to certain analogies between the immune system and the central nervous system. Both systems develop during the lifetime of an individual: each antigene that is encountered by the immune system irreversibly changes it just as the state of the CNS is changed by each experience of the individual. Both systems learn: the immune system "learns" to respond to antigenes, the CNS "learns" to respond to sensory stimuli. Both systems, as a consequence, have memory which cannot be transferred to the individual's progeny.

The immune system clearly is a selective system (Edelman 1975), where selection refers to a mechanism which recognizes and amplifies a certain property that was already present in the system prior to the arrival of the signal. In an instruction type of learning one assumes that the informational structure of the incoming signal is the primary and necessary event for the shaping of the system's structure and function (Fig. 13.2). Many biological processes, from evolution to immunization, have been found to depend upon selection among groups, rather than upon shaping by instruction.

The selection theory of brain function is elaborated upon by Edelman (1978, 1987) and Edelman and Finkel (1984). One of the main assumptions is that after ontogeny and early development, the brain contains cellular configurations (groups) that can already respond discriminably to sensory stimuli (e.g., the orientation selectivity in the visual system of newborn monkeys), because of their genetically determined structures or because of epigenetic alterations that have occurred independently of the structure of these sensory signals. This prespecified collection of neuronal groups is called a *primary repertoire* and consists of a large number of groups (of the order of 10^6), each with a modest number (50–10000) of cells. The primary repertoire is degenerate, i.e., it contains multiple neuronal groups with different internal structures, that are capable of carrying out the same function. The primary repertoire should contain enough neuronal groups, such that sensory signals have a high probability to find matching groups; and finally it must have provisions for amplifying a selective recognition event, probably by synaptic alterations, either through the formation of new synapses or through changes at already existing contacts. All these properties are very much the same as in the classical perceptron (Chap. 10). In addition, the theory requires a *secondary repertoire* as a collection of different, higher-order, neuronal groups whose internal and external synaptic connectivity can be altered by selection during experience. This cell group selection can occur in two stages, first by filtering – selecting all groups that react more or less well to the spatiotemporal input pattern – and second by an inhibition process that eliminates those selected groups from stage one that have an insufficient response (a threshold mechanism).

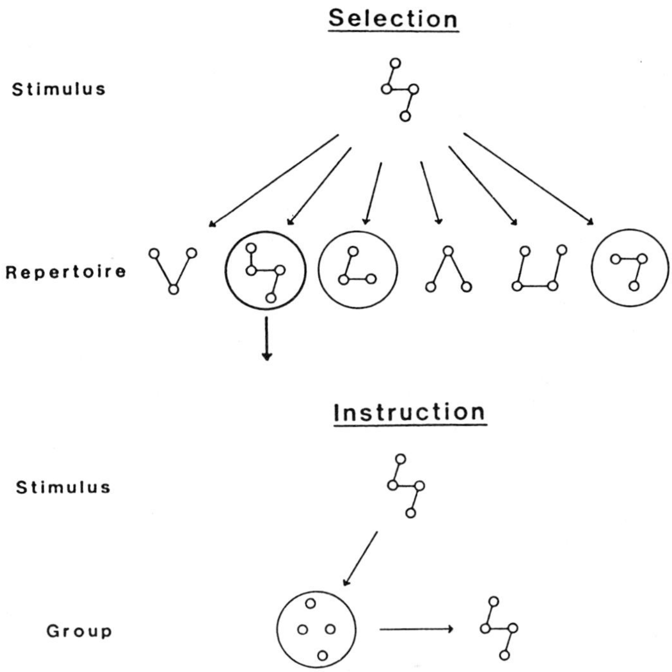

Fig. 13.2. Illustration of some principles in learning by selection and by instruction. It is crucial for a selection theory that there are broadly tuned neural elements that are capable of responding to a large variety of sensory stimuli. A certain stimulus will find certain units that are capable of responding, these will be selected and as a result more finely tuned to this particular stimulus. After a series of presentations the selected unit may become responsive only to that stimulus. In an instruction theory the first stimulus presentation commits a neuronal group to that particular stimulus, thus a correct sequence of stimuli is needed for a harmonic development. (Edelman 1978)

An important aspect of the theory is the *re-entrance* of signals at the level of the secondary repertoire. As we have seen, the pyramidal cells in the cortex receive far more collaterals from other pyramidal cells than from specific afferents. For the pyramidal cells there is probably no way to distinguish these inputs. Thus, internally generated signals are re-entered as if they were external signals; this re-entrance might be able to guarantee a continuity in the neural construct as well as a succession of temporal order of associated secondary repertoire events. Moreover, re-entrant signals in the secondary repertoire might be of different modality than the more direct input from first repertoire groups. Thereby the secondary repertoire might be able to relate multimodal activity patterns. Re-entrance is also a mechanism that allows ongoing cross-correlations between inputs from first repertoire groups and second repertoire groups, thereby providing the possibilities of association and classification.

Experimental evidence for a degenerate fashion in which neuronal groups code sensory stimuli, together with the effects of individual experience, was provided by the mapping experiments in somato sensory cortex by Merzenich and coworkers (see Chap. 10 for a summary of these results). Indications of the initial nonselectivity of primary cortex was provided by Rebillard et al. (1977, 1980). They describe that in bilaterally deaf white cats (who have an inborn deafness), or in normal cats deafened in the first postnatal week, the auditory cortex showed consistent responses to visual stimuli. Deafferenting the auditory cortex after 3 postnatal weeks, i.e., after the critical period, did not result in responses to visual stimuli from the auditory cortex. This can be interpreted as an indication for a degenerate, very broadly tuned group of neurons in the auditory cortex, offering a modest support for the selection theory of higher brain function.

14 The Correlative Brain

In the previous chapters we have investigated those aspects of brain function that are likely to involve correlations between activity patterns of neurons or neuronal groups. In addition, a survey was given of experimental techniques to measure such correlations, and of brain theories based on correlations. In this final chapter, I will first of all present a short summary of the presumed role of correlation in the normal functioning of the brain, thereafter I will discuss two types of maps in the brain: the topographic maps and the functional maps. In this way I again take up the line of thought that lead us from the single-unit receptive field (Chap. 3) to the action of the assemblies (Chap. 13). Finally, I will try to delineate the meeting ground for bottom-up (neuroscience) and top-down (cognitive) approaches to brain function.

14.1 Correlation, the Basic Mechanism of the Brain?

The brain, in a multitude of ways, explores the sensory environment and uses the information obtained to control behavior. In doing so, its main mechanism to evaluate, to control, and to learn is that of correlation. Correlation of nervous activity can take many forms: it can be the detection of coincidences in the firing of two neighboring nerve cells, or the detection of the covariation in the firing rates of two nerve cells. It can be the covariation in the activity pattern of neuronal groups, but it can also be the covariation in the postsynaptic potential activity entering the cell through distinct dendrites. Looking for coincidences provides a means of making inferences about the environment. As we have seen in Chapter 7, the conditional probability density of the co-occurrence of two series of events A and B, $P_{AB}(t,u)$, is equal to the product of their marginal densities $P_A(t)$ and $P_B(u)$ in case the event generating processes are independent. Whenever there are correlated events we have:

$$P_{AB}(t,u) = P_A(t)P_B(u) + C_{AB}(t,u) \ . \tag{14.1}$$

Thus whenever coincidences occur more often than expected on the basis of independence, there may be a causal connection which is represented in $C_{AB}(t,u)$. It is my thesis that the cortex, and most other parts of the brain, are designed to detect such correlated events, as well as to generate them, thereby resulting in meaningful commands to the motor system. Correlation is the only way to make sense out of inherently random or noisy neural activity patterns. In the form of a summary of items discussed in previous chapters I will group the consequences of this correlative activity of the brain into a small number of categories.

14.1.1 Correlation Is Used in the Formation of Topographic Maps

Genetic information is generally insufficient to provide for detailed topographic maps, such as orientation columns, ocular dominance columns or binaural interaction columns (cf. Chap. 13). The crudely outlined genetic maps are sharpened by a self-organizing process that requires spontaneous activity, lateral inhibition, plasticity of synapses, and the ability to change the connectivity between neurons (through changes in synaptic strength) on the basis of coherence in incoming signals (e.g., Singer 1985 a). Most rules for changes in synaptic strength assume the change to be proportional to the correlation or covariance in two neural activity patterns (Chap. 10). Recently, evidence for the need of temporal correlated activity patterns in the formation of somatosensory maps has been provided by Clark et al. (1988). In many cases such changes are guided by experience.

A similar phenomenon may serve to align topographic maps of space for different sensory modalities. Knudsen (1983) has shown that auditory and visual maps of young barn owls, raised with one ear plugged, are in register although the binaural input is distorted. Removing the ear plug resulted in a shift of the auditory map of space with respect to the visual one. Apparently an alignment of the representation of auditory and visual space occurs in the optic tectum during the early periods of life. Since auditory responsiveness is mainly found in the deep layers of the tectum where also visuomotor units are found, it is suggested that localizing sound (auditory) together with the sound sources (visuomotor) results in the alignment of auditory and visual spatial maps. This requires either multimodal neurons (Meredith and Stein 1986) and an inhibitory mechanism, or separate sets of auditory and visual units with some form of neural interaction.

14.1.2 Correlation Is Used and Necessary to Detect Events in the Outside World

The appraisal of the outside world by our senses is done in a parallel fashion and results in neural activity patterns organized in topographic maps of the brain. Correlation of nervous activity takes place at many levels. First of all there is the single neuron level. Neurons that have overlapping receptive fields will show a covariance in overall firing behavior as well as a coincidence in the occurrence of spikes (Chap. 8). The topographic mappings will assure that coincidences in firing will occur predominantly between neighboring units. Coincident firings generally form a subset of the firings of both neurons and may function to extract the relevant information from the neural "noise". For example, in the case of rather broadly tuned auditory neurons, the subset of coincident firings may be representing a particular stimulus (cf. Fig. 8.8). As Eggermont and Epping (1987) have reported, such enhancement of selectivity occurred in about 20% of 92 unit pairs recorded from the auditory midbrain in the frog.

Another example, from the field of auditory pathology, is the phenomenon of tinnitus (e.g., Eggermont 1984) a sensation of sound in the absence of stimulation. From Chapter 2 we know that spontaneous neural activity can be present in the absence of stimulation. The auditory system has the highest spontaneous activity of all sensory systems, yet we normally do not "hear" our spontaneous activity. Tinnitus is usually present in hearing pathology such as that induced by noise trauma. Animal models of tinnitus usually show about the same spontaneous rate as in normal situations, and except in salycilate ototoxicity, definitely not an increase in firing rate. Tinnitus can be understood on the basis of correlations in the firings of neighboring nerve fibers, a correlation that is normally absent in the auditory nerve fibers (Chap. 8). How that correlation is introduced by the pathology is presently not clear.

In the midbrain (superior colliculus or optic tectum) maps of various sensory modalities appear to be in register (Chap. 3). Correlated activity between neurons in two maps at corresponding locations may lead to a higher-order level of correlation, namel between topographic maps. Such map correlations may serve to extract meaningful events in, e.g., the combined auditory-visual environment. These correlations are reminiscent of what Nelson (1985) calls "domain interaction" and which he considers as the basis for higher perceptual processing.

14.1.3 Correlation Is the Basis of Learning, Association, Pattern Recognition, Novelty Detection, and Memory Recall

We have already suggested that coincidence detection by rather broadly tuned neurons may result in sharper tuning or greater specificity for particular stimuli. This can either be obtained by a simple convergence of two neural activity patterns on a coincidence-detecting neuron or by strengthening the connectivity between simultaneously active neurons. The latter mechanism has been postulated for the creation of sharply tuned assemblies (Hebb 1949), synfire chains (Abeles 1982), and secondary repertoires (Edelman 1978).

Learning is a process that generates a "brain state" that is different from that prior to the learning process. The results of learning are memories. Memories are assumed to be laid down in synapses. Novelty detection is related to memory recall; the evidence for the existence of novelty detectors comes among others from evoked potential studies where the "oddball" stimulus creates activity in the latency regions beyond 100 ms (e.g., Wood et al. 1984). A novelty detector is most likely a neuronal group and not an individual neuron. Whenever information is processed through various assemblies, familiar input will create already formed connections or chains in a neuronal group. An acceptable match between incoming information and stored information is based on correlation and will either generate an inhibitory signal to arousal centers or no signal at all. Whenever there is a mismatch a signal is sent to the arousal centers whose nervous activity or activity induced by the arousal centers may result in detectable evoked potentials. Novelty detection then in this view is the result of a template matching procedure that has to be carried out in parallel with a very large amount of potential templates: in principle all information that does not match with the templates has a novelty character.

14.1.4 Correlation, Motor Coordination, and Context-Dependent Behavior

Sensory and motor maps in the brain tend to be in register. For instance, in the superior colliculus a sensory map of the visual field exists in the upper layers, while in the deeper layers a motor map, governing saccadic eye movements, is found that is in register with the sensory map. Activity of cells in a certain location of the motor map will result in a saccadic eye movement that has its end position in correspondence with the sensory representation of the visual target in the corresponding upper layers (e.g., Ottes et al. 1986). It is assumed that, in order to provide for the ultimate saccade direction, a large number of motor units is active, since these units all have very large motor fields (van Gisbergen et al. 1987). The weight of each individual contribution to the final compound saccade vector will be given by the correlation of their preferred firing direction and the desired direction; this in turn is proportional to the cosine of the angle between these two vectors (e.g., Georgopoulos et al. 1986).

In many layered brain structures, spontaneous and highly synchronized neural activity can be found. One of these areas is the hippocampus (Chap. 12), where theta-activity is found that seems to reflect the initiation of purposive movement patterns such as self-generated motion through space. There seems to be little or no correlated firing of neurons in the neocortex during active processing. However, in the sleep state, many single-unit spike trains show a high degree of correlation in their firings: cross-coincidence histograms show a symmetrical peak of approximately 5 ms width around zero lag time, being suggestive of shared input (Noda et al. 1969). This correlation disappears in REM sleep and during wakefulness. Similar phenomena have been observed in other parts of the cortex such as association cortex (Noda and Adey 1970). Thus, correlated activity does not always accompany active behavior. It could be postulated that during sleep, cells in the brain are recruited into one large assembly. Its functional significance could be to generate reverberant activity, with the purpose of stabilizing memories and storing them as long-term memory.

Braitenberg and Schuz (1983) on anatomical grounds came to the conclusion that the probability of two neighboring pyramidal cells not to be interconnected at all is quite high (about 0.9) and that this probability is about 0.1 for being connected through one synapse. Thus most of the pyramidal cells are not interconnected, and if they are the connection is very weak (only one synapse out of 5000 – 10000 on the cell in question in mammals) compared to the estimated firing threshold of a pyramidal cell (about 100 simultaneously activated synapses).

Most correlations in the cortex are through shared specific afferent input and therefore serves to functionally relate, e.g., neurons with overlapping receptive fields. What the experimenter observes as topographic maps, by his explorations with, e.g., microelectrodes, does not have meaning for the brain itself on basis of this spatial organization but only through the correlations in the neuron's firings (Koenderink 1984a). Thus activity in one cell can just be spontaneous activity and will be very hard to distinguish from stimulus-induced activity. Activity in two or more neurons signifies outside world events whenever their firings are correlated.

14.2 Topographic and Functional Brain Maps

Two types of neuronal maps are generally considered to exist in the brain: topographic (somatotopic, retinotopic, or cochleotopic) maps and functional maps. Topographic maps arise as a result of anatomical structure of the receptor surface and the innervating nerve fibers preserving this orderliness in the fiber tracts and in each interposed nucleus. Receptor surface maps may undergo quite complex transformations, resulting in map deformation. An example is the complex logarithmic transformation between retinal surface and visual cortex map (e.g., Schwartz 1977a, b) or superior colliculus map (e.g., Ottes et al. 1986). Functional maps are characterized by a "correlation structure" among neurons. They are generally diffuse in structure, mainly the outcome of brain theories, and have as yet to be visualized by electrophysiological techniques. Functional maps are supposed to be highly specialized and are identified as assemblies or repertoires.

Topographic maps are organized on the basis of their sensory receptor or effector surfaces, and are therefore likely to be found at the sensory and motor sides of the brain. Functional maps mainly have an organization or ordering through their correlation structure (Koenderink 1984a, b). Neural units with strong correlations can be considered close in the neural organization, units with weak correlations have a larger functional distance. In this ordering one could also incorporate the time delay to the dominant peak in the correlation function. It is obvious from the stimulus-dependent nature of correlations (Chap. 8) that functional maps may differ for different conditions: they are plastic. Topographic maps also have a cross-correlation structure, but this is mainly the result of the anatomical ordering and overlap of receptor and motor fields: usually their correlation structure is situation-independent. Functional maps are likely to be found in the "gray areas" between the cortical sensory and motor areas, i.e., in the association areas. However, some investigators (Diamond 1979; Vaadia et al. 1982) provide evidence against this subdivision. Diamond proposed that the neocortex be subdivided in only three regions – auditory, somatosensory, and visual – and stated that no area really can be defined as association cortex. Each field in the three sensory areas was supposed to have direct or indirect connections to motorneurons. Vaadia et al. found evidence that 23 out of 146 neurons in the primary and secondary auditory cortex of the behaving monkey showed sensorimotor association properties: besides a strict sensory component in their response, these units showed task-dependent activity as well. Since these units were found widely dispersed throughout the auditory cortex, it was suggested that association cortex overlaps with sensory cortex. The possibility exists that these neurons are the final targets of the extra-lemniscal pathway, as observed in the investigations of Weinberger et al. (1984 and Chap. 10) and therefore not neurons belonging to association areas in the strict sense. If the conjecture of Vaadia et al. is correct, this suggests the co-existence of topographic and functional maps in the same cortical structures. Neurons from one type of map may simultaneously take part in the other type of map. Another example of such mixing of maps can be found in Braitenberg's (1985) model of the visual cortex.

Some maps in the brain are highly ordered, yet are not the result of the projection of a receptor surface. An example is given by the representation of auditory space in the tectum of the barn owl. Such a map arises through interaction of neural activity from both ears, in reference to a topographic map: the retinotopic map in the tectum (Knudsen 1983). On this basis we could call the map of auditory space a functional map. However, since auditory space is mapped in an orderly way onto the neurons in the tectum, we might as well call this a topographic map. To the experimenter there is no difference from, e.g., a tonotopic map: both show a spatially ordered structure. To the animal there is no difference either: the internal order is present in the correlation structure. Thus in the brain, all maps are correlation maps and there is no real way to distinguish between the two types. Correlation maps represent signal-associated aspects, both sensory and motor, as coded by individual neurons or (small) neuronal groups. Individual neurons will code mainly those aspects that are related to the topographic map representation. When neurons are simultaneously active, this is a strong indication for the occurrence of an event in the outside world: such synchronous firing rarely happens by mere chance.

Synchronous or correlated activity is the only way in which sensory events manifest themselves to the brain; it will also be the way in which such events are remembered. Thus, in a more general sense, simultaneous activity of neurons is also related to memory recall. Activity in the representations of different sensory modalities may be correlated on a larger time scale and can be an indication of events in a more general context. The relation between topographic maps of different modalities can be maintained in a functional map. This functional map then constitutes a memory of a situation in which such a multimodal stimulus was present. Later, this map can be activated again and be used to complete a certain pattern of activity, or to predict the temporal sequence of a stimulus pattern.

How is such a functional map formed? One may assume that the correlations between the occurrence of events that together form a situation can be stored in terms of connections between the neural units that represent these events (Chaps. 10 through 13). The question can be raised whether there are a sufficient number of connections to store all such items. Since the neocortex has about 10^{10} neurons and about 10^{14} synapses, there seem to be ample space to do so. Palm (1986), under the assumption that the probability of a synapse being involved in any particular assembly is about 0.5 and that an assembly should be capable of storing 25 bits of information (i.e., 2^{25} situations), calculated that one would then need to partition the cortex into about 10^{12} assemblies with an average size of 5000 neurons to store a sufficient amount of data. Feldman and Ballard (1982) argued that such models usually require more neurons than are available in the brain, unless they are more sophisticated than just switching devices. The merits of neurons as nonlinear analog computational elements are discussed in Hopfield and Tank (1985, 1986) and their application to the visual system in Koch et al. (1986). Bienenstock (1988) envisions a dual coding both in fast "switching" synapses and in the statistical relationship between the firings in different neurons that are connected. In a given connected pair short-term correlations can still change for varying conditions, thereby increasing the storage capacity considerably.

Topographic and functional maps are based on a certain degree of ordering and they may consist of distinct sets of neurons. It is possible to distinguish two classes of neurons in the brain. Jacobson (1974) called these class I and class II neurons. Class I neurons are hardwired to each other and express their function in a genetically determined way with a highly deterministic pattern of behavior that is characteristic for the species. Class II neurons were considered to form populations with diverse potential to allow the realization of behavior characteristic for the individual. Jacobson equated class I neurons with the principal or projection neurons in the brain: the pyramidal cells and the Purkinje cells, for example. Class II neurons were proposed to comprise most interneurons. Class I neurons form and differentiate before class II neurons, and in lower vertebrates one usually finds less class II neurons than in higher vertebrates (see, e.g., the evolutionary aspects of the cerebellum as described in Llinas 1969). Class II neurons develop with little genetic control of the details in their morphology and connections. Thus class I neurons form the skeleton of the brain (Braitenberg 1974) to which class II neurons are later attached. The development of class II neurons is assumed to be largely influenced by exogenous conditions such as sensory stimulation, behavior, and experience. Their development can be considered to store relevant individual information about the outside world, whereas class I neurons have innate information particular for that species. By virtue of the fact that class I neurons form the skeleton of the brain, they will be the main constituents of topographic maps. Functional maps on the basis of only class II neurons, however, are unlikely, since then the information could not be transmitted beyond the level of the local circuits. Thus efficient functional maps will require class I as well as class II neurons. Therefore neurons that are part of topographic maps are also part of functional maps and vice versa. Thus one could say that a functional map is a topographic map reorganized through experience.

The representation of sensory events in the brain is not only dependent on the afferent flow of nervous activity, but also on the status of the brain itself. This status may be reflected through the secretion of neurotransmitters or hormones. For example, in the brains of most vertebrates steroid-concentrating cells (e.g., for estradiol or testosterone) are found in the septum, the amygdala, the preoptic areas, the hypothalamus, and in the superior and the inferior colliculus. These hormones interact with the cell proteins and may influence the genetic program of protein synthesis, which in turn is connected to neurotransmitter production and hence to synaptic connectivity (Pfaff 1980). Testosterone changes the size of the nuclei RA and HVc in the forebrain of certain songbirds (Chap. 10, Nottebohm 1984). The size increase has been linked to the formation of new interneurons that create functional maps for the storage of new song.

The number of topographic maps discovered so far is quite large and through extensive mapping studies certainly more will be found. So far at least six cochleotopic maps have been found in the temporal cortex of the macaque, while 16 retinotopic maps have been identified in the owl monkey's brain. The map approach is probably the largest scale level that can be obtained using neuroscience (bottom-up) approaches, and we must now consider what top-down approaches have to offer about the correlation structure of the brain.

14.3 Top-Down and Bottom-Up Approaches to Brain Function

It could well be that the distinction between topographic maps and functional maps is the result of a difference in approach. Topographic maps are identified by bottom-up approaches such as neuroanatomy and neurophysiology, while functional maps are usually postulated in top-down approaches such as those of cognitive science. Topographic maps play a role in microtheories about the brain (such as orientation selectivity in the visual cortex), functional maps in macrotheories (such as learning and memory). Arbib (1985) distinguishes three levels of cognitive science: artificial intelligence, cognitive psychology, and brain theory. His definitions are that artificial intelligence attempts "to program computers to do things that you would swear required intelligence until you know that a computer has been programmed to do them". Cognitive psychology uses "language of information processing to design models which can be run on computers to emulate overt behavior of a human performing such intelligent tasks". Brain theory in addition "must not only explain behavior patterns of animals and humans, but must incorporate data on brain function and neural circuitry". It is only the latter of the three cognitive science (top-down) approaches that can shed light on the functions of both the normal and the pathological brain.

A central aspect of top-down approaches is to investigate what a certain part of the brain is doing, and then to develop a computational theory that can do exactly the same (e.g., Marr 1981). The implementation level of the algorithms of such a theory is usually that of assemblies or functional maps. The highest level in the bottom-up or structural approaches is in local circuits or wiring diagrams incorporating a few of such circuits (Shepherd 1979) into microcolumns, columns or modules (Mountcastle 1978). The modules are the largest topographic units, while assemblies are the smallest functional map units. In a hierarchically structured brain, columns and assemblies should be related to each other in a relatively simple way, e.g., a one-to-one relation between column system and assembly organization. In a largely parallel processing brain, assemblies and columns need not necessarily be related in a simple way.

For the dop-down approach, the crucial part should be the implementation of the algorithm into a morphologically acceptable wiring diagram, that can be investigated with current experimental techniques. This step is rarely taken; however, for an exception see Marr (1969). In the bottom-up approach the crucial part is the assignment of a function to a neuroanatomically or electrophysiologically derived wiring diagram. As Hopfield and Tank (1986) show for a hypothetical seven-neuron network in which all units could be observed and all connection strengths computed, this is not a trivial or easy task. In their particular example, pair correlation studies would never result in identification of the network, but simultaneous observations of the complete input to the network and the output (four principal neurons) might have resulted in the conclusion that this network performed the task of a four-bit analog-to-digital converter. Whenever a function can be assigned to a neural network, and this might require the study of its effects on behavior, an algorithm can be devised and be made part of a larger computational theory. An example of the latter approach is the inference of a function for the cerebellum on basis of its morphological appearance (Braitenberg and Atwood 1958). It is not surprising that most attempts to bridge the gap between top-down and bottom-up approaches have taken place for the cerebellum, a very stereotyped structure that has been extremely well studied (Chap. 11). Since much less is known for the hippocampus (Chap. 12) or neocortex (Chap. 13), major conceptual problems still exist.

14.3.1 An "Integrated Circuit" Approach to Brain Function

For both the top-down approach and the bottom-up approach it would be much easier if one had at one's disposal equivalents of "integrated circuits" instead of single neurons as design elements. Most of the modeling approaches still use single neurons with, optionally plastic, interconnections. It is likely, however, that in the central nervous system certain combinations of neurons and their interconnections are employed over and over again (although Crick and Anasuma 1987 take a different point of view). For instance, the general design of cerebellum, hippocampus, and neocortex has certain aspects in common: an input stage of granule cells equipped with feedforward and/or feedback inhibition, which outputs to many principal cells that have inhibitory interactions (Fig. 14.1 d). A neural equivalent of a large-scale integrated circuit will most likely be the largest structural model to be proposed in the bottom-up

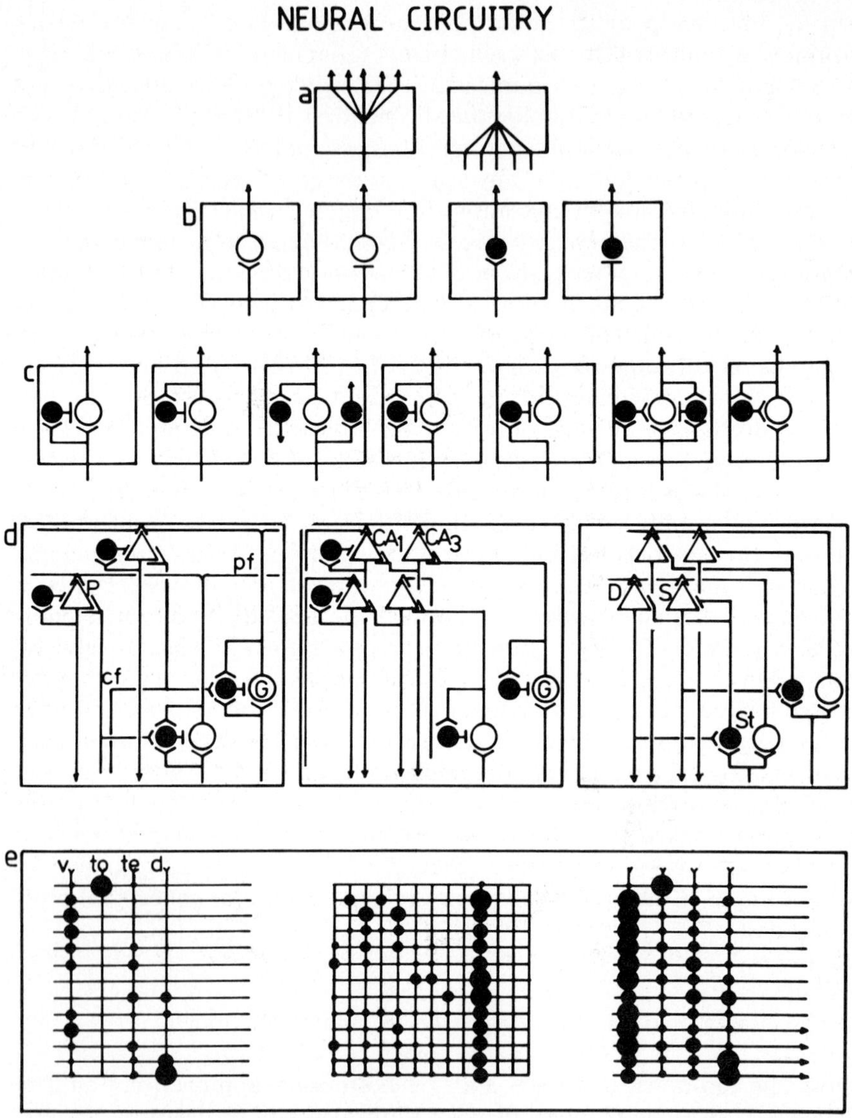

Fig. 14.1 a–e. A hierarchy of neural circuitry. **a** Divergence and convergence. **b** Excitatory and inhibitory (*black*) neurons with excitatory and inhibitory input. **c** Local circuits, *from left to right:* feedforward inhibition, feedback or Renshaw inhibition, lateral inhibition, combined feedforward and feedback inhibition, etc. **d** Similarities in neuronal groups in cerebellum (*pf* parallel fiber; *P* Purkinje cell; *G* granule cell; *cf* climbing fiber), hippocampus (CA_1 and CA_3 and granule cells are indicated), and sensory neocortex (*St* stellate cells). **e** Connectivity matrix (the size of the dot is proportional to the number of synapses between cells) for the optic tectum: vertical entries represent afferents to the optic tectum from visual (*v*), toral (*to*), telencephalon (*te*) and deep brainstem structures (*d*), *horizontal lines* represent different cell types in the tectum. The *middle diagram* shows the interconnectivity between the various cell types, the *right hand diagram* is the result of combining the two other diagrams and shows the effective connectivity between afferents and cell types in the tectum. The *lower four horizontal lines* represent output cells. It can be observed that one cell type is responsible for most of the connectivity. For references see text

approach, it will probably also be the lowest practical level of implementation in the top-down approach. Still both end products may look entirely different in their component structure although they might perform the same task. However, a comparison on this "module" level seems feasible to sort out what modifications are required to let both approaches converge. Although thinking in terms of large-scale integrated circuits is not more than a useful metaphor, it may still guide our discussions about the meeting ground of both approaches to brain function.

Small scale neural circuits – local circuits – can be inferred in all their variety from the survey by Shepherd (1979). From the peripheral ganglia up to the neocortex, Shepherd evaluated the synaptic organization and possible neural circuitry. Figure 14.1 is designed on that basis and will be used to illustrate some of the concepts of an "integrated neural circuitry". The peripheral ganglia feature axodendritic as well as dendrodendritic synapses and a basic local circuit component of most brain regions: the synaptic triade. This is a two-neuron circuit with three synapses featuring *feedforward inhibition* (Fig. 14.1 c). In the ventral horns of the spinal cord another two-neuron circuit shows up that represents *feedback* or *recurrent inhibition*; feedforward inhibition is absent in those regions. Basically the same functional structures appear in the retina, where in addition *lateral inhibition* is found. Recurrent inhibition is not found in the vertebrate retina but is present in the eye of *Limulus*. In the dorsal cochlear nucleus feedforward inhibition is likely to be present (Moore and Osen 1979). In the cerebellum triades are found between mossy fibers and granule cells with the Golgi cells acting as inhibitory interneurons. The Purkinje cells themselves are subject to recurrent inhibition through the basket cells as well as to feedforward inhibition via climbing fibers and stellate cells. In the thalamus, feedforward as well as feedback inhibition is observed on the same cell, the interneurons may in addition be activated through cortico-thalamic fibers (see Fig. 4.9). The basal ganglia show evidence of *excitatory loops* containing two, three, or four synapses, which may constitute reverberant networks. The olfactory cortex seems to be devoid of triades but direct *recurrent excitatory loops* are found together with recurrent self inhibition and lateral inhibition. The hippocampus features besides recurrent inhibition also *lateral recurrent excitation* between principal cells in CA_1 and CA_3. The sensory neocortex does not have recurrent inhibitory connections in the classical sense, but axon collaterals of the pyramidal cells, through (inhibitory) interneurons and parallel fibers, affect themselves as well as neighboring cells. This action can be excitatory as well as inhibitory; *direct collateral excitation* is also observed. Specific sensory afferents provide feedforward inhibition through shared input to excitatory and inhibitory interneurons (spiny and smooth stellate cells) that output on pyramidal cells. In the motor cortex lateral recurrent excitation, direct collateral excitation, and triadic action through the collaterals of neighboring pyramidal cells is found. The distribution of neural interaction type in motor cortex as compared to sensory cortex also reflects this more extensive excitatory connection pattern in the motor cortex (Tab. 8.1).

By now we have identified a multitude of neural elements, local circuits, and large-scale circuits on the basis of neuroanatomical and multiunit recording techniques. To further illustrate the way the bottom-up approach generally proceeds, we may construct a hierarchy of building blocks in order to build model brains. Figure 14.1 illustrates this attempt. Obviously, the smallest elements are the individual neurons – excitatory as well as inhibitory with either excitatory or inhibitory input – and their convergent and divergent connections (Fig. 14.1 a, b). The next level (Fig. 14.1 c) is formed by the various combinations of a principal neuron with an interneuron. From left to right we illustrate feedforward inhibition, feedback inhibition, lateral inhibition both through feedforward and recurrent paths, feedforward as well as recurrent inhibition onto the same principal cells, direct inhibition through an interneuron, combined excitation and inhibition through interneurons, and finally recurrent excitation. This is not an exhaustive survey, many more combinations can be found, but they are all variations on the type illustrated here.

The following level is formed by the various combinations of local circuit networks into larger-scale wiring diagrams as for, e.g., the cerebellum, hippocampus, and sensory neocortex (Fig. 14.1 d). The representations in Fig. 14.1 are highly stylized to show the parallelism in their construction. The principal neurons are indicated by triangles, open circles represent excitatory interneurons and filled circles inhibitory interneurons. For more detail see the legend of the figure and detailed neuronal circuits in Chaps. 11, 12, and 13.

Still another level of structural analysis can be found in quantitative neuroanatomy. Meek (1983) published a survey of tectal circuitry on the basis of electron microscope observation of identified Golgi-stained neurons. He studied the number and location of synapses between identified neuron types. By equating the relevance of certain connections between two types of cells with the number of synaptic contacts, he devised a functional connectivity pattern for the 13 types of neurons found in the optic tectum of the goldfish. Four of these neurons are principal neurons, the other nine are interneurons of which one type in particular outputs to nearly all other neurons. Four main input tracts were identified (Fig. 14.1 e) and are indicated as visual afferents (v), toral input (to), input from the telencephalon (te) and from deep (brainstem) structures (d). The left-hand panel illustrates the relative distribution of afferent synapses across the 13 tectal neuron types; each column is normalized to 100%. Thus the main connection of the toral afferents is to type 1, and that of the deep structure to types 12 and 13. The lowermost 4 lines represent the principal neuron types, indicating that the deep input projects dominantly on the tectal output neurons and the toral input connects with interneurons exclusively. The middle panel illustrates the interconnectivity of the tectal neuron types (now the input to each row is normalized to 100%); one observes that one particular neuron type (#10) is involved in the majority of all synaptic contacts. As a result of this interconnectivity, the effective throughput of the afferent input to the principal neurons is quite large (right-hand panel): two principal neurons receive predominantly visual input (# # 10 and 11) and the other two mainly deep structure activity (# # 12 and 13). This approach of equating effectiveness or relevance of interconnections with the number of synapses obviously has a number of shortcomings. It does not assign a weight to the synapse, e.g., on the basis of its location on the dendritic tree or cell soma, although this could easily have been done. In addition, no distinction was made between excitatory and inhibitory synapses, although, e.g., on the basis of vesicle shape some impression might have been obtained. However, such an approach begins to stress one important aspect: neurons integrate analog inputs, namely the postsynaptic potentials, and not action potentials. Neurons therefore are basically analog devices and their interconnections cannot be fully explained in terms of logical operations on incoming spike trains. The integrative action of the neurons, however, may still be revealed in the statistics of the output spike train or in the occurrence of special temporal patterns.

The goal in the above reductionist approach is to find a common ground between top-down and bottom-up approaches. Visualization of assemblies is one aspect that is needed in this approach, and the experimental technique most likely to succeed will be the optical recording technique (Chap. 6). In order to establish the detailed connection patterns between the neurons in such assemblies, standard electrophysiological multiunit recording techniques will still be required. Some combination of both the optical and the multielectrode technique is likely to be most successful. The analysis of the vast amount of data generated by such an approach will be based on one of the correlational approaches discussed in previous chapters. Unless we discover the correlational approach used by the brain, it is unlikely that we will be able to unravel the message contained in the recorded data.

Plate I. a Color representations of a four unit recording from the auditory midbrain of the grass frog using a single microelectrode. *Color* represents the individual units. The stimulus was a random frequency sequence of tone pips. In the *dot display* the dots are arranged as a function of frequency on a logarithmic scale. The horizontal axis represents time after tone pip onset. One observes that three units respond to the frequency range above 1000 Hz, which in the grass frog represents activity from the basilar papilla. The different latency ranges are clear. The other unit (*red*) responds to frequencies below 1000 Hz (amphibian papilla). (Johannesma et al. 1986). **b** Three unit recording (single electrode) of spontaneously active units in response to random tone pip stimulation. Again the responses are organized according to frequency. One observes that two of the units are highly spontaneously active (*green* and *red*), due to the limited time resolution (time axis represents 1 s of activity after tone pip onset) red and green overlapping *dots* produce a yellow dot and overlap of blue and red produces a purple/pink color. One observes three ranges of activity: the blue unit is predominantly high frequency, the green unit is sustained to the tone pip in the midfrequency range and transiently below 250 Hz. The red unit responds across the entire frequency range. One also observes the postactivation suppression for about 200 ms of the green unit in the range below 250 Hz (Epping et al. 1984)

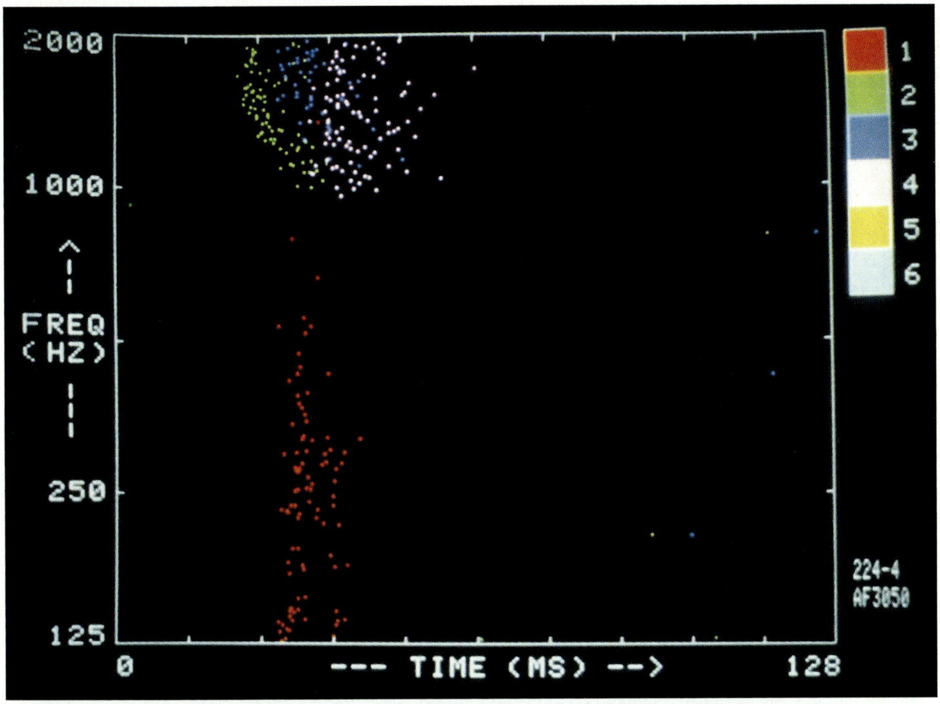
2000
1000
FREQ
(HZ)
250
125
0
--- TIME (MS) -->
128
1
2
3
4
5
6
224-4
AF3050

a

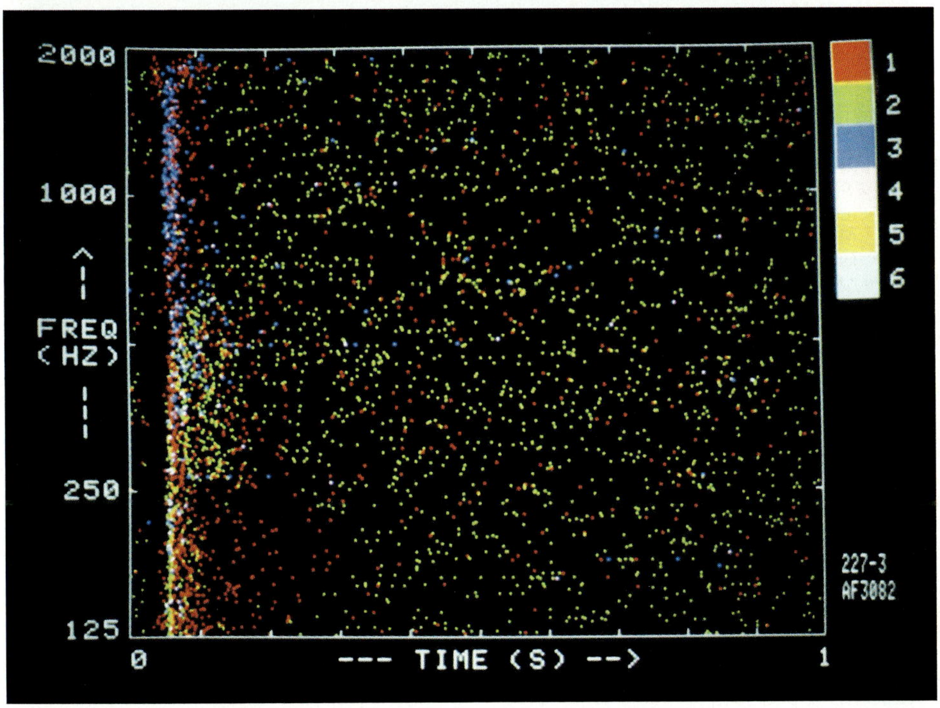
2000
1000
FREQ
(HZ)
250
125
0
--- TIME (S) -->
1
1
2
3
4
5
6
227-3
AF3082

b

References

Abeles M (1975) A journey into the brain. In: Inbar GF Signal analysis and pattern recognition in biomedical engineering. John Wiley & Son, New York, pp 41–59

Abeles M (1982) Local cortical circuits: An electrophysiological study. Springer, Berlin Heidelberg New York

Abeles M (1983) The quantification and graphic display of correlations among three spike trains. IEEE Trans BME 30:235–239

Abeles M, Goldstein MH (1977) Multispiketrain analysis. Proc IEEE 65:762–772

Abeles M, Lass Y (1975) Transmission of information in the axon: II The channel capacity. Biol Cybernet 19:121–125

Abeles M, Ribaupierre F de, Ribaupierre Y de (1983) Detection of single unit responses which are loosely time-locked to a stimulus. IEEE Trans SMC 13:683–691

Adrian ED (1928) The basis on sensation: the action of the sense organs. Christopher's, London

Adrian ED, Cattell M, Hoagland H (1931) Sensory discharges in single cutaneous nerve fibers. J Physiol 72:377–391

Aertsen AHMJ, Gerstein GL (1985) Evaluation of neuronal connectivity: sensitivity of cross-correlation. Brain Res 340:341–354

Aertsen AHMJ, Johannesma PIM (1981a) The spectro-temporal receptive field. Biol Cybernet 42:133–143

Aertsen AHMJ, Johannesma PIM (1981b) A comparison of the spectro-temporal sensitivity of auditory neurons to tonal and natural stimuli. Biol Cybernet 42:145–156

Aertsen AHMJ, Smolders JWT, Johannesma PIM (1979) Neural representation of the acoustic biotope: on the existence of stimulus-event relations for sensory neurons. Biol Cybernet 32:175–185

Aertsen A, Bonhoeffer T, Kruger J (1988) Coherent activity in neuronal populations: analysis and interpretation. In: Caianiello ER (ed) Physics of cognitive processes. World Scientific, Singapore, pp 1–34

Aitkin LM, Dunlop CW, Webster WR (1966) Click-evoked response patterns of single units in the medial geniculate body of the cat. J Neurophysiol 29:109–123

Aitkin LM, Irvine DRF, Webster WR (1984) Central neural mechanisms of hearing. In: Darian-Smith I (ed) Handbook of physiology, sec 1. The nervous system, vol 3. Sensory processes. Am Physiol Soc, Bethesda, MA, pp 675–737

Albus JS (1971) A theory of cerebellar function. Math Biosci 10:25–61

Allum JHJ, Hepp-Reymond M-C, Gysin R (1982) Cross-correlation analysis of interneuronal connectivity in the motor cortex of the monkey. Brain Res 231:325–334

Amari S-I (1977) Neural theory of association and concept formation. Biol Cybernet 26:175–185

Amari S-I (1980) Topographic organization of nerve fields. Bull Math Biol 42:339–364

Amari S-I (1982) Competitive and cooperative aspects in dynamics of neural excitation and self organization. In: Amari S, Arbib M (eds) Competition and cooperation in neural nets. Springer, Berlin Heidelberg New York, pp 1–28

Amari S-I (1983) Field theory of self organizing neural nets. IEEE Trans SMC 13:741–748

Amassian VE, Macy J, Walter HJ (1959) Patterns of activity of simultaneously recorded neurons in midbrain reticular formation. Ann NY Acad Sci 89:883–895

Amassian VE, Macy J, Waller HJ, Leader HS, Swift M (1964) Transformation of afferent activity at the cuneate nucleus. In: Gerard RW, Duyff JW (eds) Information processing in the nervous system. Exerpta Medica, Amsterdam, pp 235–254

Andersen P (1975) Organization of hippocampal neurons and their interconnections. In: Isaacson RL, Pribram KH (eds) The hippocampus. I. Structure and development. Plenum, New York, pp 155–175

Andersen P (1983) Operational principles of hippocampal neurons. A summary of synaptic physiology. In: Seifert W (ed) Neurobiology of the hippocampus. Academic Press, New York London, pp 81–86

Andersen P, Eccles JC (1965) Locating and identifying postsynaptic inhibitory synapses by the correlation of physiological and histological data. Symp Biol Hung 5:219–242

Anderson JA (1970) Two models for memory organization using interacting traces. Math Biosci 8:137–160

Anderson JA (1972) A simple neural network generating an interactive memory. Math Biosci 14:197–220

Anderson JA, Silverstein JW, Ritz SA, Jones RS (1977) Distinctive features, categorical perception, and probability learning. Some applications of a neural model. Psychol Rev 84:413–451

Andreassen S (1979) Automatic classification of muscle action potentials. In: 12th Int Conf Medical and biological engineering, Jerusalem, p 93.5

Andreassen S, Stein RB, Oguztoreli MN (1979) Application of optimal multichannel filtering to simulated nerve inputs. Biol Cybernet 32:25–33

Anninos PA, Beek B, Csermely TJ, Harth EM, Pertile G (1970) Dynamics of neural structures. J Theor Biol 26:121–148

Arbib MA (1985) Brain theory and cooperative computation. Human Neurobiol 4:201–218

Arbib MA, Kilmer WL, Spinelli DN (1976) Neural models and memory. In: Rosenzweig MR, Bennett EL (eds) Neural mechanisms of learning and memory. MIT, Cambridge, MA, pp 109–132

Arnett DW (1975) Correlation analyses of units recorded in the cat dorsal lateral geniculate nucleus. Exp Brain Res 24:11–130

Arnett DW (1978) Statistical dependence between neighboring retinal ganglion cells in the goldfish. Exp Brain Res 32:49–53

Arnett DW, Spraker TE (1981) Cross-correlation analysis of the maintained discharge of rabbit retinal ganglion cells. J Physiol 317:29–47

Bach M, Krüger J (1986) Correlated neuronal variability in monkey visual cortex revealed by a multimicroelectrode. Exp Brain Res 61:451–456

Ballard DH (1986) Cortical connections and parallel processing: structure and function. Behav Brain Sci 9:67–119

Barlow HB (1969) Scaling and refractoriness in pulse trains. J Optic Soc Am 59:1500

Barlow HB (1979) Three theories of cortical function. In: Freeman RD (ed) Developmental neurobiology of vision. Plenum, New York, pp 1–16

Barlow HB (1985) The twelfth Bartlett Memorial Lecture: the role of single units in the psychology of perception. Q J Exp Psychol 37:121–145

Barlow HB, Levick WR (1969) Changes in the maintained discharge with adaptation level in the cat retina. J Physiol 202:699–718

Barto AG (1985) Learning by statistical cooperation of self interested neuron-like computing elements. Human Neurobiol 4:229–256

Bates JAV (1973) Electrical recording from the thalamus in human subjects. In: Iggo A (ed) Handbook of sensory physiology, vol 2. Springer, Berlin Heidelberg New York, pp 561–578

Bell CC, Grimm RJ (1969) Discharge properties of Purkinje cells recorded on single and double microelectrodes. J Neurophysiol 32:1044–1055

Bell CC, Kawasaki T (1972) Relations among climbing fiber responses of nearby Purkinje cells. J Neurophysiol 35:155–169

Bendat JS, Piersol AG (1971) Random data: analysis and measurement procedures. John Wiley & Son, New York

Berger TW, Alger B, Thompson RF (1976) Neuronal substrate of classical conditioning in the hippocampus. Science 192:483–485

Bessou P, Perl ER (1969) Response of cutaneous sensory units with unmyelinated fibers to noxious stimuli. J Neurophysiol 32:1025–1043

Beurle RL (1956) Properties of a mass of cells capable of regenerating pulses. Philos Trans R Soc London Ser B240:55–94

Bienenstock E (1988) Neural-like graph-matching techniques for image processing. In: von Seelen W, Leinhos U, Shaw G (eds) Organization of neural networks: structures and models. VCH, Weinheim, pp 211–235

Bishop PO, Levick WR, Williams WO (1964) Statistical analysis of the dark discharge of lateral geniculate neurons. J Physiol 170:598–612

Bland BH (1986) The physiology and pharmacology of hippocampal formation theta rhythms. Prog Neurobiol 26:1–54

Blinchikoff HJ, Zverev AI (1976) Filtering in the time and frequency domain. John Wiley & Son, New York

Braitenberg V (1963) Histology, histonomy, histologic. In: Wiener N, Schade JP (eds) Progress in brain research. Elsevier, Amsterdam, pp 160–176

Braitenberg V (1973) Gehirngespinste. Neuroanatomie für kybernetisch Interessierte. Springer, Berlin Heidelberg New York

Braitenberg V (1974) Thoughts on the cerebral cortex. J Theor Biol 46:421–447

Braitenberg V (1978a) Cell assemblies in the cerebral cortex. In: Heim R, Palm G (eds) Theoretical approaches to complex systems. Springer, Berlin Heidelberg New York, pp 171–188

Braitenberg V (1978b) Cortical architectonics: general and areal. In: Brazier MAB, Petsche H (eds) Architectonics of the cerebral cortex. Raven, New York, pp 443–465

Braitenberg V (1982) Outline of a theory of the cerebral cortex. In: Ricciardi LM, Scott AC (eds) Biomathematics in 1980. Elsevier/North-Holland Biomedical Press, Amsterdam New York, pp 127–132

Braitenberg V (1985) An isotropic network which implicitly defines orientation columns: Discussion of an hypothesis. In: Rose D, Dobson VG (eds) Models of the visual cortex. John Wiley & Son, New York, pp 479–484

Braitenberg V, Atwood RP (1958) Morphological observations on the cerebellar cortex. J Comp Neurol 109:1–34

Braitenberg V, Schuz A (1983) Some anatomical comments on the hippocampus. In: Seifert W (ed) Neurobiology of the hippocampus. Academic Press, New York London, pp 21–37

Braitenberg V, Gambardella G, Ghigo G, Vota U (1965) Observations on spike sequences from spontaneously active cells. Kybernetik 2:197–205

Brannan JR, Boyce WE (1981) Spatially localized interactive neural populations. I. A mathematical model. Bull Math Biol 43:427–446

Brillinger DR, Segundo JP (1979) Empirical examination of the threshold model neuron firing. Biol Cybernet 35:213–220

Brillinger DR, Bryant HL, Segundo JP (1976) Identification of synaptic interactions. Biol Cybernet 35:213–228

Brindley GS (1967) The classification of modifiable synapses and their use in models for conditioning. Proc R Soc London Ser B 168:361–376

Brindley GS (1969) Nerve net models of plausible size that perform many simple learning tasks. Proc R Soc London Ser B 174:173–191

Bruce C, Desimone R, Gross CG (1981) Visual properties of neurons in a polysensory area in superior temporal sulcus of the macaque. J Neurophysiol 46:369–384

Bryant HL, Segundo JP (1976) Spike initiation by transmembrane current: a white noise analysis. J Physiol 260:279–314

Burgess PR, Perl ER (1973) Cutaneous mechanoreceptors and nocireceptors. In: Iggo A (ed) Handbook of sensory physiology, vol 2. Springer, Berlin Heidelberg New York, pp 29–78

Burns BD, Webb AC (1976) The spontaneous activity of neurons in the cat cerebral cortex. Proc R Soc London Ser B 194:211–223

Burns BD, Webb AC (1979) The correlation between discharge times of neighboring neurons in isolated cerebral cortex. Proc R Soc London Ser B 203:347–360

Burns BD, Webb AC (1982) Mechanisms governing the relation between activity in the cerebral cortex and level of arousal. Proc R Soc London Ser B 214:325–334

Caianiello ER (1961) Outline of a theory of thought-processes and thinking machines. J Theor Biol 2:204–235

Calvin WH (1973) Some simple spike separation techniques of simultaneously recorded neurons. Electroencephal Clin Neurophysiol 34:94–96

Camp C, Pinsker H (1979) Computer separation of unitary spikes from whole-nerve recordings. Brain Res 169:455–479

Cattaneo A, Maffei L, Morrone C (1981) Patterns in the discharge of single and complex visual cortical cells. Proc R Soc London Ser B 212:279–297

Changeux J-P, Danchin A (1976) Selective stabilization of developing synapses as a mechanism for the specification of neuronal networks. Nature (London) 264:705–712

Changeux JP, Heidmann T (1987) Allosteric receptors and molecular models of learning. In: Edelman GM, Gall WE, Cowan WM (eds) Synaptic function. John Wiley & Son, New York, pp 549–601

Changeux J-P, Heidmann T, Patte P (1984) Learning by selection. In: Marler P, Terrace HS (eds) The biology of learning. Springer, Berlin, Heidelberg New York, pp 115–133

Chronister RB, White LE Jr (1975) Fiber architecture of the hippocampal formation: anatomy, projections, and structural significance. In: Isaacson RL, Pribram KH (eds) The hippocampus, vol 1. Structure and development. Plenum, New York, pp 9–39

Clark SA, Allard T, Jenkins WM, Merzenich MM (1988) Receptive fields in the body-surface map in adult cortex defined by temporally correlated inputs. Nature (London) 332:444–445

Clay JR, Goel NS (1973) Diffusion models for the firing of a neuron with varying thresholds. J Theor Biol 39:633–644

Cleland BG, Dubin MW, Levick WR (1971) Simultaneous recording of input and output of lateral geniculate neurons. Nature New Biol 231:191–192

Coggshall JC (1973) Linear models for biological transducers and impulse train spectra: general formulation and review. Kybernetik 13:30–37

Cohen LB, Salzberg BM (1978) Optical measurement of membrane potential. Rev Physiol Biochem Pharmacol 83:35–88

Conel JL, (1959) The postnatal development of the human cerebral cortex. Harvard Univ Press, Cambridge

Cooper LN (1981) Distributed memory in the central nervous system: possible test of assumptions of the cerebral cortex. In: Schmitt FO, Worden FG, Adelman G, Dennis SG (eds) The organization of the cerebral cortex. MIT, Cambridge, MA, pp 479–503

Cotman CW, Nieto-Sampedro M (1982) Brain function, synapse renewal, and plasticity. Annu Rev Psychol 33:371–401

Cotman CW, Nieto-Sampedro M (1984) Cell biology of synaptic plasticity. Science 225:1287–1294

Cotman CW, Monaghan DT, Ganong AH (1988) Excitatory amino acid neurotransmission: NMDA receptors and Hebb-type synaptic plasticity. Ann Rev Neurosci 11:61–80

Cowan JD, (1968) Statistical mechanics of neural nets. In: Caianiello ER (ed) Neural networks. Springer, Berlin Heidelberg New York, pp 181–188

Cowan JD, von der Malsburg C (1985) A proposed mechanism for the origin of iso-orientation columns. In: Rose D, Dobsen VG (eds) Models of the visual cortex. John Wiley & Son, New York Chichester, pp 462–472

Cowan WM, Stanfield BB, Amaral DG (1981) Further observations on the development of the dentate gyros. In: Cowan, WM (ed) Studies in developmental neurobiology. Oxford University Press pp 395–435

Cox DR (1962) Renewal theory. Methuen, London

Cox DR, Isham V (1980) Point processes. Chapman & Hall, London

Cox DR, Miller HD (1970) The theory of stochastic processes. Methuen, London

Creutzfeldt OD (1978) The neocortical link: thoughts on the generality of structure and function of the neocortex. In: Brazier MAB, Petsche H (eds) Architectonics of the cerebral cortex. Raven, New York, pp 357–383

Creutzfeldt OD, Innocenti GM, Brooks D (1974) Vertical organization in the visual cortex (Area 17) in the cat. Exp Brain Res 21:315–336

Creutzfeldt OD, Hellweg F-C, Schreiner Chr (1980) Thalamocortical transformation of responses to complex auditory stimuli. Exp Brain Res 39:87–104

Crick F (1984) Function of the thalamic reticular complex: the searchlight hypothesis. Proc Natl Acad Sci USA 81:4586–4590

Crick F, Asanuma C (1987) Certain aspects of the anatomy and physiology of the cerebral cortex. In: Rummelhart DE, McCleland JL (eds) Parallel distributed processing, vol. 2. MIT Cambridge, MA, pp 333–371

Crunelli V, Assat SY, Kelly JS (1983) Intracellular recordings from granule cells of the dentate gyrus in vitro. In: Seifert W (ed) Neurobiology of the hippocampus. Academic Press, New York London, pp 197–213

Dallos P, Harris D (1978) Properties of auditory nerve responses in absense of outer hair cells. J Neurophysiol 41:365–383

Dayhoff JE, Gerstein GL (1983a) Favored patterns in spike trains. I. Detection. J Neurophysiol 49:1334–1348

Dayhoff JE, Gerstein GL (1983b) Favored patterns in spike trains. II. Application. J Neurophysiol 49:1349–1363

De Boer E (1967) Correlation studies applied to the frequency resolution of the cochlea. J Audit Res 7:209–217

De Boer E (1968) Reverse correlation. I. A heuristic introduction to the technique of triggered correlation with application to the analysis of compound systems. Proc Kon Ned Acad Wetensch 71:472–486

De Boer E (1969) Reverse correlation. II. Initiation of nerve impulses in the inner ear. Proc Kon Ned Acad Wetensch 72:129–151

De Boer E (1973) On the principle of specific coding. J Dyn Syst Measure Control 95:265–273

De Boer E, Kuyper P (1968) Triggered correlation. IEEE Trans BME 15:169–179

De Weerd JPCM, Martens WLJ (1978) Theory and practice of a posteriori Wiener filtering of average evoked potentials. Biol Cybernet 30:81–94

D'Hollander E, Orban GA (1979) Spike recognition and on-line classification by unsupervised learning system. IEEE Trans BME 26:279–284

Diamond DM, Weinberger NM (1984) Physiological plasticity of single neurons in auditory cortex of the cat during acquisition of the pupillary conditionned response. II. Secondary field. Behav Neurosci 98:189–210

Diamond I (1979) The subdivision of the neocortex: a proposal to revise the traditional view of sensory, motor and association areas. Prog Psychobiol Physiol Psychol 8:1–43

Dickson JW, Gerstein GL (1974) Interaction between neurons in auditory cortex of the cat. J Neurophysiol 37:1239–1261

Dill JC, Lockemann PC, Naka K (1970) An attempt to analyze multi-unit recordings. Electroencephalogr Clin Neurophysiol 28:79–82

Dinning GJ, Sanderson AC (1981) Real time classification of multiunit neural signals using reduced feature sets. IEEE Trans BME 28:804–811

Di Prisco GV (1984) Hebb synaptic plasticity. Prog Neurobiol 22:89–102

Dirac PAM (1958) The principles of quantum mechanics. Univ Press, Oxford

Dobie RA, Wilson MJ (1984) Short-latency auditory responses obtained by cross correlation. J Acoust Soc Am 76:1411–1421

Dormont JF (1972) Patterns of spontaneous activity in the ventrolateral thalamic nucleus of cat. Brain Res 37:223–239

Dormont JF, Schmied A, Conde H (1982) Motor command in the ventrolateral thalamic nucleus: neuronal variability can be overcome by ensemble average. Exp Brain Res 48:315–322

Doty RW (1984) Some thoughts and some experiments on memory. In: Squire LR, Butters N (eds) Neuropsychology of memory. Guilford, New York, pp 330–339

Dubin MW, Cleland BG (1977) Organization of visual inputs to interneurons of lateral geniculate nucleus of the cat. J Neurophysiol 40:410–427

Duda RO, Hart PE (1973) Pattern classification and scene analysis. John Wiley & Son, New York

Duffy FH (1982) Topographic display of evoked potentials: clinical application of brain electrical activity mapping (BEAM). Ann NY Acad Sci 388:183–196

Dupont JL, Crepel F (1979) Correlations among climbing fiber responses of nearby cerebral Purkinje cells in the immature cat. Exp Brain Res 37:525–535

Easton P, Gordon PE (1984) Stabilization of Hebbian neural nets by inhibitory learning. Biol Cybernet 51:1–9

Eatock RA, Manley GA, Dawson L (1981) Auditory nerve activity in the Tockay gecko. J Comp Physiol 142:203–218

Ebner TJ, Bloedel JR (1981 a) Temporal patterning in simple spike discharge of Purkinje cells and its relationship to climbing fiber activity. J Neurophysiol 45:933–947

Ebner TJ, Bloedel JR (1981 b) Correlation between activity of Purkinje cells and its modification by natural peripheral stimuli. J Neurophysiol 45:948–961

Ebner TJ, Bloedel JR (1981 c) Role of climbing fiber afferent input in determining responsiveness of Purkinje cells to mossy fiber input. J Neurophysiol 45:962–970

Eccles JC (1957) The physiology of nerve cells. Johns Hopkins Press

Eccles JC (1978) An instruction-selection hypothesis of cerebral learning. In: Buser P, Rongenul-Buser A (eds) Cerebral correlates of conscious experience. Elsevier/North-Holland Biomedical Press, Amsterdam New York, pp 155–175

Eccles JC (1981) The modular operation of the cerebral neocortex considered as the material basis of mental events. Neuroscience 6:1839–1856

Eccles JC (1983) Foreword. In: Seifert W (ed) Neurobiology of the hippocampus. Academic Press, New York London, pp XIII–XVI

Eccles JC, Ito M, Szentagothai J (1967) The cerebellum as a neuronal machine. Springer, Berlin Heidelberg New York

Eckhorn R, Pöpel B (1972) A contribution to neuronal network analysis in cat lateral geniculate body: simultaneous recordings of maintained activity. Int J Neurosci 4:137–142

Eckhorn R, Pöpel B (1974) Rigorous and extended application of information theory to the afferent visual system of the cat. I. Basic concepts. Biol Cybernet 16:191–200

Eckhorn R, Grusser O-J, Kroller J, Pellnitz K, Pöpel B (1976) Efficiency of different neuronal codes: Information transfer calculations for three different neuronal systems. Biol Cybernet 22:49–60

Edelman GM (1975) Molecular recognition in the immune and nervous systems. In: Worden FG, Swazey JP, Adelman G (eds) The neurosciences: paths of discovery. MIT, Cambridge, MA, pp 65–74

Edelman GM (1978) Group selection and phasic reentrant signaling. A theory of higher brain function. In: Edelman GM, Mountcastle VB (eds) The mindful brain. MIT, Cambridge, MA, pp 51–100

Edelmann GM (1987) Neural darwinism. The theory of neuronal group selection. Basic Books, New York

Edelman GM, Finkel LH (1984) Neuronal group selection in the cerebral cortex. In: Edelman GM, Gall WE, Cowan WM (eds) Dynamic aspects of neocortical function. John Wiley & Son, New York, pp 653–695

Edelman GM, Reeke GN (1982) Selective networks capable of representative transformations, limited generalizations, and associative memory. Proc Natl Acad Sci USA 79:2091–2095

Eggermont JJ (1973) Analog modelling of cochlear adaptation. Kybernetik 14:117–126

Eggermont JJ (1984) Tinnitus: some thoughts about its origin. J Laryngol Otol Suppl 9:31–37

Eggermont JJ (1985 a) Peripheral auditory adaptation and fatigue: a model oriented review. Hearing Res 18:57–71

Eggermont JJ (1985 b) Physiology of the developing auditory system. In: Trehub SE, Schneider B (eds) Auditory development in infancy. Plenum, New York, pp 21–45

Eggermont JJ (1985 c) Evoked potentials as indicators of auditory maturation. Acta Otolaryngol S 421:41–47

Eggermont JJ (1985 d) Mathematical models for developmental changes. Acta Otolaryngol S 421:102–107

Eggermont JJ (1986) Evoked potentials as indicators of the maturation of the auditory system. In: Gallai V (ed) Maturation of the CNS and evoked potentials. Elsevier, Amsterdam, pp 177–182

Eggermont JJ (1988) On the rate of maturation of sensory evoked potentials. Electroencephal Clin Neurophysiol 70:293–305

Eggermont JJ, Aertsen AHMJ, Hermes DJ, Johannesma PIM (1981) Spectrotemporal representation of auditory neurons: redundant or necessary. Hearing Res 5:109–121

Eggermont JJ, Epping WJM, Aertsen AHMJ (1983a) Stimulus dependent neural correlations in the auditory midbrain of the grassfrog. Biol Cybernet 47:103–117

Eggermont JJ, Johannesma PIM, Aertsen AHMJ (1983b) Reverse correlation methods in auditory research. Q Rev Biophys 16:341–414

Eggermont JJ, Epping WJM, Aertsen AHMJ (1983c) Binaural hearing and neural interaction. In: Klinke R, Hartmann R (eds) Hearing: physiological bases an psychophysics. Springer, Berlin Heidelberg New York

Eggermont JJ, Epping WJM (1987) Coincidence detection in auditory neurons: a possible mechanism to enhance stimulus specificity in the grassfrog. Hearing Res 30:219–230

Ellaway PH, Murthy KSK (1985a) The origins and characteristics of cross-correlated activity between Y-motoneurons in the cat. Q J Exp Physiol 70:219–232

Ellaway PH, Murthy KSK (1985b) The source and distribution of short term synchrony between Y-motoneurones in the cat. Q J Exp Physiol 70:233–247

Epping WJM, Eggermont JJ (1985) Single unit characteristics in the auditory midbrain of the immobilized grassfrog. Hearing Res 18:223–243

Epping WJM, Eggermont JJ (1986) Sensitivity of neurons in the auditory midbrain of the grassfrog to temporal characteristics of sound. I. Stimulation with acoustic clicks. Hearing Res 24:37–54

Epping WJM, Eggermont JJ (1987) Coherent neural activity in the auditory midbrain of the grassfrog. J Neurophysiol 57:1464–1483

Epping WJM, van den Boogaard H, Aertsen A, Eggermont J, Johannesma PIM (1984) The neurochrome. An identity preserving representation of activity patterns from neural populations. Biol Cybernet 50:235–240

Erickson RP (1974) Parallel 'population' neural coding in feature extraction. In: Schmitt FO, Worden FG (eds) The neurosciences. MIT, Cambridge, MA, pp 155–169

Erickson RP (1982) The across-fiber pattern theory: an organizing principle for molar neural function. In: Neff WD (ed) Contributions to sensory physiology. Academic Press, New York London, pp 79–110

Ervin FR, Anders TR (1970) Normal and pathological memory: data and conceptual scheme. In: Schmitt FO (ed) The neurosciences. Second study program. Rockefeller Univ Press, New York, pp 163–176

Evans EF (1975) Cochlear nerve and cochlear nucleus. In: Keidel WD, Neff WD (eds) Handbook of sensory physiology, vol 512. Springer, Berlin Heidelberg New York, pp 1–108

Evans EF, Nelson PG (1973) The response of single neurons in the cochlear nucleus of the cat as a function of their location and the anaesthetic state. Exp Brain Res 17:402–427

Evarts EV, Bental E, Bihari B, Huttenlocher PR (1962) Spontaneous discharge of single neurons during sleep and waking. Science 135:726–728

Eysel UT, Gaedt C (1971) Maintained activity in the lateral geniculate body of the cat and the effects of visual deprivation. Pflüger's Arch 327:68–81

Fawcett JW, O'Leary DDM (1985) The role of electrical activity in the formation of topographic maps in the nervous system. Trends Neurosci 8:201–206

Fay RR (1978) Coding of information in single auditory-nerve fibers of the goldfish. J Acoust Soc Am 63:136–146

Feldman JA, Ballard DH (1982) Connectionist models and their properties. Cognit Sci 6:205–254

Feldman JL, Sommer D, Cohen MI (1980) Short time scale correlations between discharges of medullary respiratory neurons. J Neurophysiol 43:1284–1295

Fetz EE, Gustafson B (1983) Relation between shapes of postsynaptic potentials and changes in firing probability of cat motoneurones. J Physiol 341:387–410

Fienberg SE (1974) Stochastic models for single neuron firing trains: a survey. Biometrics 30:399–427

Fischer B (1973a) Overlap of receptive field centers and representation of the visual field in the cats optic tract. Vision Res 13:2113–2120

Fischer B (1973b) A neural field theory: mathematical approaches to the problem of large numbers of interacting nerve cells. Bull Math Biol 35:345–357

Fitzhugh R (1957) The statistical detection of threshold signals in the retina. J Gen Physiol 40:925–948

Fitzhugh R (1958) A statistical analyser for optic nerve messages. J Gen Physiol 41:675–692
Flohr H (1983) Control of plastic processes. In: Basar E, Flohr H, Haken H, Mandell AJ (eds) Synergetics of the brain. Springer, Berlin Heidelberg New York, pp 60–74
Freeman J (1969) The cerebellum as a timing device: an experimental study in the frog. In: Llinas R (ed) Neurobiology of cerebellar evolution and development. Am Med Assoc, Chicago, pp 397–420
Freeman WJ (1975) Mass action in the nervous system. Academic Press, New York London
Freeman WJ (1980) Use of spatial convolution to compensate for distortion of EEG by volume conduction. IEEE Trans BME 27:421–429
Freeman WJ, Viana di Prisco G (1986) EEG spatial pattern differences with discriminated odors manifest chaotic and limit cycle attractors in olfactory bulb of rabbits. In: Palm G, Aertsen A (eds) Brain theory. Springer, Berlin Heidelberg New York Tokyo, pp 97–119
Frishman LJ, Levine MW (1983) Statistics of the maintained discharge of the cat retinal ganglion cells. J Physiol 339:475–494
Frostig RD, Gottlieb Y, Vaadia E, Abeles M (1983) The effects of stimuli on the activity and functional connectivity of local neuronal groups in the cat auditory cortex. Brain Res 272:211–221
Frostig RD, Frostig Z, Harper RM (1984) Information trains. The technique and its uses in spike train and network analysis, with examples taken from the nucleus parabrachialis medialis during sleep-waking states. Brain Res 322:67–74
Fujita M (1982) Adaptive filter model of the cerebellum. Biol Cybernet 45:195–206
Furukawa T, Hyashida Y, Matsuura S (1978) Quantal analysis of the size of excitatory post-synaptic potentials at synapses between hair cells and afferent nerve fibers in goldfish. J Physiol 276:211–226
Fuster JM (1984) Behavioral electrophysiology of the prefrontal cortex. Trends Neurosci 408–414
Fusters JM, Herz A, Creutzfeldt OD (1965) Interval analysis of cell discharge in spontaneous and optically modulated activity in the visual system. Arch Ital Biol 103:159–177
Gahwiler BH (1983) The action of neuropeptides on the bioelectric activity of hippocampal neurons. In: Seifert W (ed) Neurobiology of the hippocampus. Academic Press, New York London, pp 157–173
Gardiner CW (1985) Handbook of stochastic methods for physics, chemistry and the natural sciences. Springer, Berlin Heidelberg New York Tokyo
Geisler CD, Goldberg JM (1966) A stochastic model of the repetitive activity of neurons. Biophys J 6:53–69
Geisler CD, Deng L, Greenberg SR (1985) Thresholds for primary auditory fibers using statistically defined criteria. J Acoust Soc Am 77:1102–1109
Georgopoulos AP, Schwarz AB, Kettner RE (1986) Neuronal population coding of movement direction. Science 233:1357–1360
Gerstein GL (1970) Functional association of neurons: detection and interpretation. In: Schmitt FO, Worden FG (eds) The neurosciences second study program. Rockefeller Univ Press, New York, pp 648–661
Gerstein GL (1988) Information flow and state in cortical neural networks: interpreting multineuron experiments. In: von Seelen W, Leinhos U, Shaw G (eds) Organization of neural networks: structures and models. VCH, Weinheim, pp 53–75
Gerstein GL, Aertsen AHMJ (1986) Representation of cooperative firing activity among simultaneously recorded neurons. J Neurophysiol 54:1513–1528
Gerstein GL, Clark WA (1964) Simultaneous studies of firing patterns in several neurons. Science 143:1325–1327
Gerstein GL, Mandelbrot B (1964) Random walk models for the spike activity of a single neuron. Biophys J 4:41–68
Gerstein GL, Michalski A (1981) Firing synchrony in a neural group: putative sensory code. Adv Physiol Sci 30:93–102
Gerstein GL, Perkel DH (1969) Simultaneously recorded trains of action potentials: analysis and functional interpretation. Science 164:828–830
Gerstein GL, Perkel DH (1972) Mutual temporal relationships among neuronal spike trains. Biophys J 12:453–473

Gerstein GL, Perkel DH, Subramanian KN (1978) Identification of functionally related neural ensembles. Brain Res 140:43–62

Gerstein GL, Renfrew DH, Pubols BH (1980) Two behavioral paradigms for study of rapid changes in functional grouping of neurons. Acta Neurobiol Exp 40:149–160

Gerstein GL, Bloom MJ, Espinosa IE, Evanczuk S, Turner MR (1983) Design of a lab for multineuron studies. IEEE Trans SMC 13:668–676

Gerstein G, Aertsen A, Bloom M, Espinosa E, Evanczuk S (1985a) Multi-neuron experiments: observations of state in neural nets. In: Haken H (ed) Complex systems. Operational approaches. Springer, Berlin Heidelberg New York Tokyo, pp 58–70

Gerstein GL, Perkel DH, Dayhoff JE (1985b) Cooperative firing activity in simultaneously recorded populations of neurons: detection and measurement. J Neurosci 5:881–889

Geschwind N (1979) Specializations of the human brain. Sci Am 241:180–199

Gevins AS (1984) Analysis of the electromagnetic signals of the human brain: milestones, obstacles, and goals. IEEE Trans BME 31:833–850

Gevins AS, Doyle JC, Cutillo BA, Schaffer RE, Tannehil RS, Ghannam JH, Gilcrease VA, Yeager CL (1981) Electric potentials in human brain during cognition: a new method reveals dynamic patterns of correlation. Science 213:918–922

Giglmayr J (1979) Modulation of point processes as a model of neuronal impulse generation. Math Biosci 46:139–149

Gilbert CD (1983) Microcircuity of the visual cortex. Annu Rev Neurosci 6:217–247

Gilles FH, Leviton A, Dooling EC (1983) The developing human brain. Growth and epidemiologic neuropathology. Wright, Boston

Ginsburg KS, Johnsen JA, Levine MW (1984) Common noise in the firing of neighbouring ganglion cells in goldfish retina. J Physiol 351:433–450

Glaser EM (1971) Separation of neuronal activity by waveform analysis. Adv Biomed Eng 1:77–136

Glaser EM, Marks WB (1968) On-line separation of interleaved neuronal pulse sequences. Data Acquisit Process Biol Med 5:137–156

Glaser EM, Ruchkin DS (1976) Principles of neurobiological signal analysis. Academic Press, New York London

Goldberg JM, Adrian HO, Smith FD (1964) Response of neurons of the superior olivary complex of the cat to acoustic stimuli of long duration. J Neurophysiol 27:706–749

Goldstein MH, Hall JL, Butterfield BO (1967) Single unit activity in the primary auditory cortex of unanesthetized cats. J Acoust Soc Am 43:444–445

Gordon G (1973) The concept of relay nuclei. In: Iggo A (ed) Handbook of sensory physiology, vol 2. Springer, Berlin Heidelberg New York, pp 137–150

Graham K, Duffin J (1981) Cross-correlation of medullary expiratory neurons in the cat. Exp Neurol 73:451–464

Griffith JS (1971) Mathematical neurobiology. Academic Press, New York London

Griffith JS, Horn G (1966) An analysis of spontaneous impulse activity of units in the striate cortex of unrestrained cats. J Physiol 186:516–534

Grinvald A (1984) Real time imaging of neuronal activity. Trends Neurosci 7:143–150

Grinvald A (1985) Real-time optical mapping of neuronal activity. From single grown cone to the intact mammalian brain. Annu Rev Neurosci 8:263–305

Grinvald A, Anglister L, Freeman JA, Hildesheim R, Manker A (1984) Real-time optical imaging of naturally evoked electrical activity in intact frog brain. Nature (London) 308:848–850

Grossberg S (1976) On the development of feature detectors in the visual cortex with applications to learning and reaction-diffusion system. Biol Cybernet 21:145–159

Grossman RG, Viernstein LJ (1961) Discharge patterns of neurons in cochlear nucleus. Science 134:99

Grover FS, Buchwald JS (1970) Correlation of cell size with amplitude of background fast activity in specific brain nuclei. J Neurophysiol 33:160–171

Grüsser OJ, Grüsser-Cornehls U (1973) Neuronal mechanisms of visual movement perception and some psychophysical and behavioral correlations. In: Iggo A (ed) Handbook of sensory physiology, vol 7/3. Springer, Berlin Heidelberg New York, pp 333–429

Gustafsson B, Wigstrom H, Abraham WC, Huang YY (1987) Longterm potentiation in the hippocampus using depolarizing current pulses as the conditioning stimulus to single volley synaptic potentials. J Neurosci 7:774–780

Haas HL (1983) Amine neurotransmitter actions in the hippocampus. In: Seifert W (ed) Neurobiology of the hippocampus. Academic Press, New York London, pp 139–153

Hagiwara S (1954) Analysis of interval fluctuations of the sensory nerve impulse. Jpn J Physiol 4:234–240

Hakan H (ed) (1983) Synergetics. Springer, Berlin Heidelberg New York

Halgren E (1984) Human hippocampal and amygdala recording and stimulation: Evidence for a neural model of recent memory. In: Squire LS, Butters N (eds) Neuropsychology of memory. Guilford, New York, pp 165–181

Hammond P (1972) Spatial organization of receptive fields of LGN neurons. J Physiol 222:53P–54P

Hammond P (1975) Receptive field mechanisms of sustained and transient retinal ganglion cells in the cat. Exp Brain Res 23:113–128

Harper RM, Sieck GC (1980) Discharge correlations between neurons in the nucleus parabrachialis medialis during sleep-waking states. Brain Res 199:353–358

Harris LR, Blakemore C, Donaghy M (1980) Integration of visual and auditory space in the mammalian superior colliculus. Nature (London) 288:56–59

Harth E, Unnikrishnam KP (1985) Brainstem control of sensory information: a mechanism for perception. Int J Psychophysiol 3:101–119

Hartline HK (1938) The response of single optic nerve fibers of the vertebrate eye to illumination of the retina. Am J Physiol 121:400–451

Hebb DO (1949) Organization of behaviour. John Wiley & Son, New York

Heetderks WJ (1978) Criteria for evaluating multi-unit spike separation techniques. Biol Cybernet 29:215–220

Heetderks WJ, Williams WJ (1975) Partition of gross peripheral nerve activity into single unit responses by correlation techniques. Science 188:373–375

Heierli P, de Ribaupierre F, Toros A, de Ribaupierre Y (1981) Functional organization of the medial geniculate body by simultaneous recordings of single unit pairs. In: Syka J, Aitkin LM (eds) Neuronal mechanisms of hearing. Plenum, New York, pp 183–186

Heierli P, de Ribaupierre F, de Ribaupierre Y (1987) Functional properties and interactions of neuron pairs simultaneously recorded in the medial geniculate body of the cat. Hearing Res 25:209–225

Heiss WD, Bornschein H (1966) Multimodale Intervalhistogramme der Daueraktivität von retinalen Neuronen der Katze. Kybernetik 3:187–191

Hernandez L (1980) Optic discriminator for single cell activity. Physiol Behav 25:607–610

Hertz A, Creutzfeldt O, Fuster J (1964) Statistische Eigenschaften der Neuronaktivität im ascendierenden visuellen System. Kybernetik 2:61–71

Hickey TL, Winters RW, Pollack JG (1973) Center-surround interactions in two types of on-center retinal ganglion cells in the cat. Vision Res 13:1511–1526

Hinton G, Anderson JA (1981) Parallel models of associative memory. Erlbaum, Hillsdale, NJ

Holden AV (1976) Models of the stochastic activity of neurones. Springer, Berlin Heidelberg New York

Holmes O, Houchin J (1966) Units in the cerebral cortex of the anesthetized rat and the correlations between their discharges. J Physiol 187:651–671

Hopfield JJ (1982) Neural networks and physical systems with emergent collective computational properties. Proc Natl Acad Sci USA 79:2554–2558

Hopfield JJ (1984) Neurons with graded responses have collective computational properties like those of two-state neurons. Proc Natl Acad Sci USA 81:3088–3092

Hopfield JJ, Tank DW (1985) "Neural" computation of decision in optimization problems. Biol Cybernet 52:141–152

Hopfield JJ, Tank DW (1986) Computing with neural circuits: a model. Science 233:625–633

Horn G (1962) Regular impulse activity of single units in the cat striate cortex. Nature (London) 194:1084–1085

Huang KS (1963) Statistical mechanics. John Wiley & Son, New York

Hubel DH, Wiesel TN (1962) Receptive fields, binocular interaction and functional architecture. J Physiol 160:106–154

Hubel DH, Wiesel TN (1968) Receptive fields and functional architecture of monkey striate cortex. J Physiol 195:215–243

Huttenlocher PR, Courten Cde, Garey LJ, Van der Loos H (1982) Synaptogenesis in the human visual cortex-evidence for synapse elimination during normal development. Neurosci Lett 33:247–252

Ito M (1984a) The modifiable neuronal network of the cerebellum. Jpn J Physiol 34:781–792

Ito M (1984b) The cerebellum and neural control. Raven, New York

Ito M, Sakurai M, Tongroach P (1982) Climbing fibre induced depression of both mossy fibre responsiveness and glutamate sensitivity of cerebellar Purkinje cells. J Physiol 324:113–134

Jacobson M (1974) A plenitude of neurons. In: Gottlieb D (ed) Aspects of neurogenesis. Academic Press, New York London, pp 151—166

Jay MF, Sparks DL (1984) Auditory receptive fields in primate superior colliculus shift with changes in eye position. Nature (London) 309:345–347

Jerne NK (1967) Antibodies and learning: selection vs instruction. In: Quarton GC, Melnechuk T, Schmitt FO (eds) The neurosciences. Rockefeller Univ Press, New York, pp 200–205

Johannesma PIM (1968) Diffusion models for the stochastic activity of neurons. In: Caianiello ER (ed) Neural networks. Springer, Berlin Heidelberg New York, pp 116–144

Johannesma PIM (1980) Functional identification of auditory neurons bases on stimulus even correlation. In: van den Brink G, Bilsen FA (eds) Psychophysical, physiological and behavioral studies in hearing. Univ Press, Delft, pp 77–84

Johannesma PIM, Eggermont JJ (1983) Receptive fields of auditory neurons in the frogs midbrain as functional elements for acoustic communication. In: Ewert JP, Capranica RR, Ingle DJ (eds) Advances in vertebrate neuroethology. Plenum, New York, pp 901–910

Johannesma PIM, van den Boogaard HFP (1985) Stochastic formulation of neural interaction. Acta Appl Math 4:201–224

Johannesma PIM, Aertsen A, Cranen B, van Erning L (1981) The phonochrome: a coherent spectro-temporal representation of sound. Hearing Res 5:123–145

Johannesma PIM, Aertsen A, van den Boogaard H, Eggermont J, Epping W (1986) From synchrony to harmony: ideas on the function of neural assemblies and the interpretation of neural synchrony. In: Palm G, Aertsen A (eds) Brain theory. Springer, Berlin Heidelberg New York Tokyo, pp 25–47

Johansson RS (1978) Tactile sensibility in the human hand: receptive field characteristics of mechanoreceptive units in the glabrous skin area. J Physiol 281:101–123

John ER (1972) Switchboard theories versus statistical theories of learning. Science 177:850–864

Johnsen JA, Levine MW (1983) Correlation of activity in neigbouring goldfish ganglion cells: relationship between latency and lag. J Physiol 345:439–449

Johnson DH, Kiang NY-S (1976) Analysis of discharges recorded simultaneously from pairs of auditory nerve fibers. Biophys J 16:719–734

Johnson DH, Swami A (1983) The transmission of signals by auditory-nerve fiber discharge patterns. J Acoust Soc Am 74:493–501

Jones EG (1984) Organization of the thalamocortical complex and its relation to sensory processes. In: Darian-Smith I (ed) Handbook of physiology, sec 1. The nervous system, vol 3/1. Am Physiol Soc, Bethesda, MD, pp 149–212

Jones K, Tubis A, Burns EM (1985) On the extraction of the signal-excitation function from a non-Poisson cochlear neural spike train. J Acoust Soc Am 78:90–94

Kaas JH, Merzenich MM, Killackey HP (1983) The reorganization of somatosensory cortex following peripheral nerve damage in adults and developing mammals. Annu Rev Neurosci 6:325–356

Kandel ER (1976) Cellular basis of behavior. Freeman, San Francisco

Kaplan MS, Hinds JW (1977) Neurogenesis in the adult rat: electron microscopic analysis of light radio autographs. Science 197:1092–1094

Katznelson RD (1981) EEG recording, electrode placement and aspects of generator localization. In: Nunez PL (ed) Electric fields of the brain. Oxford Univ Press, New York, pp 176–213

Keehn DG (1966) An iterative spike separation technique. IEEE Trans BME 13:19–28

Kent EW (1971) An educable waveform recognition program suitable for on-line discrimination of extracellular single units. Electroencephalogr Clin Neurophysiol 31:618–620

Kiang NY-S (1984) Peripheral processing of auditory information. In: Darian-Smith I (ed) Handbook of Physiology, sect 1. The nervous system, vol 3/2. Am Physiol Soc, Bethesda, MA, pp 639–674

Kiang NY-S, Watanabe T, Thomas EC, Clark LF (1965) Discharge patterns of single fibers in the cats auditory nerve. MIT, Cambridge, MA

Kirkwood PA (1979) On the use and interpretation of cross correlation measurements in the mammalian central nervous system. J Neurosci Meth 1:107–132

Kirkwood PA, Sears TA (1978) The synaptic connections to intercostal motoneurones as revealed by the average common excitation potential. J Physiol 275:103–134

Kirkwood PA, Sears TA (1982a) Excitatory post-synaptic potentials from single muscle spindle afferents in external intercostal motoneurones in the cat. J Physiol 322:287–314

Kirkwood PA, Sears TA (1982b) The effects of single afferent impulses on the probability of firing of external intercostal motoneurones in the cat. J Physiol 322:315–336

Kirkwood PA, Sears TA, Tuck DL, Westgaard RH (1982a) Variations in the time course of the synchronization of intercostal motoneurones in the cat. J Physiol 327:105–135

Kirkwood PA, Sears TA, Stagg D, Westgaard RH (1982b) The spatial distribution of synchronisation of intercostal motoneurones in the cat. J Physiol 327:137–155

Kitzes LM, Farley GR, Starr A (1978) Modulation of auditory cortex unit activity during the performance of a conditionned response. Exp Neurol 62:678–697

Klinke R, Pause M (1980) Discharge properties of primary auditory fibers in Caiman crocodillus: comparisons and contrasts to the mammalian auditory nerve. Exp Brain Res 38:137–150

Klopf AH (1982) The hedonistic neuron: a theory of memory, learning, and intelligence. Hemisphere, Washington, DC

Knight BW (1972) Dynamics of encoding in a population of neurons. J Gen Physiol 59:734–766

Knox CK (1974) Cross-correlation functions for a neuronal model. Biophys J 14:567–582

Knox CK, Poppele RE (1977) Correlation analysis of stimulus-evoked changes in excitability of spontaneously firing neurons. J Neurophysiol 40:616–625

Knox CK, Kubota S, Poppele RE (1977) A determination of excitability changes in dorsal spinocerebellar tract neurons from spike train analysis. J Neurophysiol 40:626–646

Knudsen EI (1983) Space coding in the vertebrate auditoty system. In: Lewis B (ed) Bioacoustics. A comparative approach. Academic Press, New York London, pp 311–343

Knudsen EI, Konishi M (1978) Center-surround organization of auditory receptive fields in the owl. Science 202:778–780

Knudsen EL, Knudsen PF (1985) Vision guides the adjustment of auditory localization in young barn owls. Science 230:545–548

Koch C, Poggio T (1983) A theoretical analysis of electrical properties of spines. Proc R Soc London Ser B218:455–477

Koch C, Marroquin J, Yuille A (1986) Analog "neuronal" networks in early vision. Proc Natl Acad Sci USA 83:4263–4267

Koenderink JJ (1984a) Simultaneous order in nervous nets from a functional standpoint. Biol Cybernet 50:35–41

Koenderink JJ (1984b) Geometrical structures determined by the functional order in nervous nets. Biol Cybernet 50:43–50

Koerber KC, Pfeiffer RR, Warr WB, Kiang NY-S (1966) Spontaneous spike discharges from single units in the cochlear nucleus after destruction of the cochlea. Exp Neurol 16:119–130

Kohonen T (1982) Self-organized formation of topologically correct feature maps. Biol Cybernet 43:59–69

Kohonen T (1984) Self-organization and associative memory. Springer, Berlin Heidelberg New York

Koijima GR, Bracchi F (1980) A microprocessor based instrument for neural pulse wave analysis. IEEE Trans BME 27:515–519

Kraus N, Disterhoft JF (1982) Response plasticity of single neurons in rabbit association cortex during tone-signalled learning. Brain Res 246:205–215

Krausz HI (1975) Identification of non-linear systems using random impulse train inputs. Biol Cybernet 19:217–230

Krausz HI, Friesen WO (1977) The analysis of nonlinear synaptic transmission. J Gen Physiol 70:243–265

Kroeker JP (1977) Wiener analysis of non-linear system using Poisson-Carlier cross-correlation. Biol Cybernet 27:221–227

Kron G (1959) Tensors for circuits. Dover, New York

Krüger J (1983) Simultaneous individual recordings from many cerebral neurons: techniques and results. Rev Physiol Biochem Pharmacol 98:177–233

Krüger J, Bach M (1981) Simultaneous recording with 30 microelectrodes in monkey visual cortex. Exp Brain Res 41:191–194

Kuffler SW (1953) Discharge patterns and functional organization of mammalian retina. J Neurophysiol 16:37–68

Kuffler SW, Fitzhugh R, Barlow HB (1957) Maintained activity in the cat's retina in light and darkness. J. Gen Physiol 40:683–702

Kuperstein M, Whittington DA (1981) A practical 24 channel microelectrode for neural recording in vivo. IEEE trans BME 28:288–293

Lansky P (1983) Inference for the diffusion models of neuronal activity. Math Biosci 67:247–260

Lass Y, Abeles M (1975) Transmission of information by the axon: I. Noise and memory in the myelinated nerve fiber of the frog. Biol Cybernet 19:61–67

Lee BB, Cleland BG, Creutzfeldt OD (1977) The retinal input to cells in area 17 of the cat's cortex. Exp Brain Res 30:527–538

Legatt AD, Arezzo J, Vaughan Jr HG (1980) Average multiple unit activity as an estimate of phasic changes in local neuronal activity: effects of volume conducted potentials. J Neurosci Meth 2:203–217

Legéndy CR (1978) Cortical columns and the tendency of neighboring neurons to act similarly. Brain Res 158:89–105

Legéndy CR, Salcman M (1985) Bursts and recurrences of bursts in the spike trains of spontaneously active striate cortex neurons. J Neurophysiol 53:926–939

Lesevre N (1982) Chronotopographical analysis of the human evoked potential in relation to the visual field (data from normal individuals and hemianopic patients). Ann NY Acad Sci 388:156–182

Lettvin JY, Matuurana HR, Pitts WH, McCulloch WS (1961) Two remarks on the visual system of the frog. In: Rosenblith W (ed) Sensory communication. MIT, Cambridge, MA, pp 757–776

LeVay S (1971) On the neurons and synapses of the lateral geniculate nucleus of the monkey, and the effects of eye enucleation. Z Zellforsch Mikrosk Anat 113:396–419

Levick WR (1973) Maintained discharge in the visual system and its role for informations processing. In: Jung R (ed) Handbook of sensory physiology, vol 7/3. Springer, Berlin Heidelberg New York, pp 575–598

Levick WR, Williams WD (1964) Maintained activity of lateral geniculate neurons in darkness. J Physiol 170:582–597

Levick WR, Cleland BG, Dubin MW (1972) Lateral geniculate neurons of cat: retinal inputs and physiology. Invest Ophthalmol 11:302–311

Levine DS (1983) Neural population modeling and psychology: a review. Math Biosci 66:1–86

Levine MW (1980) Firing rates of a retinal neuron are not predictable from interspike interval statistics. Biophys J 30:9–25

Levine MW, Shefner JM (1977) A model for the variability of interspike intervals during sustained firing of a retinal neuron. Biophys J 19:241–252

Li C-L (1959) Synchronization of unit activity in the cerebral cortex. Science 129:783–784

Liberman MC, Oliver ME (1984) Morphometry of intracellularly labeled neurons of the auditory nerve: correlations with functional properties. J Comp Neurol 223:163–176

Lindstrom S (1982) Synaptic organization of inhibitory pathways to principal cells in the lateral geniculate nucleus of the cat. Brain Res 234:447–453

Liberman MC, Kiang NYS (1978) Acoustic trauma in cats. Acta Otolaryngol Suppl 358:1–63

Little WA (1974) The existence of persistent states in the brain. Math Biosci 19:101–120

Little WA, Shaw GL (1978) Analytic study of the memory storage capacity of a neural network. Math Biosci 39:281–290

Llinas R (1969) Neurobiology of cerebellar evolution. Am Med Assoc, Chicago

Looft FJ, Heetderks WJ (1978) Real-time correlator for detecting single units in peripheral nerve. IEEE Trans BME 25:564–567

Luria AR (1973) The working brain. An introduction to neuropsychology. Penguin, London

Lüscher HR, Mathis J, Schaffner H (1983) A dual time-voltage window descriminator for multiunit nerve spike decomposition. J Neurosci Meth 7:99–105

MacGregor RJ, Lewis ER (1977) Neural modeling. Plenum, New York

MacGregor RJ, Prieto-Diaz R, Miller SW, Groves PM (1973) Statistical properties of neurons in the rat mesencephalic reticular formation. Brain Res 64:167–187

MacGregor RJ, Miller SW, Groves PM (1975) Slow rhythms and correlations in spike trains from midbrain neurons. Exp Neurol 47:581–598

MacKay DM, McCulloch WS (1952) The limiting information capacity of a neuronal link. Bull Math, Biophys 14:127–135

Mahut H, Moss M (1984) Consolidation of memory: the hippocampus revisited. In: Squire LR, Butters N (eds) Neuropsychology of memory. Guilford Press, New York, pp 297–315

Manley GA, Robertson D (1976) Analysis of spontaneous activity of auditory neurons in the spiral ganglion of the guinea pig cochlea. J Physiol 258:323–336

Marmarelis PZ, Marmarelis VZ (1978) Analysis of physiological systems. The white noise approach. Plenum, New York

Marr D (1969) A theory of cerebellar cortex. J Physiol 202:437–470

Marr D (1970) A theory for cerebral neocortex. Proc R Soc London Ser B 176:161–234

Marr D (1971) Simple memory: a theory for archicortex. Philos Trans R Soc London Ser B262:23–81

Marr D (1981) Vision. Freeman, San Francisco

Mastronarde DN (1983a) Correlated firing of cat retinal ganglion cells. I. Spontaneously active inputs to X- and Y-cells. J Neurophysiol 49:303–324

Mastronarde DN (1983b) Correlated firing of cat retinal ganglion cells. II. Responses of X- and Y-cells to single quantal events. J Neurophysiol 49:325–349

Mastronarde DN (1983c) Interactions between ganglion cells in cat retina. J Neurophysiol 49:350–365

McCann GD (1973) Interactive computer strategies for living nervous system research. IEEE Trans BME 20:1–11

McCulloch WS, Pitts WH (1943) A logical calcules of the ideas immanent in nervous activity. Bull Math Biophys 5:115–133

McGill WJ (1967) Neural counting mechanisms and energy detection in audition. J Math Psychol 4:351–376

McNaughton BL (1983) Activity dependent modulation of hippocampal synaptic efficacy: some implications for memory processes. In: Seifert W (ed) Neurobiology of the hippocampus. Academic Press, New York London, pp 233–251

McNaughton BL, O'Keefe J, Barnes CA (1983) The stereotrode: a new technique for simultaneous isolation of several units in the central nervous system from multiple unit records. J Neurosci Meth 8:391–397

Meek J (1983) Functional anatomy of the tectum mesencephali of the goldfish. An explorative analysis of the functional implications of the laminar structural organization of the tectum. Brain Res Rev 6:247–297

Megala AL, Capranica RR (1981) Response patterns to tone bursts in peripheral auditory system of anurans. J Neurophysiol 46:465–478

Meredith MA, Stein BE (1986) Visual, auditory, and somatosensory convergence on cells in superior colliculus results in multisensory integration. J Neurophysiol 56:640–662

Merzenich MM, Jenkins WM (1983) Dynamic maintenance and alterability of cortical maps in adults; some implications. In: Klinke R, Hartmann R (eds) Hearing – physiological bases and psychophysics. Springer, Berlin Heidelberg New York, pp 162–167

Merzenich MM, Reid MD (1974) Representation of the cochlea within the inferior colliculus of the cat. Brain Res 77:397–415

Merzenich MM, Knight PL, Roth GL (1975) Representation of cochlea within primary auditory cortex in the cat. J Neurophysiol 38:231–249

Merzenich MM, Kaas JH, Sur M, Lin S-S (1978) Double representation of the body surface within cytoarchitectonic areas 3b and 1 in "S1" in the owl monkey. J Comp Neurol 181:41–74

Merzenich MM, Kaas JH, Wall JT, Sur M, Nelson RJ (1983) Progression of change following median nerve section in the cortical representation of the hand in areas 3b and 1 in adult owl and squirrel monkey. Neuroscience 10:639–665

Merzenich MM, Nelson RJ, Stryker MP, Cynader MS, Schoppmann A, Zook JM (1984a) Somatosensory cortical map changes following digit amputation in adult monkeys. J Comp Neurol 224:591–605

Merzenich MM, Jenkins WM, Middlebrooks JC (1984b) Observations and hypotheses on special organizational features of the central auditory nervous system. In: Edelman GM, Gall WE, Cowan WM (eds) Dynamic aspects of neocortical function. John Wiley & Son, New York, pp 397–424

Metherate R, Dykes RW (1985) Simultaneous recordings from pairs of cat somatosensory cortical neurons with overlapping peripheral receptive fields. Brain Res 341:119–129

Michalski A, Gerstein GL, Czarkowska J, Tarnecki R (1983) Interactions between cat striate cortex neurons. Exp Brain Res 51:97–107

Millar J (1983) A "wavegate" spike discriminator for sorting extracellular nerve action potentials. J Neurosci Meth 7:157–164

Millecchia R, McIntyre T (1978) Automatic nerve impulse identification and separation. Comput Biomed Res 11:459–468

Mishelevich DJ (1970) On-line real-time digital computer separation of extracellular neuro-electric signals. IEEE Trans BME 17:147–150

Moller AR, Janetta PJ (1981) Compound action potentials recorded intracranially from the auditory nerve in man. Exp Neurol 74:862–874

Moore GP (1980) Mathematical techniques for studying information processing by the nervous system. In: Pinsker HM, Willis WD Jr (eds) Information processing in the nervous system. Raven, New York, pp 17–30

Moore GP, Segundo JP, Perkel DH, Levitan H (1970) Statistical signs of statistical interaction in neurons. Biophysical J 10:876–900

Moore JK, Osen KK (1979) The cochlear nuclei in man. Am J Anat 154:393–418

Mountcastle VB (1967) The problem of sensing and the neural coding of sensory events. In: Quarton GC, Melnechuk J, Schmitt FO (eds) The neurosciences. Rockefeller Univ Press, New York, pp 393–408

Mountcastle VB (1978) An organizing principle for cerebral functions: the unit module and the distributed system. In: Edelman GL, Mountcastle VB (eds) The mindful brain. MIT, Cambridge, MA, pp 7–50

Mugnaini E, Osen KK, Dahl A-L, Friedrich VL, Korte G (1980) Fine structure of granule cells and related interneurons (termed Golgi cells) in the cochlear nucleus complex or cat, rat and mouse. J Neurocytol 9:537–570

Munemori J, Hara K-I, Kimura M, Sato R (1984) Statistical features of impulse trains in cats lateral geniculate neurons. Biol Cybernet 50:167–172

Murata K, Kameda K (1963) The activity of single cortical neurons uf unrestrained cats during sleep and wakefulness. Arch Ital Biol 101:306–331

Murphy JT, Kwan HC, Wong YC (1985a) Cross correlation studies in primate motor cortex: synaptic interaction and shared input. Can J Neurol Sci 12:11–23

Murphy JT, Kwan HC, Wong YC (1985b) Cross correlation studies in primate motor cortex: event related correlation. Can J Neurol Sci 12:24–30

Nakahama H, Nishioka S, Otsuka T, Aikawa S (1966) Statistical dependency between interspike intervals of spontaneous activity in thalamic lemniscal neurons. J Neurophysiol 29:921–934

Nakahama H, Ishii N, Yamamoto M, Saito H (1971) Stochastic properties of spontaneous impulse activity in central single neurons. Tohoku J Exp Med 104:373–409

Nakahama H, Ishii N, Yamamoto M (1972) Markov process of maintained impulse activity in central single neurons. Kybernetik 11:61–72

Nass MM, Cooper LN (1975) A theory for the development of feature detecting cells in visual cortex. Biol Cybernet 19:1–18

Nelson JI (1985) The cellular basis of perception. In: Rose D, Dobson VG (eds) Models of the visual cortex. John Wiley & Son, New York Chichester, pp 108–122

Noda H, Adey WR (1970) Firing of neuron pairs in cat association cortex during sleep and wakefulness. J Neurophysiol 33:672–684

Noda H, Adey WR (1973) Neuronal activity in the association cortex of the cat during sleep, wakefulness and anaesthesia. Brain Res 54:243–260

Noda H, Manohar S, Adey WR (1969) Correlated firing of hippocampal neuron pairs in sleep and wakefulness. Exp Neurol 24:232–247

Nottebohm F (1984) Birdsong as a model in which to study brain processes related to learning. Condor 86:227–236

Nunez P (1981) Electric fields of the brain. Oxford Univ Press, New York

O'Brien JH, Fox SS (1969) Single cell activity in cat motor cortex. II Functional characteristics of the cell related to conditioning changes. J Neurophysiol 32:285–296

Ogawa T, Bishop PO, Levick WR (1966) Temporal characteristics of responses to photic stimulation by single ganglion cells in the unopened eye of the cat. J Neurophysiol 29:1–30

Oguztöreli MN, Stein RB (1977) Optimal filtering of nerve signals. Biol Cybernet 27:41–48

Oja E (1982) A simplified neuron model as a principal component analyzer. J Math Biol 15:267–273

Olton DS (1983) Memory functions and the hippocampus. In: Seifert W (ed) Neurobiology of the hippocampus. Academic Press, New York London, pp 335–372

O'Leary DP, Honrubia V (1975) On-line identification of sensory systems using pseudorandom binary noise perturbations. Biophys J 15:505–532

Orbach HS, Cohen LS, Grinvald A (1985) Optical mapping of electrical activity in rat somatosensory and visual cortex. J Neurosci 5:1886–1895

Orban GA (1984) Neuronal operations in the visual cortex. Springer, Berlin Heidelberg New York

Ottes FP, van Gisbergen JAM, Eggermont JJ (1985a) Metrics of saccade responses to visual double stimuli: Two different modes. Vision Res 24:1169–1179

Ottes FP, van Gisbergen JAM, Eggermont JJ (1985b) Latency dependence of colour-based target vs non-target discrimination by the saccadic system. Vision Res 25:849–862

Ottes FP, van Gisbergen JAM, Eggermont JJ (1986) Visuomotor fields of the superior colliculus: a quantitative model. Vision Res 26:857–873

Paisley AC, Summerlee AJS (1984) Relationship between behavioural states and activity in the cerebral cortex. Prog Neurobiol 22:155–184

Paliouras JD (1975) Complex variables for scientists and engineers. Macmillan Publ Cie, NY

Palm G (1981) Evidence, information, and surprise. Biol Cybernet 42:57–68

Palm G (1982a) Neural assemblies. An alternative approach to artificial intelligence. Springer, Berlin Heidelberg New York

Palm G (1982b) How useful are associative memories? In: Ricciardi LM, Scott AC (eds) Biomathematics in 1980 Elsevier/North-Holland Biomedical Press, Amsterdam New York, pp 145–153

Palm G (1986) Associative networks and cell assemblies. In: Palm G, Aertsen A (eds) Brain theory. Springer, Berlin Heidelberg New York Tokyo, pp 211–228

Palm G, Aertsen A (eds) (1986) Brain theory. Springer, Berlin Heidelberg New York Tokyo

Palm G, Aertsen AMJH, Gerstein GL (1988) On the significance of correlations among neuronal spike trains. Biol Cybernet 59:1–11

Pask G (1960) The natural history of networks. In: Yovits MC, Cameron S (eds) Self-organizing systems. Pergamon, New York Oxford, pp 232–263

Patton JA, Nottebohm FN (1984) Neurons generated in the adult brain are recruited into functional circuits. Science 225:1046–1048

Patton JA, O'Loughlin BE, Nottebohm F (1985) Cells born in adult canary forebrain are local interneurons. J Neurosci 5:3088–3093

Paulin MG (1985) Is the cerebellum an adaptive filter? J Theor Neurobiol 4:143–151

Pellionisz A, Llinas R (1979) Brain modelling by tensor network theory and computer simulation. The cerebellum: Distributed processor for predictive correlation. Neuroscience 4:323–348

Pellionisz A, Llinas R (1980) Tensorial approach to the geometry of brain function: cerebellar coordination via a metric tensor. Neuroscience 5:1125–1136

Pellionisz A, Llinas R (1982) Space-time representation in the brain. The cerebellum as a predictive space-time metric tensor. Neuroscience 7:2949–2970

Pellionisz A, Llinas R (1985) Tensor network theory of the meta organization of functional geometries in the central nervous system. Neuroscience 16:245–273

Penfield W (1975) The mystery of the mind. Univ Press, Princeton, NJ

Peretto P (1984) Collective properties of neural networks: a statistical approach. Biol Cybernet 50:51–62

Perez R, Glass L, Shlaer R (1975) Development of specificity in the cat visual cortex. J Math Biol 1:275–288

Perkel DH (1970) Spike trains as carriers of information. In: Schmidt FO (ed) The neurosciences: second study program. Rockefeller Univ Press, New York, pp 587–596

Perkel DH, Bullock TH (1968) Neural coding. Neurosci Res Prog, Brookline

Perkel DH, Gerstein GL, Moore GP (1967) Neuronal spike trains and stochastic point processes. Biophys J 7:419–440

Perkel DH, Gerstein GL, Smith MS, Tatton WC (1975) Nerve impulse patterns: a quantitative display technique for three neurons. Brain Res 100:271–296

Pernier J, Gerin P (1971) Méthode d'objectivation d'organisations temporelles privilegiées d'activités neuronales unitaires. Bull Math Biophys 33:129–151

Pernier J, Gerin P (1975) Temporal patterns analysis of spontaneous unit activity in the neocortex. Biol Cybernet 18:123–136

Perrett DI, Rolls ET, Caan W (1982) Visual neurones responsive to faces in the monkey temporal cortex. Exp Brain Res 47:329–342

Pfaff DW (1980) Estrogens and brain function. Springer, Berlin Heidelberg New York

Pfeiffer RR, Kiang NY-S (1965) Spike discharge patterns of spontaneous and continuously stimulated activity in the cochlear nucleus of anaesthetized cats. Biophys J 5:301–316

Poggio GF, Viernstein LJ (1964) Time series analysis of impulse sequences of thalamic somatosensory neurons. J Neurophysiol 27:517–545

Pompeiano O (1973) Reticular formation. In: Iggo A (ed) Handbook of sensory physiology, vol 2. Springer, Berlin Heidelberg New York, pp 381–488

Precht W (1974) Characteristics of vestibular neurons after acute and chronic labyrinthine destruction. In: Kornhuber HH (ed) Handbook of sensory physiology. Springer, Berlin Heidelberg New York, pp 451–462

Prochazka VJ, Kornhuber HH (1973) On-line multi-unit sorting with resolution of superposition potentials. Electroencephalogr Clin Neurophysiol 34:91–93

Puel JL, Uziel A (1987) Correlative development of cochlear action potential sensitivity, latency, and frequency selectivity. Dev Brain Res 37:179–188

Purves D, Lichtman JW (1980) Elimination of synapses in the developing nervous system. Science 210:153–157

Radil-Weiss T, Skvaril J (1971) A technique for recording the impulse activity of four adjacent neurons in the brain. Electroencephalogr Clin Neurophysiol 30:571–574

Radil-Weiss T, Skvaril J, Syka J (1968) Quantitative analysis of spontaneous impulse activity of neurons in the brain. In: Drischel H, Tiedt N (eds) Biokybernetik. Fischer, Leipzig, pp 244–250

Rall W (1962) Electrophysiology of a dendritic neuron model. Biophys J 2:145–167

Ramos A, Schwarz EL, John ER (1976) Stable and plastic unit discharge patterns during behavioral generalization. Science 192:393–396

Ramsey RL (1974) A device for analysing multi-unit-neural records. J Physiol 244:3–5P

Rauschecker JP, Singer W (1981) The effects of early visual experience on the cat's visual cortex and their possible explanation by Hebb synapses. J Physiol 310:215–239

Ray RH, Krüger L (1983) Spatial properties of receptive fields of mechanosensitive primary afferent nerve fibers. Fed Proc 42:2536–2541

Rebillard G, Carlier E, Rebillard M, Pujol R (1977) Enhancement of visual responses on the primary auditory cortex of the cat after an early destruction of cochlear receptors. Brain Res 129:162–164

Rebillard G, Rebillard M, Pujol R (1980) Factors affecting the recording of visual evoked potentials from the deaf cat primary auditory cortex. Brain Res 188:252–254

Reeke GN Jr, Edelman GM (1984) Selective networks and recognition automata. Ann NY Acad Sci 426:181–199

Reitböck HJP (1983) A 19-channel matrix drive with individually controllable fiber microelectrodes for neurophysiological applications. IEEE Trans SMC 13:676–683

Reitböck HJP, Werner G (1983) Multielectrode recording system for the study of spatio-temporal activity patterns of neerons in the central nervous system. Experientia 39:339–341

Rescorla RA, Wagner AR (1972) A theory of Pavlovian conditioning: variations in the effectiveness of reinforcement and nonreinforcement. In: Black AH, Prokasy WF (eds) Cassical conditioning. II. Current research and theories. Appleton-Century-Crofts, New York, pp 64–99

Ricciardi LM, Sacerdote L (1979) The Ornstein-Uhlenbeck process as a model for neuronal activity. I Mean and variance of firing time. Biol Cybernet 35:1–9

Roberts WM (1979) Optimal recognition of neuronal waveforms. Biol Cybernet 35:73–80

Roberts WM, Hartline DK (1975) Separation of multi-unit nerve impulse trains by a multi-channel linear filter algorithm. Brain Res 94:141–149

Rodieck RW (1965) Quantitative analysis of cat retinal ganglion cell response to visual stimuli. Vision Res 5:583–601

Rodieck RW (1968) Maintained activity of cat retinal ganglion cells. J Neurophysiol 30: 1043–1071

Rodieck RW, Kiang NY-S, Gerstein GL (1962) Some quantitative methods for the study of spontaneous activity of single cells. Biophysical J 2:351–368

Roney KJ, Shaw GL (1980) Analytic study of assemblies of neurons in memory storage. Math Biosci 51:25–41

Roney KJ, Scheibel AB, Shaw GL (1979) Dendritic bundles: survey of anatomical experiments and physiological theories. Brain Res Rev 1:225–271

Rose G (1983) Physiology and behavioral characteristics of dentate granule cells. In: Seifert W (ed) Neurobiology of the hippocampus. Academic Press, New York London, pp 449–472

Rose D, Dobson VG (eds) (1985) Models of the visual cortex. John Wiley & Son, New York Chichester

Rose JE, Gross NB, Geisler CD, Hind JE (1966) Some neural mechanisms in the inferior colliculus of the cat which may be relevant to localization of a sound source. J Neurophysiol 29:288–314

Rosenblatt F (1962) Principles of neurodynamics. Spartan, New York

Routtenberg A (1984) The CA3 pyramidal cell in the hippocampus: site of intrinsic expression and extrinsic control of memory formation. In: Squire LS, Butters N (eds) Neuropsychology of memory. Guilford, New York, pp 536–546

Rowe MH, Stone J (1980) Parametric and feature extraction analyses of the receptive fields of visual neurons. Brain Behav 17:103–122

Rummelhart DE, McCleland JL (1987) Parallel distributed processing, vol 1. MIT, Cambridge, MA

Sachs MB, Woolf NK, Sinnott JM (1980) Response properties of neurons in the avian auditory system: comparisons with mammalian homologues and consideration of the neural encoding of complex stimuli. In: Popper AN, Fay RR (eds) Comparative studies of hearing in vertebrates. Springer, Berlin Heidelberg New York, pp 323–353

Salzberg BM, Obaid AL, Senseman DM, Gainer H (1983) Optical recording of action potentials from vertebrate nerve terminals using potentiometric probes providing evidence for sodium and calcium components. Nature (London) 306:36–40

Sanderson AC, Kozak WM, Calvert TW (1973) Distribution coding in the visual pathway. Biophys J 13:218–244

Sanseverino ER, Agnati LF, Maioli MG, Galletti C (1973) Maintained activity of single neurons in striate and non-striate areas of the cat visual cortex. Brain Res 54:225–242

Schellart NAM, Spekreijse H (1972) Dynamic characteristics of retinal ganglion cell responses in goldfish. J Gen Physiol 59:1–21

Scherg M, von Cramon D (1985a) A new interpretation of the generation of BAEP waves I–V: Results of a spatio-temporal dipole model. Electroencephalogr Clin Neurophysiol 62:290–299

Scherg M, von Cramon D (1985b) Evoked dipole source potentials of the human auditory cortex. Electroencephalogr Clin Neurophysiol 65:344–360

Schiller PH, Finlay BL, Volman SF (1976) Quantitative studies of single cell properties in monkey striate cortex. V. Multivariate statistical analyses and model. J Neurophysiol 39:1362–1374

Schmidt EM (1971) An instrument for separation of multi-unit neuroelectric signals. IEEE Trans BME 18:155–157

Schmidt EM (1984a) Instruments for sorting neuroelectric data: a review. J Neurosci Meth 12:1–24

Schmidt EM (1984b) Computer separation of multi-unit neuroelectric data: a review: J Neurosci Meth 12:95–111

Schmidt EM, Stromberg WM (1969) Computer dissection of peripheral nerve bundle activity. Comput Biomed Res 2:446–455

Schmidt RF (1973) Control of the access of afferent activity to somatosensory pathways. In: Iggo A (ed) Handbook of sensory physiology, vol 2. Springer, Berlin Heidelberg New York, pp 151–206

Schneider J, Eckhorn R, Reitbock H (1983) Evaluation of neural coupling dynamics. Biol Cybernet 46:129–134

Schuz A (1978) Some facts and hypotheses concerning dendritic spines and learning. In: Brazier MAB, Petsche H (eds) Architecture of the cerebral cortex. Raven, New York, pp 129–135

Schwartz EL (1977a) Spatial mapping in the primate sensory projection: analytic structure and relevance to perception. Biol Cybernet 25:181–194

Schwartz EL (1977b) Afferent geometry in the primate visual cortex and the generation of neuronal trigger features. Biol Cybernet 28:1–14

Schwartz EL (1980) Computational anatomy and functional architecture of striate cortex: a spatial mapping approach to perceptual coding. Vision Res 20:645–669

Schwartz EL (1984) Anatomical and physiological correlates of visual computation from striate cortex to infero-temporal cortex. IEEE Trans SMC 14:257–271

Schwartz EL, Ramos A, John ER (1976) Single cell activity in chronic unit recording: A quantitative study of the unit amplitude spectrum. Brain Res Bull 1:57–68

Sears TA, Stagg D (1976) Short-term synchronization of intercostal motoneurone activity. J Physiol 263:357–381

Segundo JP (1970) Communication and coding by nerve cells. In: Schmitt FO (ed) The Neurosciences: second study program. Rockefeller Univ Press, New York, pp 569–586

Seifert W (ed) (1983) Neurobiology of the hippocampus. Academic Press, New York London

Sejnowski TJ (1976a) On global properties of neuronal interaction. Biol Cybernet 22:85–95

Sejnowski TJ (1976b) On the stochastic dynamics of neuronal interaction. Biol Cybernet 22:203–211

Sejnowski TJ (1977) Storing covariance with nonlinearly interacting neurons. J Math Biol 4:303–321

Sewell WF (1984) The relation between the endocochlear potential and spontaneous activity in auditory nerve fibers in the cat. J Physiol 347:685–696

Shannon CE (1949) Communication in the presence of noise. Proc IRE 37:10

Shaw GL (1978) Space-time correlations of neuronal firing related to memory storage capacity. Brain Res Bull 3:107–113

Shaw GL, Vasudevan R (1974) Persistent states of neural networks and the random nature of synaptic transmission. Math Biosci 21:207–218

Shaw GL, Harth E, Scheibel AB (1982) Cooperativity in brain function: assemblies of approximately 30 neurons. Exp Neurol 77:324–358

Shaw GL, Silverman DJ, Pearson JC (1985) Model of cortical organization embodying a basis for a theory of information processing and memory recall. Proc Natl Acad Sci USA 82:2364–2368

Shepherd GM (1979) The synaptic organization of the brain. Oxford Univ Press, New York

Sherry CJ, Klemm WR (1980) An information coding system that seems to be independent of noise and variability. Brain Res Bull 5:297–300

Shiavi R, Negin M (1975) Multivariate analysis of simultaneously active motor units in human skeletal muscle. Biol Cybernet 20:9–16

Sholl DA (1956) The organization of the cerebral cortex. Methuen, London

Siebert WM (1965) Some implications of the stochastic behavior of primary auditory neurons. Kybernetik 2:206–215

Siegel JM, Nienhuis R, Wheeler RL, McGinty DJ, Harper RM (1981) Discharge patterns of reticular formation unit pairs in waking and REM sleep. Exp Neurol 74:875–891

Simon S (1965) The real time sorting of neuroelectric action potentials in multiple unit studies. Electroencephal Clin Neurophysiol 18:192–195

Singer W (1983) Neuronal activity as a shaping factor in the self organization of neuron assemblies. In: Basar E, Flohr H, Haken H, Mandell AJ (eds) Synergetics of the brain. Springer, Berlin Heidelberg New York, pp 89–101

Singer W (1985a) Activity dependent self-organization of the mammalian visual cortex. In: Rose D, Dobson VG (eds) Models of the visual cortex. John Wiley & Son, New York Chichester, pp 123–136

Singer W (1985b) Central control of developmental plasticity in the mammalian visual cortex. Vision Res 25:389–396

Skvaril J, Radil-Weiss T, Bohdanecky Z, Syka J (1971) Spontaneous discharge patterns of mesencephalic neurons: interval histograms and mean interval relationship. Kybernetik 9:11–15

Smith CE (1979) A comment on a retinal neuron model. Biophys J 25:385–386

Smith DR, Smith GK (1965) A statistical analysis of the continual activity of single cortical neurones in the cat unanaesthetized isolated forebrain. Biophys J 5:47–74

Sparks DL (1986) Translation of sensory signals into commands for control of saccadic eye movements: role of primate superior collicuous. Physiol Res 66:118–171

Spiegel MS (1965) Statistics. McGraw-Hill, New York

Squire LR (1986) Mechanisms of memory. Science 232:1612–1619

Squire LR (1987) Memory and brains. Univ Press, Oxford

Squire LS, Butters N (1984) Neuropsychology of memory. Guilford, New York

Stagg D (1973) Computer acquisition of multi-unit nerve-spike signals. Med Biol Eng 11:340–347

Stark L, Semmlov J, Terdiman J (1968) Anatomical transfer function. Math Biosci 2:425–433

Stein BE, Clamann HP (1981) Control of pinna movements and sensorimotor register in cat superior colliculus. Brain Behav Evol 19:180–192

Stein BE, Gaither NS (1981) Sensory representation in reptilian optic tectum: some comparisons with mammals. J Comp Neurol 202:69–87

Stein BE, Magalhaes-Castro B, Kruger L (1976) Relationship between visual and tactile representation in cat superior colliculus. Neurophysiol 39:401–419

Stein RB (1965) A theoretical analysis of neuronal variability. Biophys J 5:173–196

Stein RB (1967) The information capacity of nerve cells using a frequency code. Biophys J 7:797–826

Stein RB, Andreassen S, Oguztoreli MN (1979) Mathematical analysis of optimal multi-channel filtering for nerve signals. Biol Cybernet 32:19–24

Steinbuch K (1961) Die Lernmatrix. Kybernetik 1:36

Stent GS (1973) A physiological mechanism for Hebb's postulate of learning. Proc Natl Acad Sci USA 70:997–1001

Sterling P (1983) Microcircuitry of the cat retina. Annu Rev Neurosci 6:149–185

Stevens JK, Gerstein GL (1976) Interactions between cat lateral geniculate neurons. J Neurophysiol 39:239–256

Stone J, Dreher B (1982) Parallel processing of information in the visual pathways. A general principle of sensory coding. Trends Neurosci 5:441–446

Stryker MP (1982) Role of visual afferent activity in the development of ocular dominance columns. In: Rakic P, Goldman-Rakic PS (eds) Development and modifiability of the cerebral cortex. MIT, Cambridge, MA, pp 541–549

Suga N (1973) Feature extraction in the auditory system of bats. In: Moller A (ed) Basic mechanisms in hearing. Academic Press, New York London, pp 675–742

Suga N, Kuzirai K, O'Neill WE (1981) How biosonar information is represented in the bat cerebral cortex. In: Syka J, Aitkin LM (eds) Neuronal mechanisms of hearing. Plenum, New York, pp 197–219

Surmeier DJ, Weinberg RJ (1985) The relationship between cross-correlation measurements and underlying synaptic events. Brain Res 331:180–184

Sutton RS, Barto AG (1981) Toward a modern theory of adaptive networks: expectation and prediction. Psychol Rev 88:135–170

Swindale NV (1980) A model for the formation of ocular dominance stripes. Proc R Soc London Ser B 208:243–264

Swindale NV (1982) The development of columnar systems in the mammalian visual cortex. Trends Neurosci 235–241

Szentagothai J (1968) Structuro-functional considerations of the cerebellar neuron network. Proc IEEE 56:960–967

Szentagothai J (1978a) The neuron network of the cerebral cortex: a functional interpretation. Proc R Soc London Ser B 201:219–248

Szentagothai J (1978b) The local neuronal apparatus of the cerebral cortex. In: Buser P, Roungeul-Buser A (eds) Cerebral correlates of conscious experience. Elsevier/North-Holland Biomedical Press, Amsterdam, pp 131–138

Szentagothai J (1979) Local neuron circuits of the neocortex. In: Schmitt FO (ed) The neurosciences: fourth study program. MIT, Cambridge, MA, pp 399–415

Szilard L (1964) On memory and recall. Proc Natl Acad Sci USA 51:1092–1099

Takeuchi A, Amari S-I (1979) Formation of topographic maps and columnar microstructures. Biol Cybernet 35:63–72

Tam DC, Ebner TJ, Knox CK (1988) Conditional cross-interval correlation analyses with applications to simultaneously recorded cerebellar Purkinje neurons. J Neurosci Meth 23:23–33

Tanaka K (1983) Cross-correlation analysis of geniculostriate neuronal relationships in cats. J Neurophysiol 49:1303–1318

Tanaka K (1985) Organization of geniculate inputs to visual cells in the cat. Vision Res 25:357–364

Teich MC, Diament P (1980) Relative refractoriness in visual information processing. Biol Cybernet 38:187–191

Teich MC, Khanna SM (1985) Pulse-number distribution for the neural spike trains in the cat's auditory nerve. J Acoust Soc Am 77:1110–1128

Ten Hoopen M (1966a) Impulse sequences of thalamic neurons-an attempted theoretical interpretation. Brain Res 3:123–140

Ten Hoopen M (1966b) Multimodal interval distributions. Kybernetik 3:17–24

Ten Hoopen M (1966c) Probabilistic firing of neurons considered as a first passage problem. Biophys J 6:435–451

Ten Hoopen M, Reuver HA (1965) On a waiting time problem in physiology. Statist Neerl 19:27–34

Ten Hoopen M, Reuver HA (1968) Recurrent point processes with dependent interference with reference to neuronal spike trains. Math Biosci 2:1–10

Thompson RS, Gibson WG (1981a) Neural model with probabilistic firing behavior. I. General considerations. Math Biosci 56:239–253

Thompson RS, Gibson WG (1981b) Neural model with probabilistic firing behavior. II. One- and two-neuron networks. Math Biosci 56:255–285

Thompson RF, Berger TW, Madden J (1983) Cellular processes of learning and memory in the mammalian CNS. Annu Rev Neurosci 6:447–491

Toyama K (1985) Neuronal circuitry in the cat visual cortex studied by cross-correlation analysis. In: Rose D, Dobson VG (eds) Models of the visual cortex. Wiley & Son, New York, pp 366–373

Toyama K, Tanaka K (1984) Visual cortical function studied by cross correlation analysis. In: Edelman GM, Gall WE, Cowan WM (eds) Dynamic aspects of neocortical function. John Wiley & Son, New York, pp 67–86

Toyama K, Kimura M, Tanaka K (1981a) Cross-correlation analysis of interneuronal connectivity in cat visual cortex. J Neurophysiol 46:191–201

Toyama K, Kimura M, Tanaka K (1981b) Organization of cat visual cortex as investigated by cross-correlation technique. J Neurophysiol 46:202–214

Tsukahara N (1982) Brain plasticity: the themes and case studies of neuro-biophysics. Adv Biophys 15:131–172

Tsumoto T, Suda K (1984) Differences in the rate of postnatal neuronal maturation among cortical layers: early- and late-maturing neurons in the kitten visual cortex. In: (ed) Motoy Kuno, Neuronal growth and plasticity, VNU Science Press, Utrecht, pp 293–311

Tsumoto T, Creutzfeldt OD, Legendy CR (1978) Functional organization of the corticofugal system from visual cortex to lateral geniculate nucleus in the cat. Exp Brain Res 32:345–364

Vaadia E, Gottlieg Y, Abeles M (1982) Single-unit activity related to sensorimotor association in auditory cortex of a monkey. J Neurophysiol 48:1201–1213

van den Boogaard HFP, Johannesma PIM (1984) The master equation for neuronal interaction. IMA J Math Appl Med Biol 1:365–389

van Gisbergen JHM, van Opstal AJ, Tax AAM (1981) Collicular ensemble coding of saccades based on vector summation. Neuroscience 21:541–555

van Kampen NG (1981) Stochastic processes in physics and chemistry. North Holland, Amsterdam

van Kan PLE, Scobey RP, Gabor AJ (1985) Response covariance in cat visual cortex. Exp Brain Res 60:559–563

van Stokkum IHM, Johannesma PIM, Eggermont JJ (1986) Representation of time dependent correlation functions and recurrence time functions. Biol Cybernet 55:17–24

Vibert JF, Costa J (1979) Spike separation in multi-unit records: A multivariate analysis of spike descriptive parameters. Electroencephalogr Clin Neurophysiol 47:172–182

Voigt HF, Young ED (1980) Evidence of inhibitory interactions between neurons in dorsal cochlear nucleus. J Neurophysiol 44:76–96

von Baumgarten R, Schaeffer K-P (1957) Kopplungsvorgänge an benachbarten Nervenzellen im Hirnstamm und im motorischen Cortex der Katze. Pflüger's Arch 265:264–275

von der Malsburg C (1973) Self-organization of orientation sensitive cells in the striate cortex. Kybernetik 14:85–100

von der Malsburg C (1981) The correlation theory of brain function. Internal Rep 81-2. Dep Neurobiol MPI Göttingen

von der Malsburg C (1983) How are nervous structures organised? In: Basar E, Flohr H, Haken H, Mandell AJ (eds) Synergetics of the brain. Springer, Berlin Heidelberg New York, pp 238–249

von der Malsburg C (1986) Am I thinking assemblies. In: Palm G, Aertsen A (eds) Brain theory. Springer, Berlin Heidelberg New York, pp 161–176

von der Malsburg C, Cowan JD (1982) Outline of a theory for the ontogenesis of iso-orientation columns in visual cortex. Biol Cybernet 45:49–56

von der Malsburg C, Willshaw D (1981) Co-operativity and brain organization. Trends Neurosci 80–83

von Seelen W, Leinhos U, Shaw G (1988) Organization of neural networks: structures and models. VCH, Weinheim

Voronin LL, Gerstein GL, Kudryashov IE, Ioffe SV (1975) Elaboration of a conditioned reflex in a single experiment with simultaneous recording of neural activity. Brain Res 92:385–403

Vrensen G, Cardozo JN (1981) Changes in size and shape of synaptic connections after visual training: an ultra structural approach to synaptic plasticity. Brain Res 218:79–97

Walsh BT, Miller JB, Gacek RR, Kiang NY-S (1972) Spontaneous activity in the eighth cranial nerve of the cat. Int J Neurosci 3:221–236

Walsh EJ, McGee J, Javel E (1986a) Development of auditory evoked potentials in the cat. I. Onset of response and development of sensitivity. J Acoust Soc Am 79:712–724

Walsh EJ, McGee J, Javel E (1986b) Development of auditory evoked potentials in the cat. II. Wave latencies. J Acoust Soc Am 79:725–744

Walsh EJ, McGee J, Javel E (1986c) Development of auditory evoked potentials in the cat. III. Wave amplitude. J Acoust Soc Am 79:745–754

Webb AC (1976a) The effects of changing levels of arousal on the spontaneous activity of cortical neurones. I. Sleep and wakefulness. Proc R Soc London Ser B 194:211–223

Webb AC (1976b) The effects of changing levels of arousal on the spontaneous activity of cortical neurons. II. Relaxation and alarm. Proc R Soc London Ser B 194:239–251

Weinberger NM, Hopkins W, Diamond DM (1984) Physiological plasticity of single neurons in auditory cortex of the cat during acquisition of the pupillary conditionned response: I Primary field. Behav Neurosci 98:171–188

Werner G, Mountcastle VB (1963) The variability of central neural activity in a sensory system and its implications for the central reflection of sensory events. J Neurophysiol 26:958–977

Wheeler BC, Heetderks WJ (1982) A comparison of techniques for classification of multiple neural signals. IEEE Trans BME 29:752–759

Widrow B (1962) Generalization and information storage in networks of adaline 'neurons'. In: Yovitz M, Jacobi G, Goldstein G (eds) Self organizing systems. Spartan, New York, pp 435–452

Wieraszko A (1983) Glutamic and aspartic acid as putative neurotransmitters: release and uptake studies on hippocampal slices. In: Seifert W (ed) Neurobiology of the hippocampus. Academic Press, New York, pp 175–191

Wiesel TN (1982) Postnatal development of the visual cortex and the influence of environment. Nature (London) 299:583–591

Willshaw DJ, Buneman OP, Longuet-Higgins HC (1969) Non-holographic associative memory. Nature (London) 222:960–962

Wilson HR, Cowan JD (1972) Excitatory and inhibitory interactions in localized populations of model neurons. Biophys J 12:1–24

Wilson HR, Cowan JC (1973) A mathematical theory of the functional dynamics of cortical and thalamic nervous tissue. Kybernetik 13:55–80

Wilson MJ, Dobie RA (1987) Human short-latency auditory responses obtained by cross-correlation. Electroencephalogr Clin Neurophysiol 66:529–538

Winocur G (1984) Memory localization the brain. In: Squire LS, Butters N (eds) Neuropsychology of memory. Guilford, New York, pp 122–133

Wise ME (1981) Spike interval distributions for neurons and random walks with drift to a fluctuating threshold. In: Taillie C (ed) Statistical distributions in scientific work. Reidel, Dordrecht, pp 211–231

Wood CC, McCarthy G, Squires NK, Vaughan HG, Woods DL, McCallum WC (1984) Anatomical and physiological substrates of event related potentials. Ann NY Acad Sci 425:681–721

Woolsey CN (1982) Cortical sensory organization: multiple auditory areas. Humana, Clifton, NJ

Wyler AR (1985) Synchrony between cortical neurons during operant conditioning. Brain Res 341:66–72

Yang GL, Chen TC (1978) On statistical methods in neuronal spike train analysis. Math Biosci 38:1–34

Young ED, Voigt HF (1981) The internal organization of the dorsal cochlear nucleus. In: Syka J, Aitkin LM (eds) Neuronal mechanisms of hearing. Plenum, New York, pp 127–133

Young JZ (1975) Sources of discovery in neuroscience. In: Worden FG, Swazey JP, Adelman G (eds) The neurosciences: paths of discovery. MIT, Cambridge, MA, pp 15–46

Subject Index

Studies of Brain Function

Volumes already published in the series:

1 *W. Heiligenberg*
Principles of Electrolocation and Jamming Avoidance in Electric Fish

2 *W. Precht*
Neuronal Operations in the Vestibular System

3 *J.T. Enright*
The Timing of Sleep and Wakefulness

4 *H. Braak*
Architectonics of the Human Telencephalic Cortex

5 *H. Collewijn*
The Oculomotor System of the Rabbit and Its Plasticity

6 *M. Abeles*
Local Cortical Circuits

7 *G. Palm*
Neural Assemblies

8 *J. Hyvärinen*
The Parietal Cortex of Monkey and Man

9 *E. Zrenner*
Neurophysiological Aspects of Color Vision in Primates

10 *U. Bässler*
Neural Basis of Elementary Behavior in Stick Insects

11 *G.A. Orban*
Neuronal Operations in the Visual Cortex

12 *M. Heisenberg, R. Wolf*
Vision in Drosophila

13 *Yu.I. Arshavsky, I.M. Gelfand, G.N. Orlovsky*
Cerebellum and Rhythmical Movements

14 *G. Roth*
Visual Behavior in Salamanders

15 *U. Windhorst*
How Brain-like is the Spinal Cord?

16 *Jos J. Eggermont*
The Correlative Brain